Particle Interactions 2018

Topical at the 8th World Congress on Particle Technology

Orlando, Florida, USA
22 - 26 April 2018

ISBN: 978-1-5108-6973-8

Printed from e-media with permission by:

Curran Associates, Inc.
57 Morehouse Lane
Red Hook, NY 12571

Some format issues inherent in the e-media version may also appear in this print version.

Printed by Curran Associates, Inc. (2018)

For permission requests, please contact AIChE
at the address below.

AIChE
120 Wall Street, FL 23
New York, NY 10005-4020

Phone: (800) 242-4363
Fax: (203) 775-5177

www.aiche.org

Additional copies of this publication are available from:

Curran Associates, Inc.
57 Morehouse Lane
Red Hook, NY 12571 USA
Phone: 845-758-0400
Fax: 845-758-2633
Email: curran@proceedings.com
Web: www.proceedings.com

TABLE OF CONTENTS

Some abstracts in this book refer to images that were not made available by the publisher

„AFM MEASUREMENTS OF ADHESION FORCE ON HYDROPHOBIC SURFACES"

Ditscherlein, L.; Knüpfer, P.; Fritzsche, J.; Peuker, U.A.; Institute of Mechanical Engineering and Mineral Processing, Agricolastrasse 1, 09599 Freiberg, Germany*

**corresponding author: lisa.ditscherlein@mvtat.tu-freiberg.de*

With the investigation of the surface force apparatus and the atomic force microscope (AFM) [1, 2], the investigation and determination of adhesive forces in a wide range of applications became possible. Disperse or LIFSHITZ-VAN DER WAALS interactions can be calculated for ideal surfaces (smooth, known geometry), but in the past years there exist extensions for nanoscale rough surfaces, too. It is seen for many water-based systems and processes that ordinary DLVO-forces do not explain the interactions between hydrophobic surfaces [3]. ISRAELACHVILI and PASHLEY were the first who provided that VAN DER WAALS forces are weaker than hydrophobic interactions [4]. This hydrophobic force decayed exponentially in the range of a few nanometers. In the last years, there have been various investigations on hydrophobic surfaces [5-11]. These surfaces include (I) methylated/chemisorbed and (II) physisorbed monolayers on surfaces, (III) water-air interfaces and (IV) other surfaces like dehydroxilated silica. Hydrophobic surfaces, which are nonpolar, tend to adhere to each other due to the lowering of the total free energy by reducing the solid-water interfacial area. They are not able to form polar, ionic or hydrogen bonds with water molecules. All investigations have in common a significant stronger interaction than predicted from LIFSHITZ-VAN DER WAALS interactions. Difficulties might occur because measuring in water is prone to pick up impurities such as excess surfactants, small bubbles or polymers and therefore there must be great attention during preparation and measurement [12].

The origin of hydrophobic forces is still under concept [13-15]. Although it is more a surface force than a body force, it is unclear whether it is a solvation, a long-ranged electrostatic, a VAN DER WAALS interaction or something else. Promising approaches divide hydrophobic forces into two parts, a short- and a long-range attractive interaction. The true hydrophobic, short-range interaction (1-2 nm) is caused by a structuring of water molecules due to the strong nonpolar properties of the hydrophobic surfaces. Here, the dipoles are re-orientated parallel to the surface, because it is energetically more favourable to interact with themselves than with the surface. In consequence, a water depletion layer is generated. The long-range part of the force that can act over several tens and hundreds of nm is also subject of discussion. In the last years, the theory of small bridging bubbles has become established [16-18]: Such nanobubbles are formed during immersion with liquid on poor wetted rough surfaces, after ethanol-water exchange or after a temperature increase of liquid or substrate (lowered gas solubility) or by perturbation during contact. Widely investigated on very smooth surfaces, there are only a few studies about nanobubbles on rough surfaces. In nearly all published work there exist only a small number of force spectroscopic data (less than 20 force distance curves and calculated adhesion forces/energies etc. per publication), although there is so much diversity of surface properties due to physical and chemical heterogeneity. Our research group has introduced normalized force and energy distributions for overall adhesion forces on highly rough hydrophobic surfaces and split the data into distributions with and without capillary interactions.

Contact angle measurements (sessile drop) were done using the G10 from Kruess. The atomic force microscope XE-100 from Park Systems is used for all topographical and force spectroscopy investigations. In this study we focus on technical rough ceramic samples, an example of a sample and colloidal particle is shown in Figure 1.

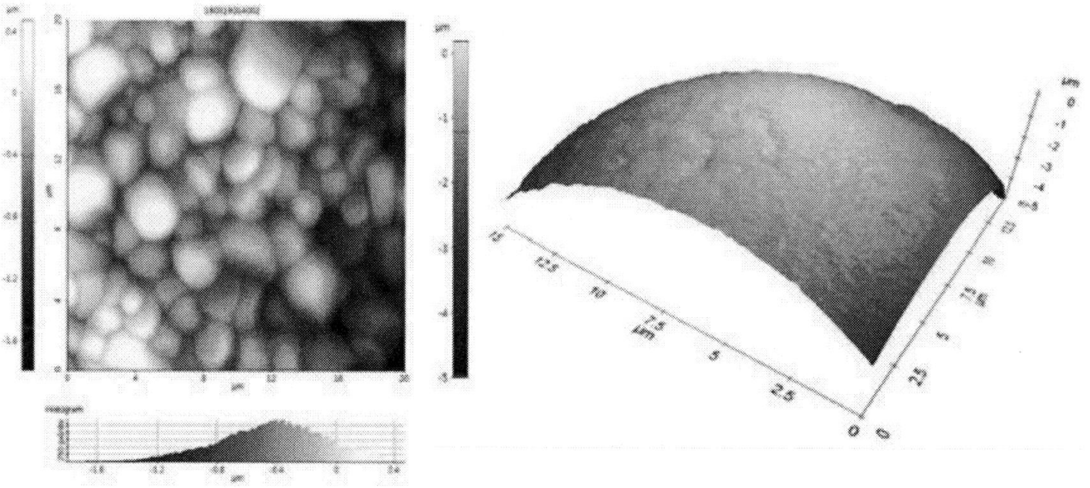

Figure 1: example of a rough ceramic surface (left) and a colloidal ceramic probe particle (right)

All measurements were carried out in a water-based model system. Ceramic samples and particle were coated with the silane Dynasilan F8261 of Evonik Industries Germany as it is described in [19]. The silanization process leads to an apparent water contact angle of 104° on the sample, respectively the colloidal particle. This is because of a decrease of surface energy. By the addition of pure ethanol to tap-filtered water, the wetting can be improved and adhesion in good and poor wetted systems can be compared. The alumina particles with sizes between 30-40 µm were glued on tipless cantilevers using a water-insoluble epoxy resin. Cantilevers that are used in poor wetted systems should be stiffer due to the larger adhesive force observed; for total hydrophobic systems AlOAl-TL B cantilevers from Budget Sensors were used (calibrated spring constants lie between 1.55-4.50 N/m) in our study. For small adhesive interactions, soft AlOAl-TL A cantilevers were used with spring constants between 0.15-0.30 N/m. For each sample 64 force distance curves on a 15 x 15 µm² area on at least 5 different parts of the substrate surface were measured to get statistical valid results. A MatLAB® routine recalculates the raw data (e.g. baseline correction etc.) and determines the adhesion force that is the minimum of the retrace for each force distance curve. These pulls off forces are normalized on the particle diameter because of the linear relationship between adhesion force and particle size. Topographical scans are done using a ContAl-G cantilever from Budget Sensors with a spring constant of 0.20 N/m.

In Figure 3 the impact of wetting by different ethanol-water solutions on a hydrophobic ceramic sample is shown. With larger amount of ethanol, the surface tension of the mixture is decreased as well as the contact angle of the mixture onto the ceramic surface (Figure 3, left). With decreased contact angle, respectively better wetted surface, the probability of capillary interaction is lowered. During immersion of the rough hydrophobic surface with water, nanobubbles are generated inside of the pores. Due to an improved wettability, the nanobubble amount is decreased. Furthermore, ethanol has higher gas solubility than water and air inside of bubbles might be dissolved in the mixture. Measured adhesion energies, determined from force distance curves, are much stronger if nanobubbles act as capillary bridges between the surfaces (Figure 3, right). Together with a decreased bubble amount at higher ethanol saturation, a size reduction of the bubbles occurs, because measured adhesion forces/energies are reduced. Also for non-capillary interactions, the attraction between the surfaces is reduced with a higher amount of ethanol in the liquid mixture that corresponds to theoretical approaches.

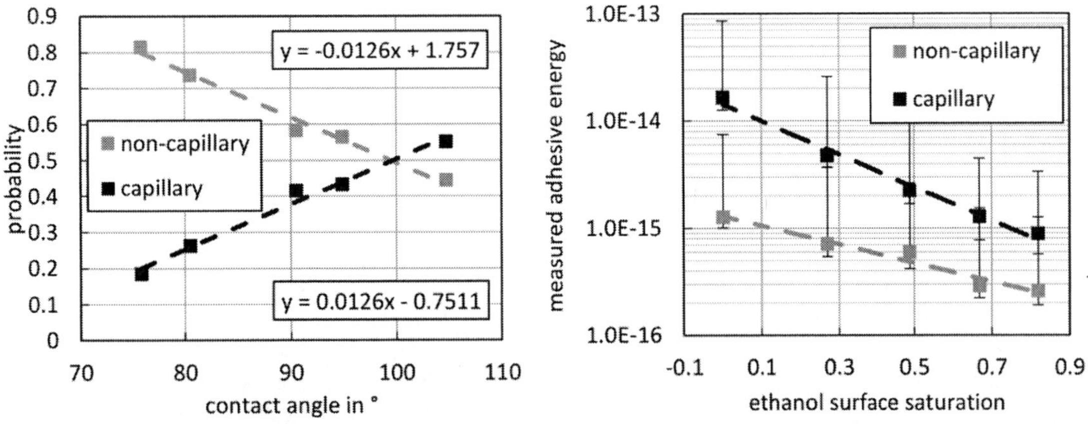

Figure 2: probability of measured capillary, respectively non-capillary interactions depending on the contact angle of water-ethanol solutions on a hydrophobic ceramic sample (left) and corresponding measured adhesive energies (right). Shown adhesive energies are median values

In Figure 3 force distributions for ceramic surfaces with different wetting behaviour in pure water are shown. Clearly visible is an increase in adhesion if wetting is decreased.

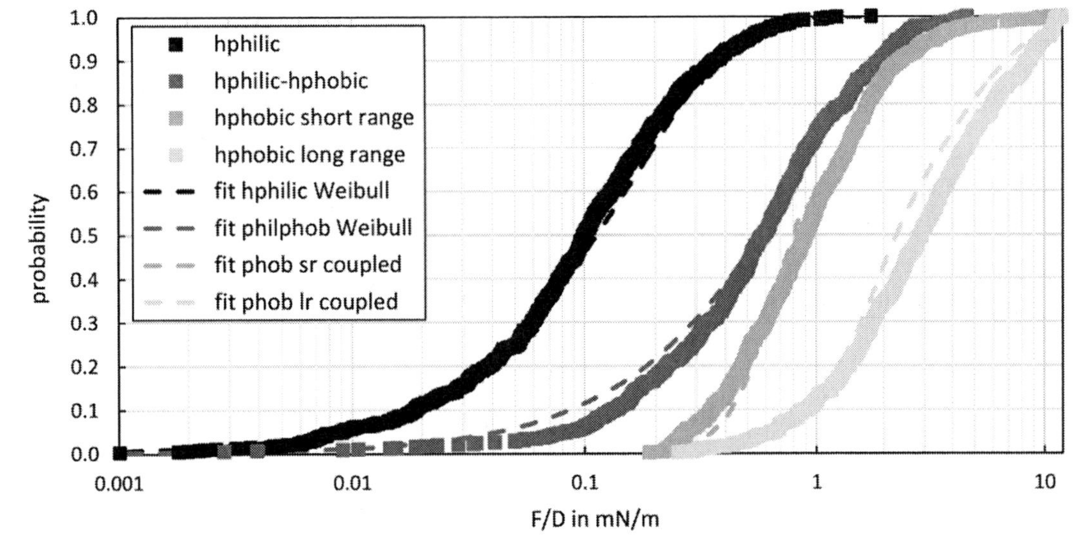

Figure 3: normalized force distributions of ceramic substrates under variation of wetting due to changed surface energies of the sample, respectively colloidal probe

For two hydrophilic surfaces, the distribution of VAN DER WAALS interactions is determined by the changing interaction area due to variations in surface roughness. No nanobubbles are detected on the surface (Figure 4, left). Because of the high roughness of the sample (rms = 0.2 µm), existing theoretical approaches that describe van der Waals interactions on rough surfaces are not applicable. Here, a Weibull distribution fits the data very well. Another Weibull distribution is used to fit the data for the hydrophilic-hydrophobic system. Here, the particle is hydrophobic, that is hydrophobic interactions occur on the particle surface and increase the overall adhesion. For two hydrophobic surfaces, the data is split into a distribution with and without capillary interactions. Again, clearly visible is an increase in adhesion if capillary interactions take place. For fitting the data, a coupling of Weibull and lognormal distribution is needed, because at small measured adhesive forces the contact area of the surface determines the interactions (Weibull) and large attraction is caused by lognormal distributed nanobubbles. A scan of the hydrophobic surface clearly shows the existence of such bubbles (Figure 4, right).

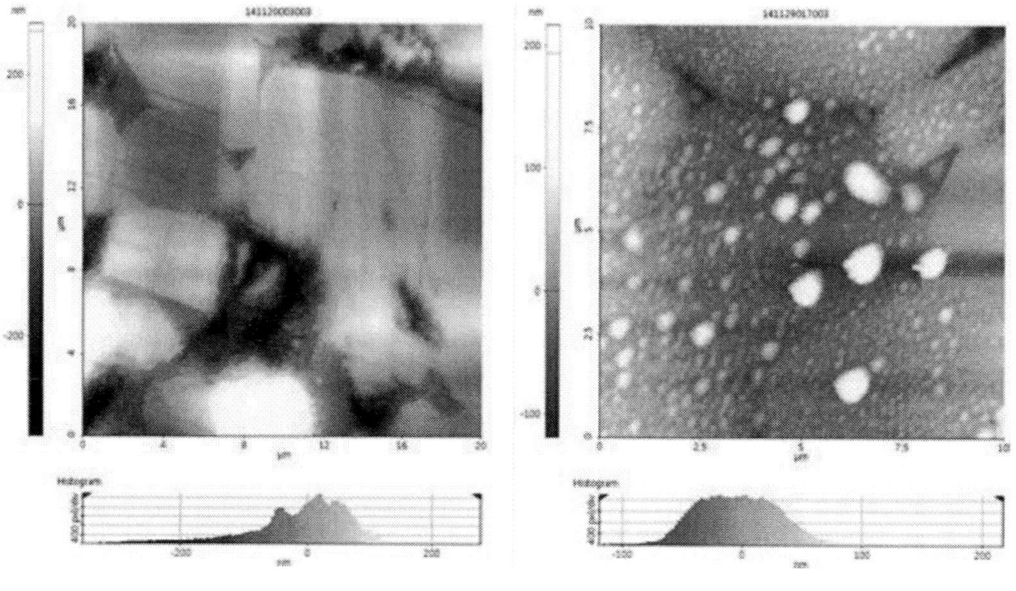

Figure 4: intermittent mode scan of a hydrophilic surface without nanobubbles (left) and a hydrophobic surface with nanobubbles (right)

In this study, the impact of wetting on adhesion and capillary interactions is shown. Improved wetting results in smaller adhesion forces and energies due to a smaller number of capillary interactions. Capillary interactions are caused by small bridging bubbles, so-called nanobubbles.

The authors would like to thank the German Research Foundation (DFG) for supporting the studies in subproject B01 which is part of the Collaborative Research Center CRC 920.

[1] D. Tabor and R. H. S. Winterton, "DIRECT MEASUREMENT OF NORMAL AND RETARDED VAN DER WAALS FORCES," vol. 312, pp. 435-50 plate, 1969.
[2] G. Binnig, C. F. Quate, and C. Gerber, "Atomic force microscope," *Physical Review Letters,* vol. 56, pp. 930-933, 1986.
[3] H. K. Christenson and P. M. Claesson, "Direct measurements of the force between hydrophobic surfaces in water," *Advances in Colloid and Interface Science,* vol. 91, pp. 391-436, 2001.
[4] J. Israelachvili and R. Pashley, "The hydrophobic interaction is long range, decaying exponentially with distance," *Nature,* vol. 300, pp. 341-342, 1982.
[5] P. Attard, "Nanobubbles and the hydrophobic attraction," *Advances in Colloid and Interface Science,* vol. 104, pp. 75-91, 2003.
[6] T. D. Blake and J. A. Kitchener, "Stability of aqueous films on hydrophobic methylated silica," *Journal of the Chemical Society, Faraday Transactions 1: Physical Chemistry in Condensed Phases,* vol. 68, pp. 1435-1442, 1972.
[7] A. Carambassis, L. C. Jonker, P. Attard, and M. W. Rutland, "Forces measured between hydrophobic surfaces due to a submicroscopic bridging bubble," *Physical Review Letters,* vol. 80, pp. 5357-5360, 1998.
[8] Y. Wang and B. Bhushan, "Boundary slip and nanobubble study in micro/nanofluidics using atomic force microscopy," *Soft Matter,* vol. 6, pp. 29-66, 2009.
[9] X. Cui, C. Shi, L. Xie, J. Liu, and H. Zeng, "Probing Interactions between Air Bubble and Hydrophobic Polymer Surface: Impact of Solution Salinity and Interfacial Nanobubbles," *Langmuir,* 2016/06/13 2016.
[10] W. A. Ducker, Z. Xu, and J. N. Isrealachvili, "Measurements of hydrophobic and DLVO forces in bubble-surface interactions in aqueous solutions," *Langmuir,* vol. 10, pp. 3279-3289, 1994.
[11] M. L. Fielden, R. A. Hayes, and J. Ralston, "Surface and capillary forces affecting air bubble-particle interactions in aqueous electrolyte," *Langmuir,* vol. 12, pp. 3721-3727, 1996.
[12] J. N. Israelachvili, *Intermolecular and Surface Forces*, 3 ed. Oxford: Elsevier LTD., 2011.
[13] E. Ruckenstein and N. Churaev, "A possible hydrodynamic origin of the forces of hydrophobic attraction," *Journal of Colloid And Interface Science,* vol. 147, pp. 535-538, 1991.
[14] Y. I. Rabinovich and B. V. Deryagin, "DIRECT MEASUREMENT OF FORCES OF ATTRACTION OF HYDROPHOBIZED QUARTZ FIBERS IN AQUEOUS KCl SOLUTIONS," *Colloid journal of the USSR,* vol. 49, pp. 598-602, 1987.
[15] J. L. Parker, P. M. Claesson, and P. Attard, "Bubbles, cavities, and the long-ranged attraction between hydrophobic surfaces," *Journal of Physical Chemistry®,* vol. 98, pp. 8468-8480, 1994.
[16] M. A. Hampton and A. V. Nguyen, "Nanobubbles and the nanobubble bridging capillary force," *Advances in Colloid and Interface Science,* vol. 154, pp. 30-55, 2010.
[17] D. Lohse and X. Zhang, "Surface nanobubbles and nanodroplets," *Reviews of Modern Physics,* vol. 87, 2015.
[18] J. R. T. Seddon and D. Lohse, "Nanobubbles and micropancakes: Gaseous domains on immersed substrates," *Journal of Physics Condensed Matter,* vol. 23, 2011.
[19] J. Fritzsche, J. Teichmann, F. Heuzeroth, K. G. Van Den Boogaart, and U. A. Peuker, "Impact of wetting to the agglomeration of dispersed particles in an aqueous medium," *Advanced Engineering Materials,* vol. 15, pp. 1299-1306, 2013.

508578 Flow Optimization in Cohesive Powder Blends: Identifying and Modeling Impact of Shape and Size on Observed Cohesivity Minimum

Abstract:

Through analysis of particle morphology, particle size, and compositional influences, we present experimental case studies revealing unexpected transitions in flowability and cohesion of two binary powder blend systems containing components with elongated (needle-like) and more spherical particle morphologies. We explore interactions between the needle-like and the more spherical components in the blend to explain these transitions, and optimal concentrations to improve flowability are identified. We observe, for cohesive binary mixtures, a local minimum phenomenon, in which a blend with a small amount of needle-like morphologies exhibits improved flowability compared to the single component system. A surface-area based model considering particle geometry, surface coverage, and blend composition is implemented to understand and predict trends in flowability. Discrete Element Method (DEM) modeling is employed to study the structure and particle interactions in simulated binary powder blends containing spherical and elongated particle morphologies. These results should translate both to the improved understanding of mixed component morphology systems and to a novel approach towards pharmaceutical product formulation optimization.

508167 Influence of Droplet Size on Particle-Particle Adhesion of Colliding Particles through Droplet: Directnumerical Simulation Study

Abstract:

In a spray wet granulation, a particle-particle adhesion by a liquid bridge, which shows dynamic motion due to moving particles, is the most fundamental phenomenon at the primary particle scale. Therefore, understanding of the particle-particle adhesion by such a dynamic liquid bridge is very important to elucidate mechanisms of particle agglomeration phenomenon in the wet granulation. In this study, the particle-particle adhesion phenomenon at the individual particle scale was analyzed using a direct numerical simulation. Collision of two particles mediated by binder droplets on a particle surface was simulated. In particular, the effect of droplet size under constant total liquid volume on adhesiveness of two colliding particles was investigated. In the present simulation, multiple liquid bridges were simultaneously formed, and these liquid bridges coalesced into single liquid bridge. The simulation results exhibited that the adhesiveness of particles increased with a decrease in the droplet diameter under a constant total liquid volume. In an initial stage of the particle growth in a fluidized bed spray granulation, the tendency of the calculations was consistent with experimental results. We revealed that the capillary pressure force and shapes of the liquid bridge are key factors for particle-particle adhesion at different droplet sizes under constant total liquid volume.

507221 Influence of Filter Cake Thickness during Backwash Regeneration

Abstract:

In industrial solid/liquid separation, the removal of fine particles (≤ 10 µm) in a suspension with a low solids content (0.01-1 v/v %) is a challenging and demanding task. This separation occurs in a wide array of industrial production applications such as metal processing and treatment industries. In order to achieve the required purity, different types of filters, such as candle and leaf filters, can be used. The selected filter will be used (semi-)discontinuously and the actual "filtration" process step is just as important as the subsequent "regeneration" of the filter cloth. In case of an improperly regenerated filter, the next filtration process becomes shorter until a subsequent regeneration is required. This behaviour is presented in the well-known filtrate flow – time – plot in figure 1. In the case of a properly regenerated filter (a), the filtrate flow at the beginning of each step is still equal whereas the initial filtrate flow decreases in the case of an improperly regenerated filter (b).

With respect to the field of applications, the removal procedure has revealed that there is potential for improvement. A properly regenerated filter cloth highly depends on the interaction between the particle system and the filter cloth (weave type, thread thickness, material). In order to guarantee the removal of the filter cake, the applied removal force has to be larger than the adhesive forces. Furthermore, to remove the filter cake in large fragments, the removal force has to be lower than the cohesive force. These two effects also depend on the porosity of the filter, pH-value, and further properties.

Currently, the boundary conditions for a properly regenerated filter cloth has not been investigated effectively. Hence, further research must be carried out to improve the backwash treatment with respect to the process conditions in an economic and process-technical context.

This presentation shows the influence of cake thickness on the discharge during backwash filtration. For this, different types of cake discharge will be described. Furthermore, the project and experimental facility will be presented.

Figure 1: Filtrate-Time-Diagrams and schematic representation of process management with properly regenerated filter cloth (a) and improperly regenerated filter cloth (b). In case of the properly regenerated filter, the filtrate flow at the beginning of each filtration step is equal. In case of an improperly regenerated filter, the filtrate flow decreases as a result of the remaining filter cake fragments on the filter cloth surface.

507883 Understanding Adhesive Mixing Via Energy-Based Stick/Bounce Model

Abstract:

An energy-based stick/bounce model is presented to explain relationships between particle interactions dominated by *Van-der-Waals* force during mixing of fine and coarse particles at different mixing intensities. The investigation concerns examination of the adhesive mixing efficiency (Kq) based on relative magnitudes of *Van-der-Waals* energy, dissipated energy, de-agglomeration energy and relative kinetic energy during impacts. $Kq>1$ indicates good adhesive mixing efficiency, while $Kq<1$ indicates poor adhesive mixing efficiency. Effect of size, material properties and mixing intensity of fine and coarse particles are investigated to attain a deep understanding of the mechanisms of adhesive mixing. Most interestingly, larger coarse particle size leads to larger agglomerate size of fine particles. Larger fine particle size and high intensity both lead to poor adhesive mixing efficiency. Size of fine particle agglomerates was also considered in the model. De-agglomeration energy was found to be highest amongst other energy terms during processing, indicating the de-agglomeration step was the most energy consuming step of adhesive mixing. The experimental results using a vibrational mixer and two sizes of fine (silica and cornstarch) and several sizes of coarse particles (KCl and cornstarch) validate the modeling as it explains various mechanisms. The guest silica particles tended to attach to the host KCl particles with a good adhesive mixing efficiency, while the guest cornstarch particles tended to detach from the host KCl particles beyond certain mixing intensity. Also, larger KCl particles would lead to the larger agglomerate size of cornstarch, indicating the poorer mixing efficiency. Finally, higher intensity leads to the worse mixing efficiency.

508289 CFD-DEM Modelling of Dilute-Phase Pneumatic Transport of Cohesive Dairy Powder in a Stainless Steel Pipe

Abstract:

The design and operation of efficient pneumatic transport systems for cohesive dairy powders depend on the type of dairy powder, its properties, and the conveying parameters of the fluid phase. In this study, a CFD-DEM model is used to investigate the effects of powder properties and conveying parameters on the dilute-phase pneumatic transport of cohesive milk powders in a 50mm diameter stainless steel pipe with horizontal and vertical sections connected by a 90^{0} bend of 200 mm radius.

The effects of milk powder properties (i.e. particle size, shape, density, cohesion, etc.) and/or conveying parameters (i.e. solid loading ratio, air velocity, bend configuration and orientation, etc.) on the flowability of the milk powder were investigated through the CFD-DEM model.

A thorough analysis of the effects of the various DEM parameters on particle transport was performed. These parameters (such as the surface energy of the milk powder) model the degree of cohesiveness of our dairy powders.

By modelling the particle-wall interactions, the effects of particle-geometry adhesion phenomena (i.e. particle deposition, resuspension, agglomeration, and clogging/blockage) can be used to analyse the critical conditions that must be satisfied to avoid complete pipe blockage during transport of cohesive powders.

DEM Simulation of Wet Particulate Flow over a Circular Cylinder

Hongsheng Chen, Ran Tao, Huang Zhang, and Shuiqing Li*

Key Laboratory for Thermal Science and Power Engineering of Ministry of Education, Department of Energy and Power Engineering, Tsinghua University, Beijing 100084, P. R. China

Abstract: Flow of micron-sized particle in humid system have received extensive attentions over past decades in energy, environment, chemical and pharmaceutical processes. In this paper, the three dimensional simulations of both deposition and aggregation of wet fine particles over a circular cylinder are performed by means of the adhesive discrete element method (adhesive DEM), taking into account the effects of sliding, rolling and twisting between particles. The particles involved in this work are of particle size ranging from 10 to 100 μm, and the capillary force introduced by the surface tension and the lubrication effect introduced by the viscosity of the interstitial liquid between adjacent particles are considered, with which the morphology, collision probability and capture efficiency of wet fine particles on the circular fiber are further studied in comparison with that of adhesive dry particles, which is described by the Johnson-Kendall-Roberts (JKR) model with the short-range van der Waals adhesion. Moreover, the dimensionless adhesion parameter (Ad) proposed by Li and Marshall [1-3] and the dimensionless Weber number are extended to interpret the filtration mechanism of fine particles in humid circumstances.

Keywords: fine particles, fibrous filtration, liquid bridge, discrete element method

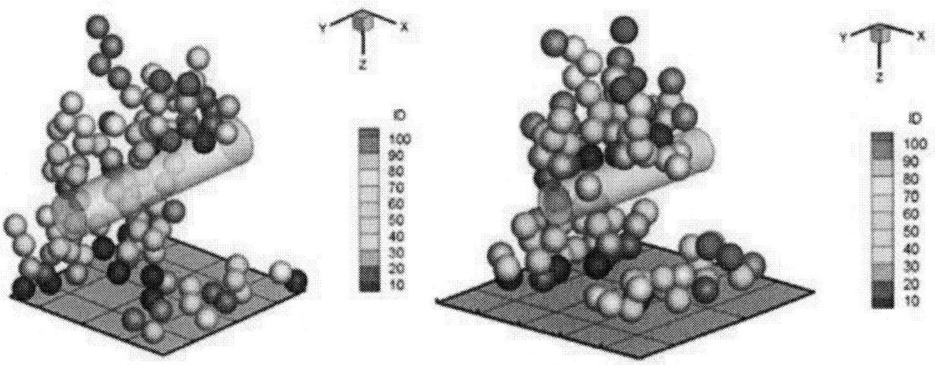

Fig. 1 Particle capture by a circular fiber, (a) wet fine particles using nonlinear spring-dashpot model (Hertz theory) combined with liquid bridge forces, and (b) adhesive dry particles using JKR model.

References

[1] S. Li, J.S. Marshall, Discrete element simulation of micro-particle deposition on a cylindrical fiber in an array, Aerosol Science, 38 (2007) 1031-1046.

[2] S. Li, J.S. Marshall, G. Liu, Q. Yao, Adhesive particulate flow: The discrete-element method and its application in energy and environmental engineering, Progress in Energy and Combustion Science, 37 (2011) 633-668.

[3] J. S. Marshell, S. Li, Adhesive particle flow: A discrete-element approach Cambridge University Press, New York, 2014.

509464 Review of Contact Models for Cohesive-Frictional Materials and Applications

Abstract:

The interaction between visco-elasto-plastic and adhesive particles is the subject of this review, where also â€œmeso-particlesâ€ are introduced, i.e., simplified particles, whose contact mechanics is not taken into account in all details. A few examples of meso-particles include agglomerates or groups of primary particles, or inhomogeneous particles with micro-structures of the scale of the contact deformation, such as core-shell materials.

A simple, flexible contact model for meso-particles is proposed, which allows to model the bulk behavior of assemblies of many particles in both rapid and slow, quasi-static flow situations. An attempt is made to categorize existing contact models for the normal force, discuss all the essential mechanical ingredients that must enter the model (qualitatively) and finally solve it analytically. The model combines a short-ranged, non-contact part (resembling either dry or wet materials) with an elaborate, visco-elasto-plastic and adhesive contact law. Using energy conservation arguments, an analytical expression for the coefficient of restitution is derived in terms of the impact velocity (for pair interactions or, equivalently, without loss of generality, for quasi-static situations in terms of the maximum overlap or confining stress).

Adhesive particles (or meso-particles) stick to each other at very low impact velocity, while they rebound less dissipatively with increasing velocity, in agree- ment with previous studies. For even higher impact velocities an interesting sec- ond sticking and rebound regime is reported. The low velocity sticking is due to non-contact adhesive forces, the first rebound regime is due to stronger elastic and kinetic energies with little dissipation, while the high velocity sticking is generated by the non-linearly increasing, history dependent plastic dissipation and adhesive contact force. As the model allows also for a stiff, more elastic core material, this causes the second rebound regime at even higher velocities.

506835 High Speed Video Imaging of Jet Cup Attrition Hydrodynamics

Abstract:

High speed video imaging has been performed to better understand the particle hydrodynamics and kinetic energy distribution of Geldart Group A and B particles in a conical jet cup [4]. Particle-particle and particle-wall interactions lead to the appearance of particle attrition. Particle tracking velocimetry (PTV) was used to measure the particle velocity and change in the particle velocity before, during, and after collisions. By measuring the kinematics of particle-wall interactions and how particles are recycled back into the jet, a mechanistic attrition model was developed. DEM simulations were used to bridge the gap between 2D and 3D hydrodynamics. Once this mechanistic model is in place, quantitative comparisons between jet cup results and commercial fluidized bed cyclone operations can be made.

507371 The Stress, Orientation, Displacement Deviation, and Solid Fraction Predictions from a Modified Attrition Cell Containing Needle-Shaped Particles

Abstract:

A discrete element model (DEM) is used to predict the stress distribution, the orientation distribution, the displacement deviation, and the solid fraction in a modified attrition cell containing needle-shaped particles. The effects of particle aspect ratio, particle-particle friction coefficient, agitation blade height, pressure on the top plate, and particle-wall friction coefficient were investigated in this study. Prior studies have shown that particle breakage in an attrition cell mainly occurs at the base of the cell and in front of the blades near the cell circumference. Therefore, instead of simulating the entire attrition cell, a smaller control volume in the form of a periodic box containing an agitation blade is used in this study. Average particle stresses increase with an increase in particle aspect ratio, particle-particle friction coefficient, pressure on the top plate, and agitation blade height, indicating a higher probability of particle breakage. Particles orientation distributions indicate that as particle aspect ratio increases, particles orient such that their major axes are nearly aligned with the flow streamlines. Time-averaged displacement deviation, which is a measure of particle mixing, increases with increasing agitation blade height and decreases with increasing particle-particle friction coefficient. Increasing particle aspect ratio increases displacement deviation for smaller aspect ratios but displacement deviation becomes approximately constant for larger aspect ratios. Other factors such as, particle-wall friction coefficient and top plate pressure do not have any significant effect on either particle orientation or displacement deviation.

507800 Attrition Prediction of Grid Jets in Fluidized Bed Systems

Abstract:

The development of novel carbon capture and commodity chemical production implementing fluidized bed and circulating fluidized bed (CFB) systems has garnered much attention in recent years. These emerging technologies include chemical looping combustion (CLC), continuous temperature swing adsorption, and transport gasifiers. However, a major uncertainty for determining the economic feasibility and system operations of these technologies is the attrition of the solid bed materials. Attrition in these systems is caused by the mechanical attrition of the particulate material, undergoing high-velocity impacts with walls and with other particles, and low-velocity wear of the surface of particles, . While mechanical attrition can occur in a variety of areas in CFB's such as cyclones, L-valves, impactors, standpipes, risers, etc., the fluidized bed has shown to be an area where attrition is severe. A fluidized bed has two regions where attrition occurs, the jetting region and bubbling region. The bubbling region tends to cause particles to rub together and attrit through a wearing mechanism, whereas the jetting region entrains particles in a high velocity jet and impacts them against a bed of particles at the top of the jet. Experimental data and models predict the jetting region to have a much more severe impact of attrition than the bubbling region.

A common grid-jet attrition model in the literature is that by Werther and Xi, where they show that the attrition rate is a function of orifice velocity, orifice diameter, gas density, and an unknown proportionality constant based on material properties. The issue in using this model is that it requires extensive experimental testing to obtain this constant. The goal of this work is to determine the value of this constant based on fundamental material properties by developing an attrition model for the jetting region in the fluidized bed based on Ghadiri's impact model (Eq. 1):

where α is a proportionality constant, ρ_p is particle density, U_p is particle velocity, d_p
particle, H is hardness, and K_c is the fracture toughness of the material. Since Eq.1 only considers the fractional impact attrition of a single particle, a term accounting for the mass of particles entrained by the jet must also be considered. This paper presents the model development and compares predictions to data from the literature.

507873 "Perfect Particles" - 3D Printing of Tuneable Agglomerates for Validation of DEM Breakage Models

Abstract:

One of the long term barriers to advanced and accurate DEM modelling of agglomerate breakage is the lack of a suitable set of test particles that can be used to validate DEM models. Generally the approach has been to take a specific, simplified particle system, measure the relevant properties as accurately as possible, and compare and fit the model to experimental measurements. There main limitation of this is approach is that agglomerate breakage is a destructive experimental tests, and so the experiment represents the date from a single individual agglomerate in a single test condition. Ideally, what we want is the ability to create hundreds of identical agglomerates, with systematic strength and structure variations, and to test each set of agglomerates under controlled variations. This has historically not been possible via experimental agglomeration production techniques.

3D printing offers the ability to design and produce test agglomerates with controlled properties to replicate simple or complex agglomerate structures, and to use these agglomerates to validate DEM models of agglomerate breakage (Ge et al, Powder Technology 2017). In this work, agglomerates with well-defined structure (tetrahedral and random structures, with varied porosity) and tuneable bond strength properties were designed in CAD and EDEM. A Stratasys Object500 polyjet printer was used to 3D print multiple, identical copies of each agglomerate design, and quasi-static breakage tests were then conducted to vary the strain rate and examine the breakage behaviour. The macroscopic agglomerate breakage experiments were compared with DEM simulations, which used Timoshenko Beam Bond Model (TBBM) contact model recently implemented by Brown et al (Granular Matter, 2014).

Qualitative comparisons of the failure mode between experimental tests and DEM simulations (using TBBM bond model) were successful. The quantitative load-displacement results shown that DEM simulation was an excellent match for the cubic tetrahedral agglomerate structure. However, DEM underestimated compressive loads for spherical random agglomerates. However, the results show that the approach is a significant step forward in the ability to "calibrate" and "validate" DEM models of agglomerate breakage.

508432 Particle Attrition: Mechanisms and Jet Cup Attrition Method

Abstract:

Particle attrition can be categorized into the two mechanisms of abrasion and fragmentation. In the abrasion attrition mechanism, particles are attriting off the surface of the parent particles. The resulting size distribution of parent particles shows almost no change in cut sizes, showing only a peak of finer constituent particles under this attrition mechanism. With the fragmentation attrition mechanism, the parent particles are breaking down to intermediate cut sizes as well as generating constituent particles. This results in a shift in the size distribution of parent particles in addition to producing finer constituent particles.

Several methods are available in characterizing attrition characteristics of catalyst particles. Jet cup attrition testing is a common method for ranking of relative attrition characteristics by comparing their attrition indices. PSRI jet cup attrition testing is a two-step process. The first step in jet cup attrition testing is to identify the threshold velocity and to determine the optimum jet velocity for evaluating particle attrition. The threshold velocity represents the gas jet velocity at which the transition of the attrition mechanism from abrasion to fragmentation takes place.

It is crucial to use a more reliable jet velocity in the jet cup attrition test since abrasion type attrition tends to be the dominant source of attrition for a well-designed process of fixed fluidized beds and circulating fluidized beds. PSRI jet cup data demonstrates how to determine the mechanisms of particle attrition purely by looking at the particle size distributions, and how the sizes of parent particles shift over the extended period for each relevant mechanism.

507133 Dispersion of Magnetite Nanoparticles in Brine Solutions Via Coating with Sulfonated Phenolic Resin

Abstract:

It has been proposed to use superparamagnetic magnetite nanoparticles (MNP) as contrast agents for electromagnetic surveys of underground oil reservoir. However, underground oil reservoir has high salinity condition that may lead to particle agglomeration, preventing long-distance propagation of magnetite particles through the subsurface porous rocks. One method for preventing agglomeration is to increase repulsive forces between particles by introducing polymers or polyelectrolytes onto the particle surfaces. In this study, a surface modification was conducted by coating magnetite nanoparticles with sulfonated phenolic resin (SPR). SPR has numerous sulfonic groups in its molecular structure that can induce highly negative charge to the surface of the magnetite particles. SPR was synthesized from 4-hydroxybenzenesulfonic acid and formaldehyde. Coating magnetite nanoparticles with was implemented by mixing the SPR solution with the iron precursor solution and then adding the sodium hydroxide solution that induced in-situ co-precipitation. In order to further enhance the stability of the particle the various types of polymer such as polyethylene glycol (PEG), polyvinyl alcohol (PVA) and polyacrylic acid (PAA) was also introduced. As a result, the PAA treated SPR coated MNP formed stable dispersion in highly salt solutions containing NaCl up to 10 wt.% for a wide range of pH.

509334 Dispersibility of Carbon Black As Function of Dispersing Intensity Using Hansen Dispersibility Parameters – Evaluation of Nonideal Mixing Effects Based on Normalized Relative Sedimentation Time

Abstract:

Dispersibility of agglomerated/aggregated, i.e. complex solids is of key interest for application in various dispersion-based formulations. To make an economic use of raw materials in suspension-based formulations, a high degree of dispersion, ideally down to the level of primary particles has to be obtained. This is related also to practical aspects like achieving high hiding power at lower pigment concentration and of outstanding material properties in composites. Dispersibility and the resulting quality of dispersion are determined by several factors including structure/quality of the raw materials and dispersion media, dispersion process and additives. Hansen Solubility Parameters (HSP) have gained growing interest for selection, optimization and prediction of suitable dispersion media, e.g. [1]. HSP concept builds on Hildebrandt solubility parameter (square root of cohesive energy between molecules, related to vaporization) but is subdivided into a disperse, a polar and a H-bond contribution to account for different kinds of interaction. Due to its root (cohesive energy), HSP describe thermodynamic (equilibrium) not kinetic properties. Extending HSP concept to dispersions of particulates in liquids has to depart from this solid basis, however, it has been proven to be widely? applicable in various fields [2]. This is why we suggest using the term Hansen Dispersibility Parameters (HDP) whenever the HSP concept is applied to dispersions and for characterization of surface properties. Extending the use of solubility parameters to dispersibility and dispersion stability, i.e. going from solutions to dispersions, a boundary is crossed. Instead of only interactions between molecules, interactions between molecules, between solvent molecules and particle surface, as well as interparticle forces have to be considered. While solutions are thermodynamically stable, dispersions (interface dominated) are inherently thermodynamically unstable systems. Experimental evaluation of dispersion stability of agglomerated/aggregated solids goes with the necessity of dispersing solids first hand, i.e. dispersion stability and dispersibility are usually studied in one experiment. Nevertheless, media for optimum dispersibility and optimum dispersion stability may be different. In case of complex solid particulates, intensity of the dispersion process must have an effect on HDP characteristic for the solid phase. It is often suggested [2] to use blends of 'green' solvents instead of harmful solvents to match optimal HDP values assuming ideal mixing. However, beside prevalent nonideality of mixing solvents, potential preferential adsorption of one solvent compound should be considered. Deviation of solvent composition in bulk and at interface should be rather the rule than the exception similar to preferential solvation in solvent mixtures [3].

The paper intends to investigate this complexity for dispersibility of carbon black with the focus on effect of the dispersion process. HDP were determined based on experimental sedimentation behavior by means of automatically monitoring the integral extinction (IE) over sample height by STEP-Technology [4] for carbon black dispersed in different individual liquids of the 3D-Hansen space (chemical mapping). A modified relative sedimentation (standard) time (RST) taking into account acceleration and optical path was introduced [5]. Higher RST values indicate higher dispersibility. Obviously RST values will be affected in case of fast agglomeration when particles are not properly stabilized by the solvent. Degree of dispersion processing intensity was varied between simple vortex mixing and sonication with varied input of energy. As predicted, best liquid(s) and HDP exhibit remarkable dependence on dispersion intensity. H-bond contribution reduces markedly in case of sonication with highest intensity. Noteworthy, liquids with H-bond acceptor properties proved to be especially effective in dispersion and dispersion stabilization.

To investigate and quantify nonideality for dispersion in binary liquid mixtures (experiments with low mixing intensity) it proved convenient to directly use RST values determined as function of mixing ratio. Synergistic and antagonistic effects were clearly identified and dispersibility improved.

Acknowledgements

The authors want to thank the funding of Deutsche Forschungsgemeinschaft (DFG) through the Cluster of Excellence "Engineering of Advanced Materials". Moreover, we acknowledge the Federal Ministry of Economic Affairs through the Arbeitsgemeinschaft industrieller Forschungsvereinigungen "Otto von Guericke" e.V. (AiF, project no. KF 2347922UW4).

References

[1] J. Jang, V.H. Pham, S.H. Hur, J.S. Chung, Dispersibility of reduced alkylamine-functionalized graphene oxides in organic solvents, J. Colloid Interface Sci. 424 (2014) 62–66.

[2] [2] S. Abbott, C.M. Hansen, H. Yamamoto, R.S. Valpey, Hansen Solubility Parameters in Practice, Complete with eBook, software and data, 4th Edition version 4.1, Hansen-Solubility.com, 2013, ISBN 978-0-9551220-2-6.

[3] Y. Marcus, Pharm. Anal. Acta, Preferential Solvation of Drugs in Binary Solvent Mixtures, 8 (2017) 537. doi:10.4172/2153-2435.1000537

[4] D. Lerche, S. Horvat, T. Sobisch, Efficient instrument based determination of the Hansen Solubility Parameters for talc-based pigment particles by multisample analytical centrifugation: Zero to One Scoring, Dispersion Letters (2015) 13–18.

[5] S. Süß, T. Sobisch, W. Peukert, D. Lerche, D. Segets, Determination of Hansen parameters for particles: a standardized routine based on analytical centrifugation, to be submitted.

508519 The Role of Surfactant Structures at the Metal/Electrolyte Interface for Corrosion Inhibition

Abstract:

The Role of Surfactant Structures at the Metal/Electrolyte Interface for Corrosion Inhibition

By

Akshay Rajopadhye, Anoop Nautiyal, Premal Shukla, and Brij Moudgil

NSF I/UCRC Center for Particulate and Surfactant Systems

University of Florida, Gainesville, Florida 32611

Abstract: The Center for Particulate and Surfactant Systems (CPaSS) at the University of Florida focuses on resolving issues that are important to energy, oil and gas and other industry sectors. The aim of this study is to understand the fundamental mechanism of corrosion inhibition of surfactant based inhibitors for oil transportation application.

Self-assembled surfactant structures formed on the surface of metal pipeline can impede the electrochemical reactions leading to the oxidation of steel (corrosion). We are using several characterization tools such as Atomic Force Microscopy, Electrochemical, and Spectroscopic techniques to examine the structure-property relationship of these surfactant films. The results of these investigations will enable us to gain deeper understanding of various factors, such as surfactant packing density, their molecular architecture, their mechanical strength, synergy with other additives etc., that govern the ultimate corrosion inhibition performance. We are developing correlations between the mechanical and electrochemical properties of surfactant films to improve their performance as corrosion inhibitors.

508566 Particulate and Surfactant Systems for Industrial Applications: Challenges and Opportunities

Abstract:

The Center for Particulate and Surfactant Systems (CPaSS â€" NSF Industry/University Cooperative Research Center) promotes sustainable use of surfactants and particulate systems. Center research focusses on generating the knowledge and technology platforms for industry to develop greener and more sustainable particle and surfactant applications. Current research projects include dispersion of high solids suspensions, corrosion inhibition, antiscaling agents, microbe removal from surfaces, engineered clays for enhanced performance in ceramics. Process efficiencies in some of these systems are determined by the nanostructures formed at the interfaces. For example, molecular structures on surfaces can impede the diffusion/electrochemical reactions that give rise to surface oxidation leading to corrosion, or create steric repulsion between particles leading to dispersion.

We are using atomic force microscopy, electrochemical impedance spectroscopy, and other spectroscopic techniques to understand the structure of surface films/coatings at the molecular/atomic scale. Experimental and, simulation and modeling techniques are being employed to gain deeper understanding of factors that govern molecular structures, their packing density, yield strength, elastic modulus, strength of adhesion, and synergistic/antagonistic behavior in mixed reagent systems.

In this presentation, select project developments will be discussed, and challenges and opportunities for developing greener and more sustainable processes and products will be outlined.

507254 Evaluating the Influence of Dry Granulation Processing Variables on the Rheological Properties of Granules

Abstract:

Granulation is often employed in the production of pharmaceutical formulations for oral solid dose delivery. It is most often required when constituent materials in a formulation have poor flow properties, enabling these components to be combined to improve process performance and ensure content uniformity.

Dry granulation, based around the use of a roller compactor and integral mill/screen, can have significant advantages over more common wet granulation techniques. Dry granulation can reduce processing costs and can also be used with moisture/solvent/heat sensitive materials. Unlike a wet granulate, the dry granulate does not require additional drying and milling before encapsulation or tabletting.

Dry granulation is not suitable for all types of material, however, there is little research into which properties make a formulation suitable for this process. In addition, there is little to indicate which process parameters produce optimal granulate quality in order to achieve interruption free processing and high quality final products.

In this study an FT4 Powder Rheometer® (Freeman Technology, Gloucestershire, UK) was used to demonstrate how processing parameters influence the properties of a placebo formulation granulate. The granulate was produced using a roller compactor (MiniPactor, Gerteis Maschinen+Processengineering AG, Switzerland) on which the roll gap and compaction force were varied. The dynamic flow, bulk and shear properties of the resulting granulates were then evaluated using the FT4 Powder Rheometer.

The data clearly demonstrates that several rheological properties, known to influence die filling efficiency and Critical Quality Attributes (CQAs) of the resulting tablets, could be controlled by changes in roller compactor settings.

This study emphasises the importance of assessing powder properties using techniques that closely simulate process conditions and how it is possible to identify Critical Process Parameters to optimise a dry granulation process and ensure high final product quality.

507489 Assessment of the Homogeneity of an Alumina Powders Mixture Using Rheological Parameters

Abstract:

This study has been carried out in order to improve the comprehension of powder mixing in the nuclear fuel manufacturing process. Thus, a binary mixture of two Alumina (Al_2O_3) powders, GE15 and CR6, was used to simulate the oxides powders actually involved in the process. Different mixing technics and parameters were tried and compared in terms of homogeneity of the mixtures produced. The concept of intensity of segregation [1] was used for assessing mixture homogeneity through a sampling protocol and classical statistical analysis tools [2]. One of the difficulties of working with these compounds is that, due to the similarities between CR6 and GE15 powders (same chemical composition, same colour, etc.), the direct measure of the composition of a given sample is not possible. Since homogeneity of a mixture is assessed from the variation of the composition between samples, it is important to find an indirect way to measure the composition of an Alumina powders mixture. Therefore, different correlations between measurable parameters (specific surface area, Sauter diameter, cohesion and flow function coefficient) and the composition of the mixtures were found empirically allowing indirect measurement of the composition of a given sample.

The four studied parameters: specific surface area (, measured by nitrogen adsorption using BET method), Sauter diameter (, measured by laser granulometry, dry transportation with air at 0.1 bar), cohesion and flow function coefficient (and , measured by FT4 shear test at 9 kPa on a 10 mL cell), were compared in terms of precision, relevance and feasibility.

It appears that the homogeneity assessment methods involving rheological parameters (cohesion and flow function coefficient) show many advantages compared to the other methods. Not only are they the most accurate correlations found but both parameters are measured simultaneously, meaning that it is possible to apply both methods on the same samples while all the tests (N_2 adsorption, laser granulometry and FT4) are destructive. Those two methods also allow the estimation of the homogeneity at a more interesting scale (around 7 g of powder), indeed, specific surface area and Sauter diameter methods give a homogeneity at a scale of less than one gram which is not relevant for the end-use properties of the final product.

[1] P. Danckwerts, The definition, measurement and some characteristics of mixtures, Applied scientific research. Section A, 3 (1952) 279-296

[2] S. Massol-Chaudeur, H. Berthiaux, S. Muerza, et J. Dodds, A numerical model to identify the structure of a high-dilution powder mixture, *Powder Technol.*, 128 (2002) 131-138

508516 Characterising the Grain Shape: In Search of Size Independent Shape Descriptors

Abstract:

Granular materials contain a large number of grains that differ in size and shape. While the characterisation of the size distribution is typically done using sieve analysis, the characterisation of the shape of the grains remains a difficult task and is one of the major bottlenecks in the characterisation of the granular material. In this study, we attempt to characterise the shape of three different types of grains commonly used in iron and steel making industry, namely pellet, sinter and iron ore lump. While the pellet particles can be considered nearly spherical in shape, imposing the spherical shape on the iron ore lumps and highly irregular sinter particles seems a very poor choice. We consider two different size ranges of pellet particles and four different size ranges for sinter and iron ore lumps. Imaging of the grains is done using an iPhone 7 Plus camera and sufficient number of grains of each type in all the size ranges are considered. We perform the image analysis and obtain various boundary parameters (such as projected area, minor and major axis, perimeter, etc.) of the grains using MATLAB image processing toolbox. In addition, we define two additional boundary parameters to characterise the deviation of the projected area from that of a circle and calculate their values for all the different type of grains considered in this study. Using these length scales, we define a total of nine different shape descriptors and obtain the distribution of these shape descriptors for each type of grain for all the size ranges. Interestingly, for a given type of grains, the frequency distribution curves for these nine shape descriptors turn out to be nearly independent of the size range of the grains. This suggests that the chosen shape descriptors indeed represent properties related to the shape of the grains alone, irrespective of the size of the grains considered.

We investigate the correlation among these shape descriptors by evaluating the values of the correlation coefficients between each pair of the shape descriptors using the Pearson method, Kendall method and Spearman method. Our results indicate that out of the nine shape descriptors considered in this study, five shape descriptors are strongly correlated with each other and only the remaining four shape descriptors need to be considered in characterising the shape of the grains. Out of these four uncorrelated shape descriptors, two shape descriptors, namely the aspect ratio of the grains and the root of form factor, have been considered important by researchers in previous studies. The two other shape descriptors proposed in this study, which are used to characterise the deviation of the projected grain shape from that of a circle, are not measured previously to the best of our knowledge. We compare these four shape descriptors for the pellets, sinter and iron ore lumps having the same sieve size range and show that the shape descriptors for these three different types of grains are different.

We next calculate the values of these four shape descriptors for relatively larger coke particles using 2d image analysis. We also perform 3d image analysis of some of these coke particles and calculate the values of the analogous shape descriptors in 3 dimensions. We investigate if the identified 2d shape descriptors (obtained from 2d imaging) can be correlated to the analogous 3d shape descriptors (obtained from 3d imaging) and explore if it is possible to use the 2d imaging technique to estimate the 3d shape descriptors, thus eliminating the need of 3d imaging.

508543 Investigating Mechanical Properties of Surfactant Films at the Solid-Liquid Interface Using AFM

Abstract:

Corrosion is a well-known problem in the oil and gas sector and also in many other industries. Surfactant based inhibitors have been used for corrosion control in oil pipelines but their inhibition mechanism is still not fully known. Our group is studying surfactant corrosion inhibition mechanism by various characterization techniques (AFM, Electrochemical and Spectroscopic techniques). A better understanding of corrosion inhibition mechanism will lead to the design and development of not only cost-effective but also greener surfactants system for corrosion inhibition purposes.

Atomic force microscopy (AFM) is an effective tool to study the barrier properties of surfactant films at the solid-liquid interface. Determination of mechanical and topographical properties of single and mixed surfactant coatings by atomic force microscopy (AFM) can delineate the role of physical vs. electrochemical barrier by such films in the corrosion inhibition process. . So far, the AFM data, obtained using a closed cell, indicate the effect of surfactant concentration and that of hydrocarbon chains interactions on the mechanical properties of cationic surfactant films. Ultimately, integration of the AFM and electrochemical results will reveal the correlation between the surfactant structures at the solid-liquid interface and their corrosion inhibition efficacy.

508545 Deposition Rate Consequences of the Formation of Multi-Spherule Cluster Aggregates in Gases "Role of Momentum Shielding™"

Abstract:

For engineering flow systems characterized by particulate suspensions in gases, will the formation of *cluster aggregates* (CAs) in the mainstream accelerate deposition rates on immersed targets or confining walls ---or is aggregate formation more likely to suppress capture rates? Recent research on the drag properties of large cluster aggregates (e.g., stimulated in part by the Sorensen (2011) review) has opened the door to answering these important mechanism- (and pressure-) dependent questions. Sub-100 nm "primary" spherules, which may find themselves in even tenuous 'fractal-like' N-spherule clusters, can take advantage of what we call "momentum shielding"---analogous to the drag advantages experienced by birds who elect to fly in formation. We show here that the extent of this "momentum shielding" can be conveniently quantified (in a dimensionless function we call: $S_{mom}(N;Kn_1;\text{aggregate structure})$), and used to predict the deposition-rate consequences of aggregation in aerosol flow systems when the cluster deposition mechanism is dominated by either: i) isothermal convective diffusion (C-D), ii) thermo-phoresis (TP-) or iii) inertial impaction (II). As expected, we show that for any type of aggregate structure (including 'fractal-like'), momentum shielding increases with cluster size (e.g. N) and gas pressure (via the reciprocal Knudsen number based on spherule radius R_1). We also show how S_{mom} can be calculated for the limiting case of a N>>1 random close-packed-D_f=3 cluster ---which is nearly gas impermeable. For fractal-like aggregates, including DLCAs (D_f=1.8) and RLCAs (D_f=2.1) we have recently shown that the abovementioned function: $S_{mom}(N, Kn_1; D_f)$ is actually sufficient to make useful engineering estimates of CA Brownian diffusion coefficients (Rosner and Tandon (2017b), thermophoretic diffusivities (Rosner and Tandon (2017a) and aggregate "stopping times" (Rosner and Tandon (2017c)----each in terms of their previously known values (D_1, $\alpha_T D_1$, and $t_{p,1}$) for isolated individual spherules in the same environment.

This information now enables instructive engineering calculations of the deposition-rate consequences of spherule aggregation to form large cluster aggregates-----i.e. at the same mainstream spherule volume fraction we can now determine how much aggregation will increase or decrease the total spherule acquisition rate on an immersed target. We present here representative results for each of the abovementioned particle transport mechanisms----i.e.: i) isothermal convective-diffusion (C-D), ii) thermo-phoresis (TP-) or iii) inertial impaction (II)----- including a brief account of the consequences of dealing with Kn1-dependent coagulation-'aged' aggregate size distributions (ASDs) in the mainsteam. We summarize our current re-examination of the canonical problem of inertial impaction on a circular cylinder in Re>>1 crossflow (Rosner and Tandon (2017c), dealing now with CA-"projectiles" but in the absence of spherule loss from the much larger target. We conclude by briefly outlining our planned work on the efficient "inversion" of sampled cluster-ASD-data to reconstruct, using the abovementioned $S_{mom}(N,Kn_1;D_f$ information, what ASD must have existed in the mainstream---at least in the absence of appreciable supercritical aggregate break-up.

References:

1. Rosner, D. E. and P. Tandon (2017a). Knudsen Transition Effects on the Thermophoretic Properties of Fractal-like Aggregates; Implications for Thermophoretic Sampling of High-Pressure Flames, Aerosol Science and Technology (in press).

2. Rosner, D. E. and P. Tandon (2017b). Aggregation- and Knudsen Transition Rarefaction- Effects on Particle Mass Deposition Rates by Convective-Diffusion, Thermophoresis or Inertial Impaction; Consequences of Multi-spherule 'Momentum Shielding', Aerosol Science and Technology (in press).

3. Rosner, D. E. and P. Tandon (2017b). Mainstream Aggregation and Polydispersity Effects for Inertially-Dominated Deposition Rates on Cylinders (or Leading 'Edges') in Crossflow"---- Consequences of 'Momentum Shielding' for the Impaction of Large, "Compact" ($2<D_f<3$) Multi-spherule Aggregates, Aerosol Science and Technology (prepared for submission, August 2017).

4. Sorensen, C. M. (2011). The Mobility of Fractal Aggregates: A Review, Aerosol Science and Technology. 45: 765-779.

508609 Experimental and Numerical Studies on the Thermal Diffusivity of PackedÂ Powder Beds

Abstract:

Several industrial processes, such as calcination and drying, involve thermal treatment of particles. In order to achieve efficient processes and good quality products, the thermal properties of the particles and that of the bulk powder must be well defined. Previously, some models, such as the Krischer and Maxwell-Eucken models, have been proposed to relate particle thermal properties to that of the bulk powder. Though these models have been shown to work well, it is not very clear how to account for changes in the packing density, moisture content, and arrangement of particles. In addition, the particle thermal properties of most materials are not readily available, which limits the application of these models. To get a better understanding of the relationship between the particle thermal properties and that of the bulk powder and also to quantify the effects of packing density and moisture content on the bulk powder thermal properties, we run several experiments and discrete element method (DEM) based simulations. In both the experiments and simulations, a pipe is filled with particles where the temperature of the pipe is kept constant. We measured the temperature change in the powder bed and compute the actual thermal diffusivity of the powder bed. We tested several types of powders and glass beads in the experiments. The packing fraction, the moisture content, and size of particles were varied in the experiments. In the simulations, in addition to the packing fraction and particle size, the thermal properties of the particles were also varied. In this talk, we will present the details of the experimental and simulation results.

509556 Determination of Size, Size Distribution and Refractive Index of Artificial and Biological Microparticles

Abstract:

For the approval of clinical studies, it is mandatory to validate the production process of artificial oxygen carriers by reliably specifying the properties of the microparticles used like the newly developed blood substitute HbMP-700 (1-3). In particular, besides shape, size and size distribution, the amount of hemoglobin (Hb) and methemoglobin (metHb) content is essential.

We present a method based on the simultaneous measurement of collimated and diffuse transmission spectra in the visible and ultraviolet range and numerical analysis to infer the wavelength-dependent refractive index, size and size distribution of microparticles.

For the validation of the method, we used aqueous suspensions of monodisperse polystyrene microspheres of known concentration, determined by flow cytometric measurements. The related data analysis is carried out for synthetic and experimental data. For homogeneous spheres, Mie theory is applied to compute the solution for the scattering of a plane electromagnetic wave by a particle and to calculate the extinction cross section C_{ext}, i.e. the shadow cast by the particle as a function of the vacuum wavelength λ, the particle size, and the complex refractive index $n(\lambda)$. For dilute suspensions of particles, the measured quantity is an ensemble-averaged extinction cross section C_{ext}. This average is obtained by integrating over the particles' size distribution, which may not be known a priori. From the resulting spectral extinction cross sections, the complex refractive index and the mean and standard deviation of the size distribution can be inferred by means of nonlinear least-squares optimization. In order to reduce the number of free parameters of the problem to a feasible amount, $n(\lambda)$ is represented in a set of harmonic-oscillator-like resonance functions. The refractive index of polystyrene microspheres obtained is in good agreement with literature values for the bulk material, the mean and standard deviation of the distribution of particle *size* D are determined with an uncertainty in the range of a few nanometers. For example, for microspheres of mean diameter mean(D) = 2μm and distribution width std(D) = 9.4nm the uncertainty of the mean is u{mean(D)} = 1.2nm and the uncertainty of the distribution width amounts to u{std(D)} = 0.4nm.

In a second step, we applied the method to characterize HbMP-700 microparticles, approximately featuring a prolate spheroidal shape with 700nm length and 400nm width. Besides shape, size and size distribution, the amount of hemoglobin (Hb) and methemoglobin (metHb) content is essential. The relative concentrations of oxyHb, deoxyHb, metHb are accessible from spectral extinction and diffuse transmission measurements, since the different Hb variants exhibit different wavelength-dependent refractive indices. The method to infer size and refractive index described above can be applied to these non-spherical particles, since the deviation of their light scattering properties from those of spheres turns out to be small. The results based on the optical method will be compared to alternative procedures for HbMP particle characterization with respect to their size and size distribution, i.e. electron microscopy, centrifugation and dynamic light scattering. In addition, x-ray spectrometry will be used to detect the ratio between Fe^{2+} (oxy/dexyHb) and Fe^{3+} (metHb) as well as the iron concentration.

References

1) Novel hemoglobin particles--promising new-generation hemoglobin-based oxygen carriers. Bäumler H, Xiong Y, Liu ZZ, Patzak A, Georgieva R. Artif Organs. 2014 Aug;38(8):708-14. doi: 10.1111/aor.12331.

2) Nonvasoconstrictive hemoglobin particles as oxygen carriers. Xiong Y, Liu ZZ, Georgieva R, Smuda K, Steffen A, Sendeski M, Voigt A, Patzak A, Bäumler H. ACS Nano. 2013 Sep 24;7(9):7454-61. doi: 10.1021/nn402073n.

3) Hemoglobin-based oxygen carrier microparticles: synthesis, properties, and in vitro and in vivo investigations. Xiong Y, Steffen A, Andreas K, Müller S, Sternberg N, Georgieva R, Bäumler H. Biomacromolecules. 2012 Oct 8;13(10):3292-300. doi: 10.1021/bm301085x.

517131 FlowCam NanoÃ,Â® Provides Counts, Sizes and Images of Nano-and Microparticles: Application to a Therapeutic Protein Pumping Study

Abstract:

Sub-visible particle characterization is becoming a more important method in assessing drug substance and drug product stability. However, the current USP protocol does not require any characterization of particles below 10 Î¼m. This greatly hampers our ability to properly assess protein stability as a function of particulate formation and obscures most sub-visible particles and various morphological properties.

In this study, we assess the capabilities of the new FlowCam Nano from Fluid Imaging Technologies, in conjunction with a pumping study that focuses on particulate formation generated during fill operations. Initial experiments show that the new FlowCam Nano can generate highly resolved images of nanoparticulates, allowing the simultaneous assessment of particulate morphology for nano- and microparticulates. This is extremely useful, as currently unpublished data (for the pumping study) show that as a function of post-pumping agitation, a significant increase in microparticle concentration coincides with a significant depletion of the nanoparticle concentration. This suggests that agglomeration (and probably not nucleation) is the more likely mechanism for larger microparticulate formation.

The ability to characterize the nano- and microparticulate morphology simultaneously, will allow us to better assess the cause of particulate formation (such as observing the transition of nanoparticles to microparticles and seeing if specific morphological properties of larger nanoparticulates are conserved in microparticulates). In addition, the FlowCam Nano is also sensitive enough to assess differences in the colloidal stability of a drug product in different formulations. During filling operations, and accelerated degradation studies, FlowCam Nano showed the differences in size distributions of IVIG in PBS and 0.25 M glycine buffer. Since IVIG is less colloidally stable in PBS, the size distribution varies significantly more per size bin and the overall mean effective spherical diameter is also ~ 600 percent greater after 24 hours of agitation in comparison to the size distribution data of IVIG in glycine. Thus, Flow Cam Nano has capabilities that are very useful in assessing drugs in the substance (e.g. optimizing drug formulations) and product (e.g. assessing particulates post fill & finish operations) phases.

Numerical and experimental estimation on the normal and tangential capillary bridge force adhered to two spheres

Kazuo Murase*, Keisuke Arai, Takato Ootsuka, Daiki Sakamoto and Futa Egawa,
Department of Applied Chemistry, Chuo University
1-13-27, Kasuga, Bunkyo-ku, Tokyo,1128551, Japan
**hmurase@kc.chuo-u.ac.jp*

Introduction

Liquid bridges extensively occur as a binder liquid adhered to solid surfaces in a variety of engineering applications such as crystal growth, ink-jet printing, material processing and powder technology. Especially the strength of adhesion and cohesion of liquid bridges adhered to particles contribute to the consolidation of an agglomerate in wet granular materials. Estimation of the attractive or repulsive force between particles is important in controlling kinetics of agglomeration and handling of granular materials. Characteristics of a liquid bridge e.g. the equilibrium bridge geometry [1], stability [2], static and dynamic binding force [3-6], rupture energy [7,8], and wetting hysteresis [9] etc., have been experimentally and numerically investigated.

The liquid distributions within a particles bed are classified into four types of categories, i.e. pendular, funicular, capillary and slurry states, according to the liquid contents [10]. The static normal capillary force at pendular liquid state have elaborately been investigated as Fisher [11], toroidal model [8] and the exact solution of Yung-Laplace equation [12]. The dynamic normal capillary forces [13] are experimentally estimated with lubrication theory. The tangential force of a liquid bridge [14] also contributes to the primary factor for the granular consolidation with the shear stress of wet granular materials but have not yet substantially been investigated.

This work aimed to estimate experimentally and numerically the normal and tangential capillary force of a static and dynamic liquid bridge adhered to two spheres. The asymmetry static liquid bridge with the shear stress demand for the exact solution of Young-Laplace equation for the equilibrium liquid drop including the effect of gravity.

1. Theory

1.1 Formulation of a asymmetry static liquid bridge

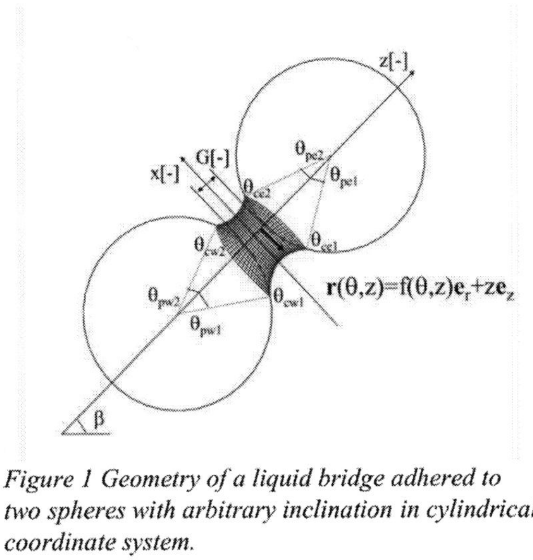

Figure 1 Geometry of a liquid bridge adhered to two spheres with arbitrary inclination in cylindrical coordinate system.

The asymmetry static bridge shape demands for the exact solution of the Young-Laplace equation due to the limit of the toroidal approximation. A prevalent geometry of a liquid bridge adhered to two spheres is drawn in Figure 1. The dimensionless position vector of a liquid surface in cylindrical coordinates is described as $\mathbf{r}(\theta,z) = f(\theta,z)\mathbf{e_r} + z\mathbf{e_z}$ divided by spherical radius, R', which $\mathbf{e_r}$, $\mathbf{e_z}$ are unit basis vectors in r-,z-directions. The symbol (') means the dimensional value. Fig.1 indicates the two dimensional front view with four half-filling (θ_{Pw1}, θ_{Pw2}, θ_{Pe1}, θ_{Pe2}) and contact angles (θ_{Cw1}, θ_{Cw2}, θ_{Ce1}, θ_{Ce2}).

The unit outward normal vector of a bridge surface, $\mathbf{n}_L$, is presented as follows [15],

$$\mathbf{n}_L \equiv \frac{r_\theta \times r_z}{|r_\theta \times r_z|} = \frac{f\mathbf{e_r} - f_\theta \mathbf{e}_\theta - ff_z\mathbf{e}_z}{[f^2(f^2+f_z^2)+f_\theta^2]^{1/2}} \quad ...(1)$$

,where low suffixes (r,θ,z) describe partial derivatives.

The dimensionless effective potential energy, $E=E'/\sigma R'^2$ with Lagrangian multiplier, $K=K'/\sigma R'^2$, on the constrained condition of a constant volume, $V=3V'/4\pi R'^3$, is described as follows [16],

$$E = \int_{-\pi/2}^{\pi/2}\int_{z_{min}}^{z_{max}} \left[f^2\left(f^2+f_z^2\right)+f_\theta^2\right]^{1/2} d\theta dz + \frac{Bo}{2}\int_{-\pi/2}^{\pi/2}\int_{z_{min}}^{z_{max}} (z\sin\beta + x\cos\beta) f^2 d\theta dz$$
$$+ K\left[V - \frac{1}{2}\int_{-\pi/2}^{\pi/2}\int_{z_{cp1}}^{z_{cp4}} f^2 d\theta dz\right] \quad ...(2)$$

Bond number (Bo) is a dimensionless number defined by $(\rho_L-\rho_G)gR'^2/\sigma$, where is the ratio of gravitational force to surface tension force equally as Etovoes number (Eoe).

The integral domain of a bridge surface is determined by the trapezoidal section of the upper and lower surface lines, [$z_{w2}<z<z_{e2}$ at $\theta=\pi/2$, $z_{w1}<z<z_{e1}$ at $-\pi/2$] and contact lines, [$z_{w1}<z<z_{w2}$, $z_{e1}<z<z_{e2}$ at $-\pi/2<\theta<\pi/2$] with four contact points settled by four half-filling angles ($\theta_{Pe1},\theta_{Pe2},\theta_{Pw1},\theta_{Pw2}$) in Fig. 1. Two contact lines on the upper and lower spheres were approximated by ellipse lines with parametric aspect ratios, which pass through respectively pairs of two contact points settled with half filling angles, ($\theta_{Pw1},\theta_{Pw2}$) and ($\theta_{Pe1},\theta_{Pe2}$). Filling angles, ($\theta_{PW}(\theta)$, $\theta_{PE}(\theta)$) on the lower and upper spheres have variable properties along the circumferential direction. The formulation of a static bridge shape at equilibrium state results in the extremum problem, which corresponds to the condition that the first order variation equals to zero, $\delta E=0$.

1.2 Estimation of the capillary force of a liquid bridge

The dimensionless mean curvature (H) and contact angles (θ_c) of a bridge surface are respectively calculated by eqs. (3),(4) once the outline of a liquid bridge is numerically evaluated.

$$-2H = \nabla_s \cdot \mathbf{n}_L = \left[(\alpha^2 + f_\theta^{\,2}) - f(f^2 + f_\theta^{\,2})f_{zz} - f(1+f_z^{\,2})f_{\theta\theta} + 2ff_z f_\theta f_{\theta z}\right]/\alpha^3 \quad \ldots(3)$$
$$,where\ \ is\ \ \alpha^2 = f^2(1+f_z^2) + f_\theta^2$$

$$\theta_c(\theta) = \cos^{-1}(\mathbf{n}_s \cdot \mathbf{n}_L) \quad \ldots(4)$$

,where contact angles also depend on the circumferential angle.

According to the Young-Laplace equation, the dimensionless pressure difference, $\Delta P= \Delta P'R'/\sigma$, across a liquid surface is related with the mean curvature of the surface.

$$\Delta P = (-2H) \quad \ldots(5)$$

The capillary force on a liquid bridge adhered to two spheres must be evaluated only on the contact line as known by the boundary method due to unspecified neck radius of a bridge and asymmetric condition. Dimensionless averaged capillary forces, $F_{Ci} = F'_{Ci}/(2\pi R'\sigma)$ on the upper and lower sphere (i=1,2 respectively correspond to lower, upper sphere) are presented as follows by three terms, which correspond to the surface tension, hydrostatic pressure and buoyancy terms along the longitudinal angle (θ). The sign of the capillary force set to be positive for attractive direction.

$$F_{ci} = \frac{1}{\pi}\int_{-\pi/2}^{\pi/2} \sin(\theta_{Pi})\sin(\theta_{Pi}+\theta_{Ci})d\theta - \frac{1}{2\pi}\int_{-\pi/2}^{\pi/2}\sin^2(\theta_{Pi})\Delta P d\theta \mp \frac{2}{3}\mathrm{Bo}\cdot V_{bi} \quad \ldots(6)$$

, which V_b is sphere volume submerged in a bridge

Finite Element Method is applied for the solution of the $\delta E=0$ equation. Isoparametric elements of a biquadratic polynomial shape function were selected for the integral domain and unknown radius. The set of non-linear algebraic equations for unknown radius at element nodes and Lagrangian multiplier was resolved by the Newton-Raphson method [9].

2. Experiment

Tensile strength of a static and a dynamic pendular liquid bridge

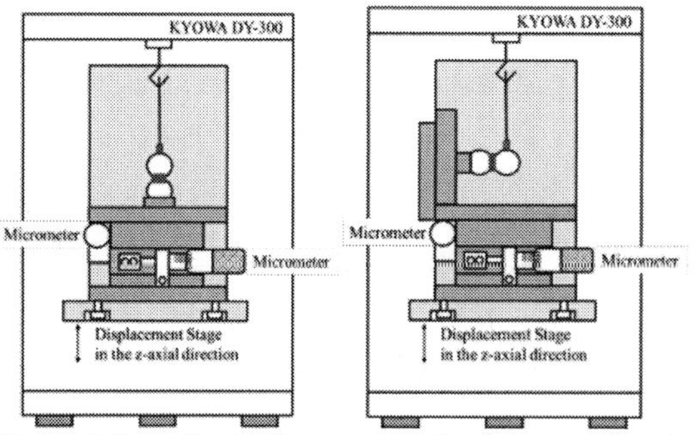

Figure 2 Experimental apparatus for (a) normal and (b) tangential tensile strength of a liquid bridge adhered to two spheres

Figure 2 shows the surface tensiometer (Kyowa interface Science Co., Ltd. DY-300) for measuring (a) the normal and (b) the tangential tensile strength of a water liquid bridge adhered to two stainless steel spheres (radius 3.0mm). The rod attached with one sphere is hanged up to the tensiometer and the other sphere fixed on the displacement stage is moved upward and downward by the tensiometer. The discrete and continuous displacement of the stage are performed for the normal and tangential deformation of a liquid bridge for the static and dynamic states. Two micrometers adjust the appropriate position of the lower spheres for the upper sphere. Water droplets are poured into the spheres aperture and substantially wetted by the displacement of the stage. The tensile strength at static and dynamic state and snapshots of liquid droplets by the digital camera and movie are recorded. The outlines of droplets are evaluated by a digitizer and estimate the half filling angles ($\theta_{Pe1}, \theta_{Pe2}, \theta_{Pw1}, \theta_{Pw2}$) and contact angles ($\theta_{Cw1}, \theta_{Cw2}, \theta_{Ce1}, \theta_{Ce2}$) at four contact points. The sphere configuration (radius, gap), the droplet volume, the half filling angles and the contact lines on spheres initiate parameters of numerical simulation.

3. Results and discussions

3.1 The normal tensile strength of a liquid bridge adhered to two vertical spheres.

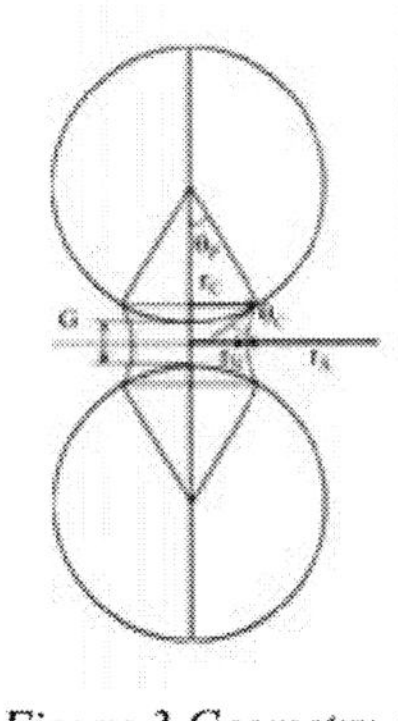

Figure 3 Geometry of toroidal model

The symmetric geometry of a liquid bridge adhered to two vertical spheres without gravity has been extensively investigated by toroidal approximation in Figure 3. The boundary method overestimates the capillary for of a liquid bridge rather than the gorge method but is feasibly and distinctly defined for the asymmetric bridge outline. Symmetric condition simplifies eq.(3) for the mean curvature and eq.(6) for the capillary force into eq.(7),(8) [7,16,17].

$$-2H = \nabla_s \cdot \mathbf{n}_L = \frac{1}{\alpha}\left(\frac{1}{f} - \frac{f_{zz}}{\alpha^2}\right), \textit{where} \quad \alpha^2 = 1 + f_z^2 \qquad \ldots(7)$$

$$F_{Boundary} = r_C \sin(\theta_{Pi} + \theta_G) - \frac{r_C^{\,2}}{2}(\Delta P) \mp \frac{2Bo}{3} V_{bi}, \;\; r_C = \sin\theta_P \qquad \ldots(8a)$$

$$F_{Borge} = r_N - \frac{r_N^{\,2}}{2} \sin\theta_{cpi}(\Delta P) \mp \frac{2Bo}{3} V_{bi} \qquad \ldots(8b)$$

,where $r_N - r_A = \dfrac{A}{\tan(\pi/2 - \theta_P - \theta_C)} + \sin\theta_P, r_A = \left[A^2 + \{\sin\theta_P - (r_N - r_A)^2\right]^{1/2} A = G/2 + 1 - \cos\theta_P$

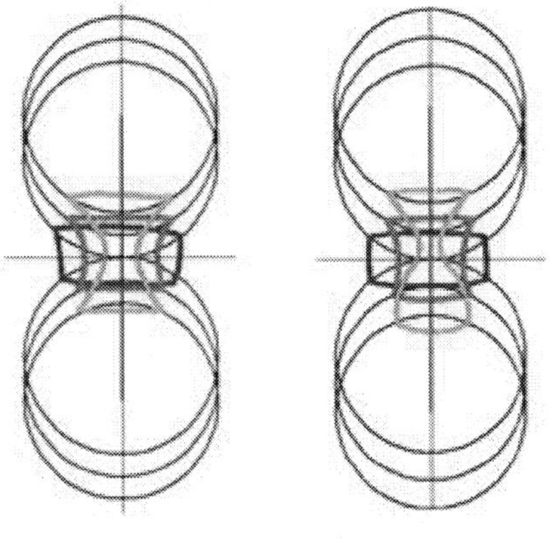

(a) Static state *G=0, 0.47, 0.91* (b) Dynamic state *G=0, 0.59, 1.16*

Figure 4 Experimental droplet outlines at static and dynamic state, Bo=1.21, V=0.088

Figures 4 show symmetric outlines of liquid bridges around the vertical axis respectively at the static and dynamic state. Static state results in the shorter rupture gap and the larger dimensionless normal tensile strength from Figures 5a. Sphere gaps at the peak static capillary force depend on liquid volume. The contact angle on the upper sphere corresponds to the advancing contact angle at the higher Bo number due to gravity effect.

Fig. 4b indicates the downward transition of a liquid immediately before the rupture due to the wettability a liquid. The toroidal approximations over-estimate experimental data at the lower sphere gap but the sphere gaps at the peak dynamic capillary force depend on the velocity of stage displacement. The velocity reduces the peak value of dynamic capillary due to the downward transition of a liquid in Fig. 5b. The stepwise profiles of capillary force are due to the movement sensitivity of tensiometer for the stage displacement.

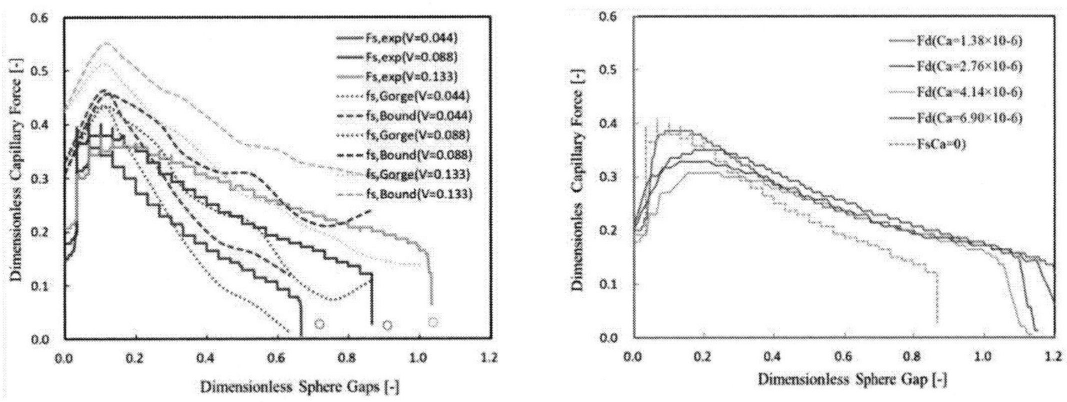

(a) at the static state (b) at the dynamic state

Figure 5 Effect of sphere gap on dimensionless capillary force

3.2 Estimation of boundary condition at contact line by LBM

The Young's equation describes the force balance among three interface tensions (solid-liquid, solid-gas and liquid gas) at three phase contact point and also defines the relationship between interface tension and contact angle. The solution of the Young-Laplace equation at the static state demands for one unknown boundary condition of either a contact angle or a contact line. The simulation for the wettability on three dimensional contact line demands for the elaborate process. Our experiment determines the contact lines on spheres under the constant liquid volume and is utilized for the boundary condition. Besides the perfect wettability the smooth transition of three phase contact lines is unexpected due to the advancing and receding contact angle. Lattice Boltzmann Method (LBM) is expected to estimate the boundary condition at static and dynamic state. The geometry of a liquid bridge is numerically estimated by FEM with half-filling angles predicted by 2 dimensional Shan & Chen model of LBM, which is used by D2Q9 model for transition of distribution function. The estimation for the liquid geometry by FEM has been verified. Figures 6 show the predicted liquid outlines by LBM and FEM. at a given liquid volume.

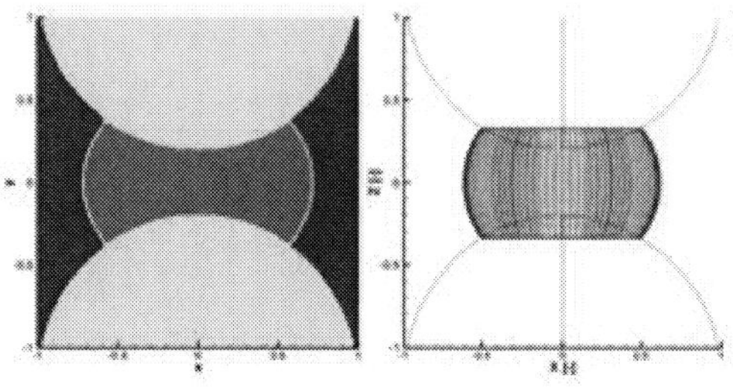

(a) Repulsive interparticle force

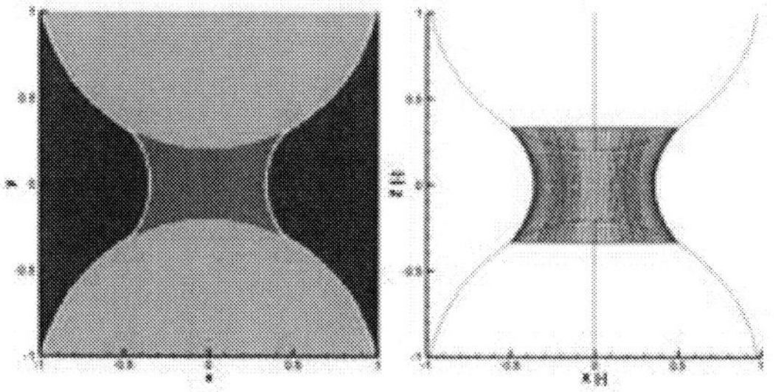

(b) Attractive interparticle force

Figure 6 Snapshots of liquid geometries at static state by LBM and FEM

Figures 7 show snapshots of dynamic liquid outlines at a given liquid volume up to the rapture state by LBM. The transition of boundary conditions, i.e. filling and contact angles, is smoothly described. Gravity and imperfect wettability, i.e. advancing and receding contact angles, are excluded in Figs. 7 but the effect of gravity on the liquid outline is avairable.

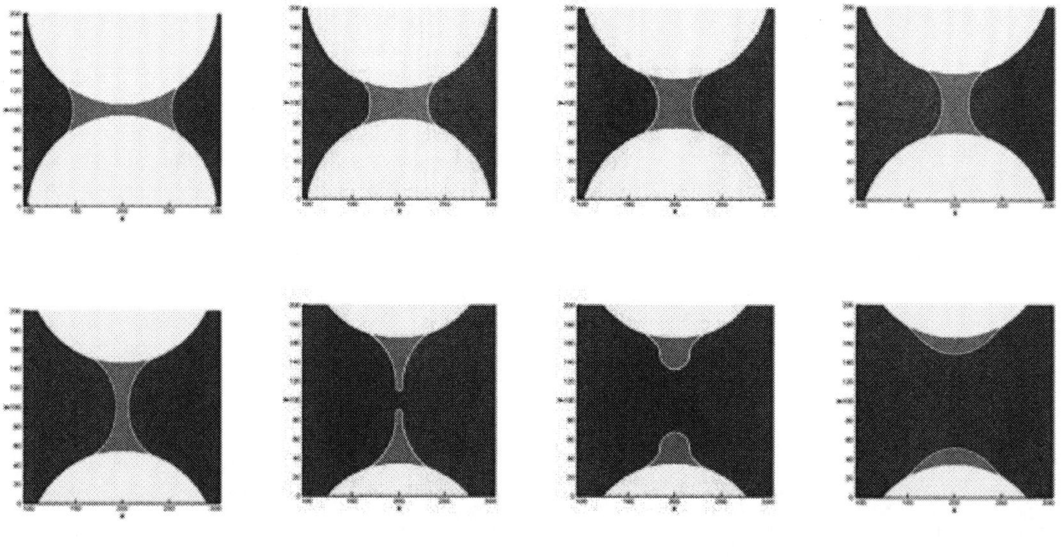

Figure 7 Snapshots of snapshots of dynamic liquid outlines at a given liquid volume

3.3 The tensile strength of a liquid bridge with shear deformation.

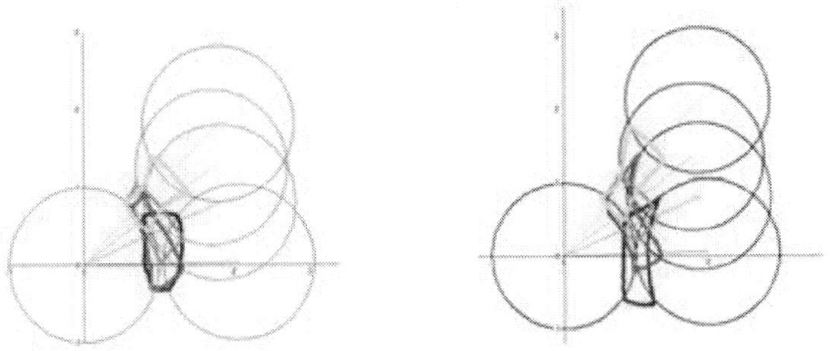

(a) Static state (G=0, 0.59, 0.98) (b) Dynamic state (G=0, 0.59, 0.98)

Figure 8 Liquid outlines at static and dynamic state (Bo=1.21, V=0.088)

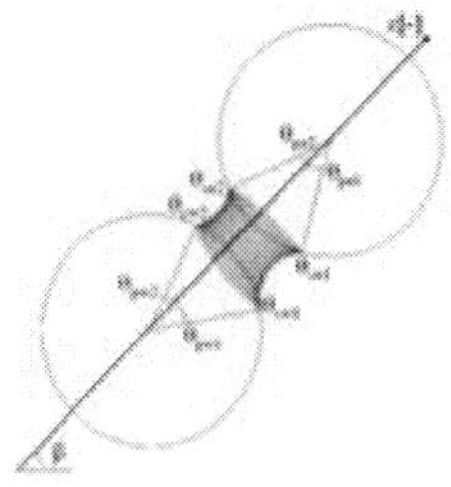

Definitions of half-filling and contact angles.

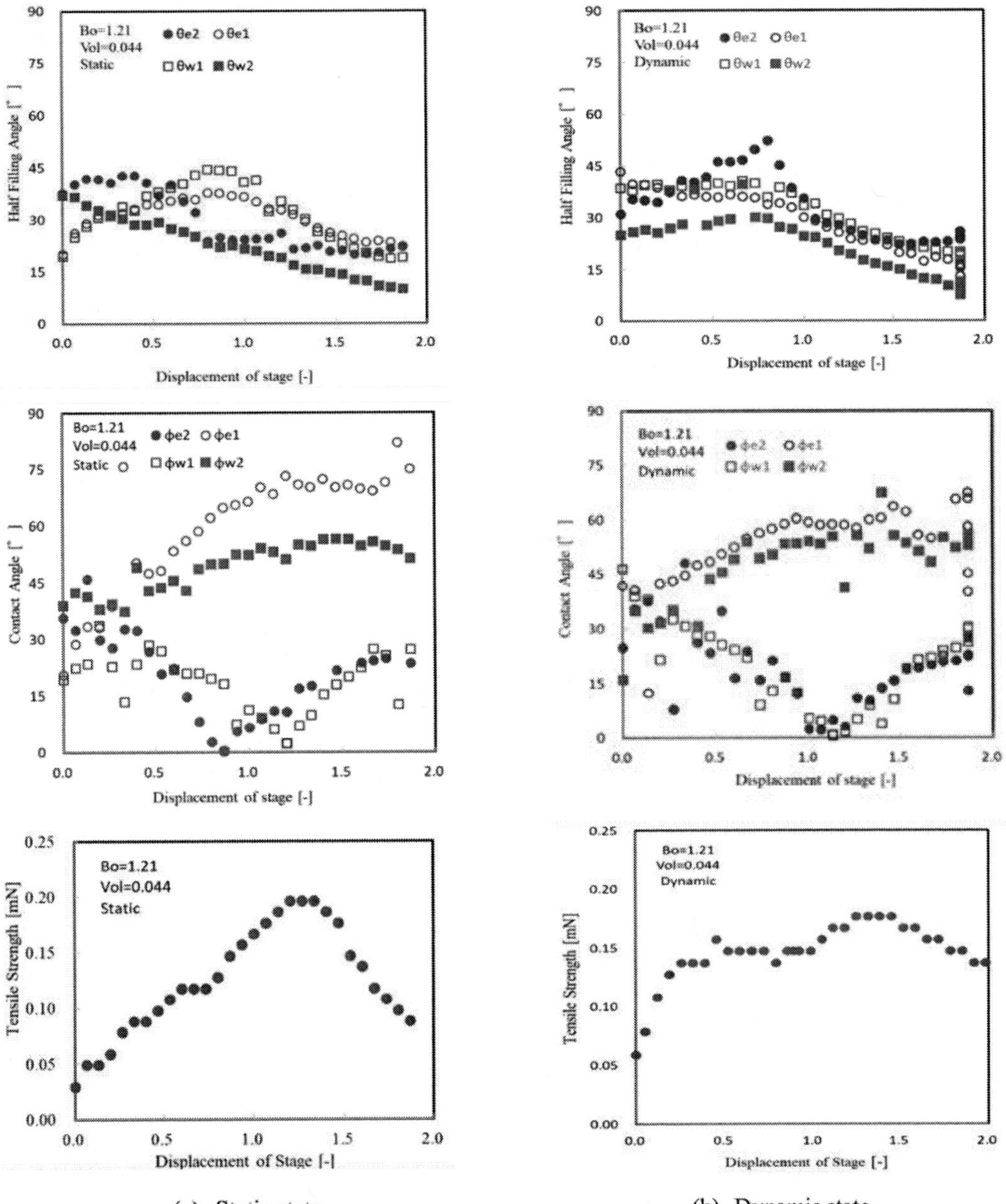

(a) Static state (b) Dynamic state

Figure 9 Profiles of half-filling angles, contact angles and tensile strength , Bo=1.21, V=0.088

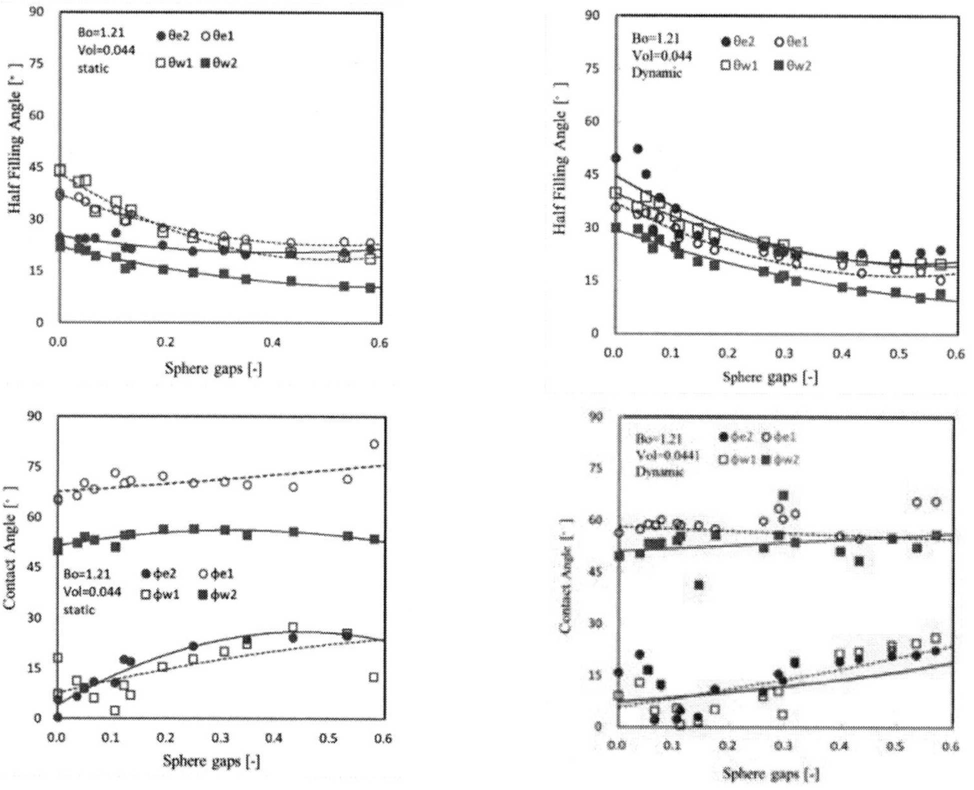

(a) Static state (b) Dynamic state

Figure 10 Profiles of half-filling angles, contact angles with sphere gaps , Bo=1.21,V=0.088

Figures 8 show four types of liquid outlines at static and dynamic state. Magenta and green plots are drawn respectively at the extreme values of half filling angles, contact angles and tensile strengths in Figures 9. During the lower stage displacement two spheres keep at the contact state and the half-filling angles reached to the peak values. Half-filling angles sifted with displacements of a stage have the peak values except the static half-filling angle at the top point of the lower sphere (θ_{Pw2}), which monotonously decrease with stage displacement. The increase and decrease of contact angle respectively corresponds to the advancing and receding contact angle. The contact angles at the bottom point of the lower sphere (θ_{Cw1}) and at the top point of the upper sphere (θ_{Ce2}) transit from the receding to the advancing contact angle due to the infinitesimal values. Figures 10 redraw the half-filling and contact angles with respect to sphere gap. Contact angles contribute to the peak tensile strength, which is measured as resultant force (F_T) of normal (F_n) and tangential (F_t) tensile strength in Figure 11. The normal tensile strength conventionally corresponds to the capillary force. The stage displacement at the peak tensile strength minimize the sum of half-filling and contact angle due to the positive hydrostatic pressure across liquid surface. The larger dynamic tangential strength occurs during the sphere contacting due to the asymmetric deformation. The adhesion force of liquid to sphere at the static state contribute to the larger peak tensile strength.

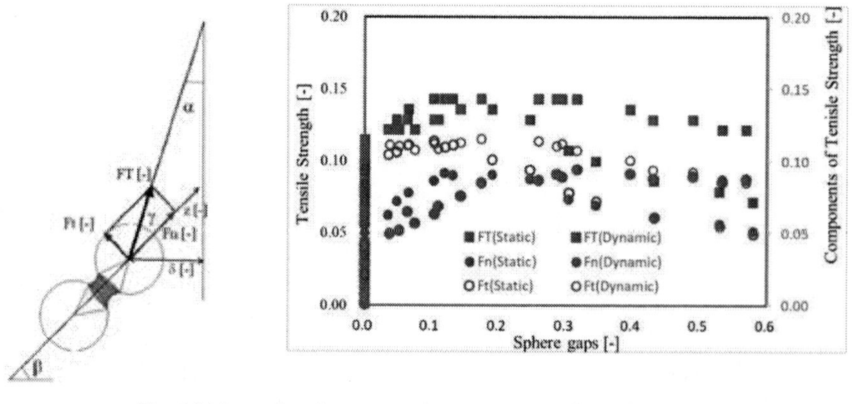

Fig.11 Normal and tangential components of tensile strength at static state

3.4 Numerical simulation at the static state

The solution of Young-Laplace equation demands for one unknown necessary condition on the contact point, i.e. half-filling or contact angle. Liquid surfaces simulated by FEM are drawn with solid lines in Figures 12. The simulation reveals the asymmetric liquid geometry with half-filling and contact angles along longitudinal angle. Disagreements of liquid interfaces results from the approximation of contact lines on spheres. The tangential strength influences depend on the mean curvature. Especially the second figure at the peak half-filling angle has the negative hydrostaic pressure in the direction of relative sphere displacement and positive pressure in the oppsite direction.

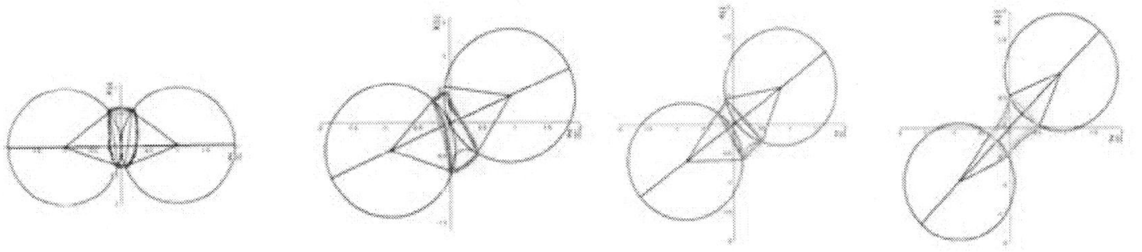

At the initial geometry At the peak halffilling angle At the peak tensile strength At the maximum shpere gap

Figure 12 Numerical estimation at the static state

Figure 13 displays transitions of measured normal tensile strengths ($F_{exp,n}$) and simulated capillary force (F_{cal}) to sphere gap. The initial droplet is attached to the above aperture in the direction of sphere movement due to gravity. The stage displacement drifts parts of liquid volume into the lower aperture of spheres. The sphere on displacement stage keep contacting with the other sphere hang up to the tensiometer up to the peak tensile strength. Fcal and FcalB are numerically estimated by outlines of liquid surfaces respectively with Young-Laplace equation and boundary toroidal model. The sign of hydrostatic pressure with deformation of liquid droplets results in the repulsive capillary force. The normal tensile strength corresponds to the capillary force during the range of symmetric liquid bridge. The angle of beta is defined by the inclination of center distance of a sphere pair. The lower normal tensile strength than the capillary force results from the orthogonal component of capillary force at the low range of beta.

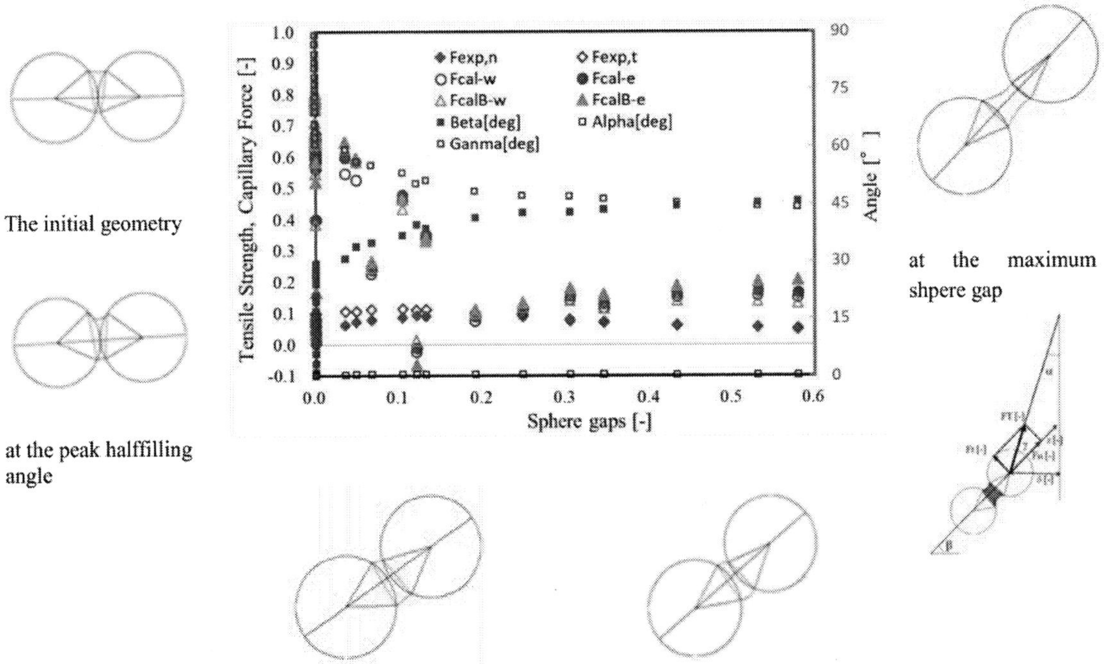

Figure 13 Normal tensile strength and numerical capillary force

Conclusions

Liquid bridges adhered to particles are deformed with the normal and tangential stress in wet granulator. The interparticle force with asymmetric droplet by shear stress must be estimated by the three dimensional solution of Young-Laplace equation. A surface tensiometer and numerical simulation contribute to the normal and tangential tensile strength of liquid bridge adhered to two spheres.

1. The normal strength of symmetric liquid bridge with a surface tensiometer is verified by the conventional toroidal model and the solution of Young-Laplace equation.
2. Sphere gaps at the peak capillary force in the vertical sphere configuration depend on liquid volume and displacement velocity of sphere.
3. The lattice Boltzmann Method is expected to numerically predict the boundary conditions on the contact lines, which are half-filling angle and contact angle at the static and dynamic state.
4. The tensile strength depends on transition of contact angles, which correspond to advancing or receding angles. The larger sum of contact and half-filling angles than the orthogonal angle corresponds to the positive hydrostatic pressure across liquid surface.
5. The normal tensile strength is related with the solution of Young-Laplace equation by FEM. The tangential tensile strength depends on the liquid transition within droplets and the variable mean curvature.

References

[1] K.Murase, H.Shibata, Y.Yasuda and S.Toyama, *Jpn J. Multiphase Flow,* **5**, 322(1991)
[2] A.E.Saez and R.G.Carbonell, *J. Colloid Interface Sci.,* **140**, 408(1990)
[3] D.N.Mazzone, G.I.Tardos and R.Peffer, *Powder Tech.*, **51**, 71(1987)
[4] C.D.Willet, M.J.Adama, S.A.Johnson and J.P.K.Seville, *Langmuir*, **16**, 9405(2000)
[5] M.J.Adama, S.A.Johnson, J.P.K.Seville and C.D.Willet, *Langmuir*, **18**, 6180(2002)
[6] R.K.Niven, *J Petroleum Sci. Eng.*, **52**, 1(2006)
[7] O.Pitois, P.Moucheront and X.Chateau, *Eur. Phys. J.B*, **23**, 79(2001)
[8] S.J.R.Simons, J.P.K. Seville and M.J. Adams, *Chem. Eng. Sci.*, **49**, 2331(1994)
[9] C.D.Willet, M.J.Adama,S.A.Johnson and J.P.K.Seville, , *Powder Tech.*, **130**, 63(2003)
[10] S.M.Iveson, J.A.Beathe and N.W.Page, *Powder Tech.*, **127**, 149(2002)
[11] R.A.Fisher, *J. Agri. Sci.*, 16,492(1926)
[12] F.M.Orr, L.E.Scriven and A.P.Rivas, *J. Fluid Mech.*, **67**, 723(1975)
[13] O.Pitois, P.Moucheront and X.Chateau, *J. Colloid Interface Sci.*, **231**, 26(2000)
[14] U.El Shamy and T.Groeger, *Int. J. Numer. Anal. Meth. Geomeh.*, **32**, 1763(2008)
[15] F.M.Orr, R.A.Brown and L.E.Scriven, *J. Colloid Interface Sci.,* **60**, 137(1977)
[16] K.Murase, T.Mochida, Y.Sagawa and H.Sugama, *Adv. PowderTech.,* **19**, 349(2008)
[17] C.D.Willet, M.J.Adama,S.A.Johnson and J.P.K.Seville, *Handbook of powder technology*, **11**, chap.28, ed. A.D.Salman, M.J.Hounslow and J.P.K.Seville,Elsevier (2007)

NUMERICAL SIMULATION OF WIRE-PLATE ELECTROSTATIC PRECIPITATOR - EFFECT OF PARTICLE CONCENTRATION

Ding Yang[1], Baoyu Guo[2], Jun Guo[1], Aibing Yu[2]*

[1] *Fujian Longking Co. Ltd, Longyan, Fujian, 364000, China*
[2] *Department of Chemical Engineering, Monash University, Clayton VIC 3800, Australia*

Introduction

Electrostatic Precipitators (ESP) have been widely used to separate fine particles from large scale industrial exhaust gases, such as coal fired power plants. In-depth knowledge and careful design optimization are critical to meeting the increasingly strict emission standard in terms of $PM_{2.5}$ emission limit, e.g., 5-10 mg/Nm^3. Numerical modelling has been employed as a powerful tool in this area. Complicated interactions among electric field, gas flow and particle transport are summarized by Adamiak [1]. To model such a system, various levels of simplifications have been made in published papers.

Among them, Lagrangian particle tracking technique is a widely accepted approach. The mass concentration of ash particles in coal-fire flue gas is typically 10-50 g/Nm^3, corresponding to volume fraction in the order of 10^{-5}. Such a small value is often regarded as sufficiently dilute particle concentration in numerical modeling of multiphase flows. For the case of ESP, complicated interactions among the e-field, gas flow and charged particles are involved, even if the deposited dust layer is ignored. Full coupling between the continuous gas phase and discrete particle phase is computationally expensive to achieve a converged steady solution. To avoid such a high cost, the model for ESP was simplified as so called one-way coupling in many previous works, so that the electric field is solved without the particle phase (Guo et al. [2,3]; Farnoosh et al. [4]). This simplification may be acceptable in the end stage of ESP, but will introduce significant error in the early collection stage. Limited studies reported in the literature have considered the particle charges as a source term in the Poisson's equation of electrostatics. For example, Choi and Fletcher [5] predicted a reduced current for the particle-loaded case (d_p = 1-5μm) compared with the particle-free case, but the corona ion charge density somehow shows an increasing trend along the stream. When fine dust particles are treated, the large specific surface area of the particle phase, combined with reduced collection efficiency, implies high space charge density in the collection region. Its effect was understood as a partial suppression of the corona current [6]. As was experimentally proved by Podlinski et al. [7], space charge of charged particles can contribute to the EHD flow pattern and particle transport. In present work, particle space charge is considered through two-way coupling and interesting results are generated in terms of ESP performance as particle loading changes.

Integrated ESP Modelling

Mathematical Equation

Coupled electric field, gas-particle flow and particle charging processes are simulated using

Eulerian-Lagrangian method in Computational Fluid Dynamics (CFD). Major governing equations of vector format are listed below for completeness. For electric field, Poisson equation of potential

$$\nabla^2 V = -(\rho_{ion} + \rho_c)/\varepsilon_0 \quad \mathbf{E} = -\nabla V \tag{1}$$

Current continuity equation

$$\nabla \bullet ((b\mathbf{E} + \mathbf{U}_f)\,\rho_{ion} - D\nabla\rho_{ion}) = S_c \tag{2}$$

where ρ_{ion} is corona charge density, b the ion mobility, D is ion diffusivity, ρ_c is particle space charge density, evaluated by averaging the charges of particles that pass through a control volume (cell),

$$\rho_c = \frac{1}{V_{cell}} \sum_{n=1}^{N} \dot{n}_p q_p \Delta t \tag{3}$$

Where $\dot{n}_p$ is the number flow rate of a parcel of particles tracked, Δt is the residence time in the cell. The source term, S_c, in Equation (2) is transfer rate of ion charges to particles, which is not considered at the moment, i.e., $S_c = 0$. For the gas flow, the momentum equation is,

$$\nabla \bullet (\rho_g \mathbf{U}_g \otimes \mathbf{U}_g) = \nabla(p + \tfrac{2}{3}\rho_g k) + \nabla \bullet [(\mu + \mu_t)(\nabla \mathbf{U}_g + (\nabla \mathbf{U}_g)^{\mathrm{T}})] + \rho_{ion}\mathbf{E} \\ - \frac{3\pi n_p d_p \mu}{\mathrm{C_C}}(1 + \mathrm{Re}_p^{0.687})(\mathbf{U}_g - \mathbf{U}_p) \tag{4}$$

For the particle flow, the force balance of a single particle is

$$m_p \frac{d\mathbf{U}_\mathrm{p}}{dt} = \frac{3\pi n_p d_p \mu}{\mathrm{C_C}}(1 + \mathrm{Re}_p^{0.687})(\mathbf{U}_g - \mathbf{U}_p) + \tfrac{1}{6}\pi d_p^3 \mathbf{g}(\rho_p - \rho_g) + n_e q_e \mathbf{E} + \mathbf{F}_{dis} \tag{5}$$

k-ω SST turbulence model is chosen for gas flow, and the dimensionless Lawless charging rate equation is used for particle charging [8].

Model Conditions

An integrated ESP model developed comprises two parallel modules or submodels, namely, electric field model and coupled gas-particle flow model. One-way coupling between the two modules is used earlier is upgraded to two-way in the present work. The electrostatic field caused by charged particles in suspension is considered in the solution of electric field. The two submodels are coupled through a small under-relaxation factor, in views of strong interaction between the electric field and particles flow, and the stochastic nature of the discrete particle phase. In this way all the volume averaged quantities related to the particle phase (such as the particle space charge density) are smoothed out over time in the simulation.

A single channel, 4-wire and flat plate configuration is considered. The dimensions are 3.5mm in wire diameter, 0.24m for wire-wire spacing, 0.2m for wire-plate spacing. The plate length is 1.2m. The simulation domain is defined as 1.74m long by 0.4m wide box with the 4 holes inside. The grid is smoothly refined at places where higher gradient in electric quantities is expected, such as the proximity of the wires and middle of the adjacent wires. Other base case conditions are as follows.

- Applied voltage on the wire electrode: Va = -60kV. The charge density on the wire surface is determined using an injection law combined with Peek equation (Guo et al.[9]).
- Gas temperature: 120°C. As the inlet boundary condition for the gas phase, the velocity is set as 1 Nm/s; relative turbulence intensity 3.7%, turbulent length scale $0.3D_h = 0.24$m.
- Dust particle size is mono-dispersed at d_p=2.5 μm. Particle density 2000 kg/m^3; ash dielectric constant of κ_p=6.6, Particle loading at the inlet, $(\rho_p)_{in}$, is varied with default value of 10 g/Nm^3.

Typical Results

Electric Field without Particles

When the inlet/outlet boundaries are moved sufficient long distance away from the electrodes in the simulation domain (referred to as "natural boundary condition"), the electric field around the boundary wires show an asymmetric distribution pattern that differ from that around the inner wires. Electric field-lines/current can go out of the limit of the collection region (Figure 1). Electric field intensity on the collection plate is of interest as it is considered a factor to determine the upper limit of operation before spark-over [10]. Normally the field intensity on the plate is highest at the point that faces exactly the wire center with the least distance. This maximum field intensity on the plate is 3.80×10^5 V/m. However, when the natural boundary condition is used, the highest value (1.06×10^6 V/m) predicted in field intensity appears at the endpoint of the plate due to its sharpness. This is a likely point that would trigger electric spark-over prematurely. Practically such a sharp edge should be avoided. Generally, the field strength at the middle wires is smaller than that at a boundary wire, due to the counteraction of adjacent wires. Then the corona charge density at the boundary wires should be greater than that at the inner wires (Figure 2), which is similar to the simulation result of Kasdi [11].

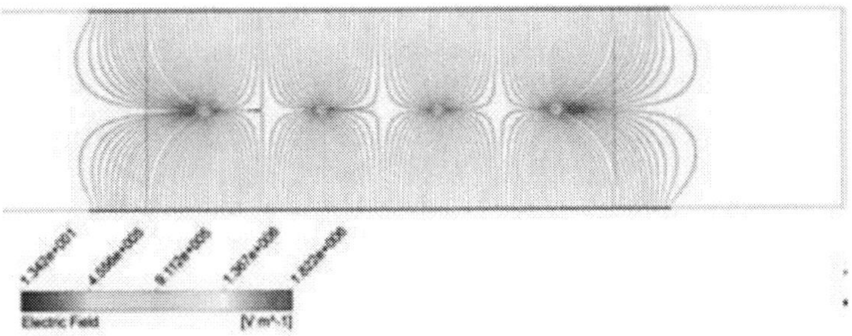

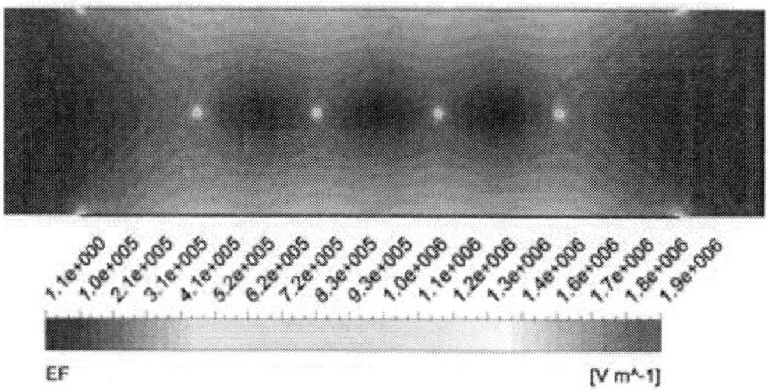

Figure 1. Field streamlines and intensity distribution (no particles).

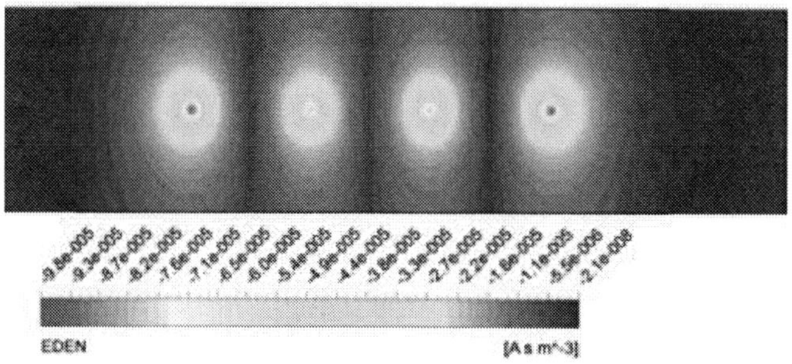

Figure 2. Corona charge density (no particles).

Effect of Particle Space Charge

Particles are randomly injected from the inlet plane with uniform probability in locations and no-slip for velocity (same as the gas velocity). In using random number generator, it is ensured that the injection points are not repeated for every coupling cycle. Figure 3 shows sampled particle trajectories with color scaled to particle charge. Figure 4 shows a pattern of particle deposition rate on the plate. It is seen that the end points of the plate, particularly the downstream one, are special spots to accumulate more dust.

Figure 3. Particle trajectories with colour scaled to particle charge.

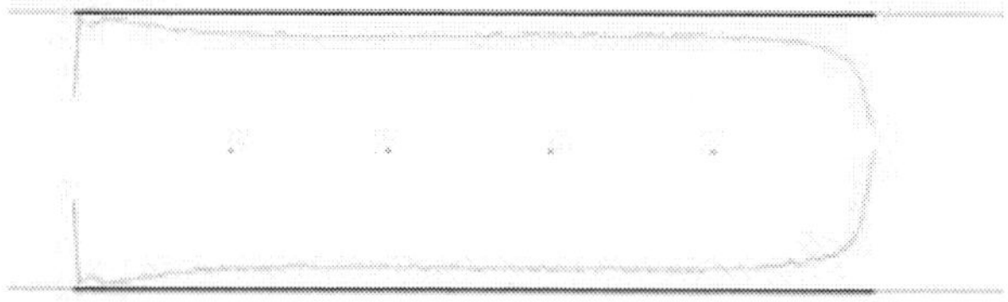

Figure 4. Indicative pattern of particle deposition rate on the plate.

Figure 5 shows the averaged space density of charges carried by particles. Highest space charge density appears near the first wire electrode, where electric current density is highest. It diffuses gradually from the centre line with particles drifting toward the plate while moving forward. Due to the radiating nature of the field-lines around each wire, the lateral component of Coulomb's force in the region close to the centre line may not be as large as that off the centre line. Not collected charged particles of negative polarity are conveyed towards the outlet, lowering the potential downstream outside the collection region (Figure 6), causing distortion of the overall electric field distribution.

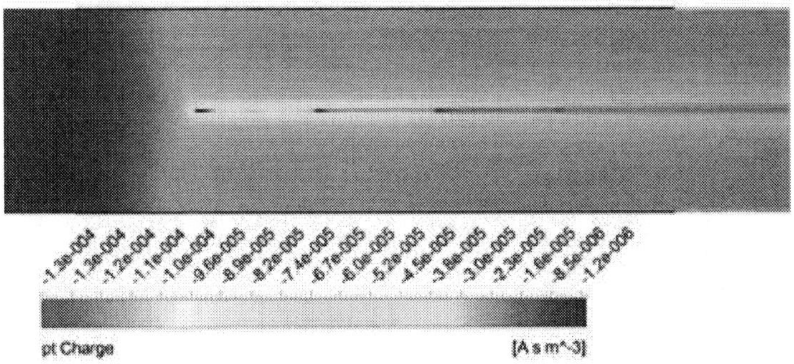

Figure 5. Distribution of space charge density of particles.

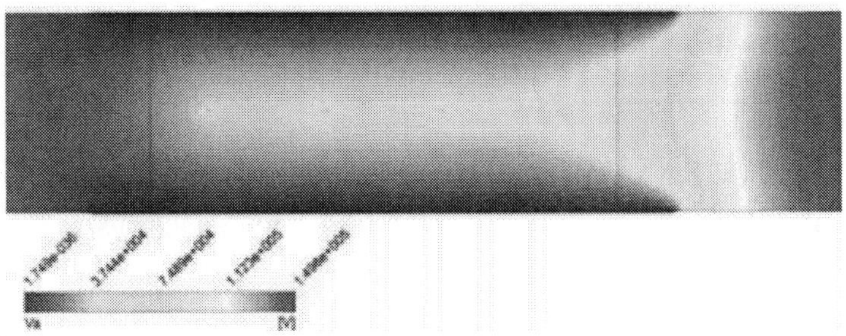

Figure 6. Distribution of electric potential (in absolute value).

Effect of Particle Loading

On Electric Current

Figure 7(a) shows the mean corona current, based on unit length of wire. As can be seen, the corona current drops quickly as particle loading increases, although a fixed voltage is applied. This is caused by the secondary electrostatic field created by the space charges of particles that cancels out the primary electric field at the wire surface and subsequently suppressed the corona generation.

On Particle Collection Efficiency

Predicted particle collection efficiency, as particle loading changes, is given in Figure 7(b). The collection efficiency, increases initially with particle loading (<3 g/m^3), peaks at particle loading of 4-5 g/m^3 and then decreases as particle loading increases further. When ignoring particle space charge, the electric field would be independent of particle loading. This is the case for very low particle loading, where the fractional collection efficiency is constant, or in other words, the quantity of dust collected as well as that penetrated increases proportionally with particle loading.

The observed enhancement of collection by a slight amount of particles is due to the redistribution of electric field so that the near-plate and volume-averaged field intensity happens to increase, which favor the charging of most particles. The asymptotic lower limit of the curve should correspond to the situation when the corona current is so small that turbulent dispersion dominates over the drift force.

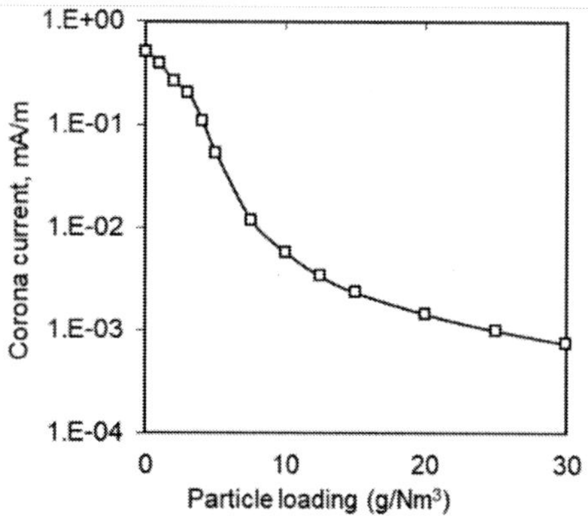

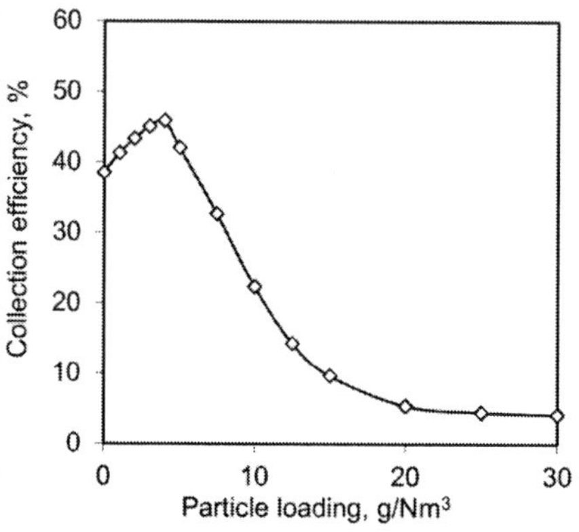

(a) (b)

Figure 7. Mean corona current (a) and collection efficiency (b) as function of particle loading.

On Electric Field Distribution

Figure 8 visualizes the predicted corona charge density and field-lines, which shows how the electric field distribution evolves with particle loading. As the electric field is distorted starting first from the end side, corona density at 3 of the 4 downstream wires is suppressed faster than the 1st wire. When the particle loading is above 7.5 g/m^3, all wires except the first one stop emitting. Meanwhile, reversed field-lines appear around the last wire, and the affected region expands as the loading increases further. Otherwise the field-lines become smoother with increasing loading. Since the field-lines are extended to the outlet, some escaped particles can be pulled back into the collection region, a factor that contributes to significantly greater deposition rate around the end-point of the plate.

Overall, the particle charges affect the particle collection adversely, as they tend to suppress the corona emission or reduce the corona current. Particularly the penetrated particles, which carried significant charges, have prolonged effect as they float along. The corona at the downstream electrodes even extinguish unless the applied voltage is raised individually. When highly charged, more particles precipitate to the plate, reducing the number of particles floating ahead (or less space charge density). The charges carried by these remaining particles exert a negative feedback to the electric field and particle charging, which compromises the collection efficiency.

Electric Quenching

Corona quenching is a common phenomenon in ESP when fine particles are treated. Riebel et al. [12] experimentally studied the corona discharge characteristics of ESP with glycerol aerosol, which found corona onset voltage increases and the corona current significantly decreases after the glycerol steam was injected into the system. Under fixed voltage in the present simulation, corona currents from all the wire electrodes drop with increasing solid loading. The downstream coronas (from the second one) will eventually extinguish at some point. An interesting phenomenon arises when the particle loading increases above 10 g/Nm^3, i.e., the electric streamlines in a small zone around the last wire reverse direction locally. This implies that a charged particle, once falls accidently within the zone, could be trapped and attracted to the same wire, as seen from the particle trajectories in Figure 9. Thus this wire is subject to heavier dust deposition.

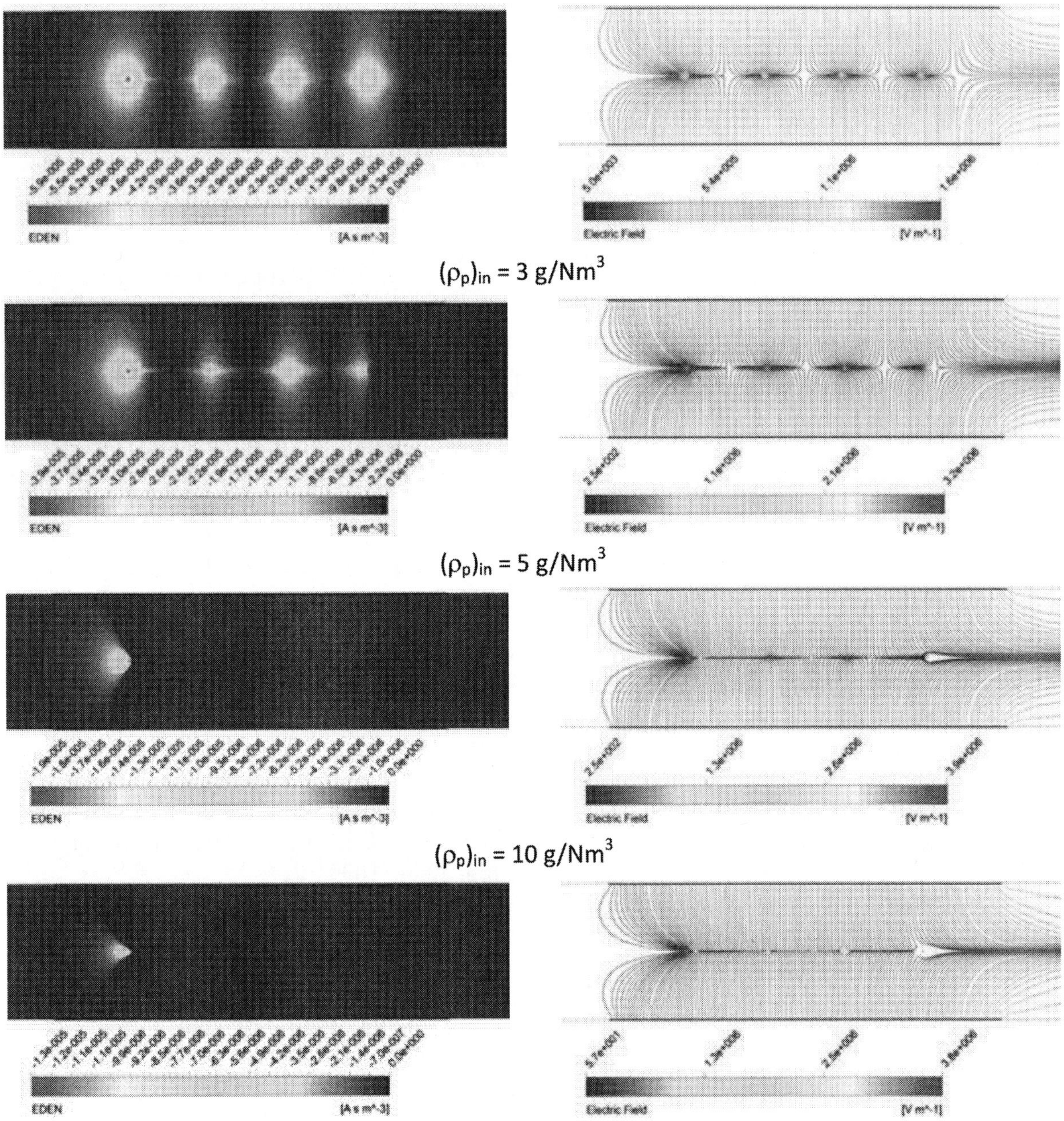

$(\rho_p)_{in}$ = 3 g/Nm^3

$(\rho_p)_{in}$ = 5 g/Nm^3

$(\rho_p)_{in}$ = 10 g/Nm^3

$(\rho_p)_{in}$ = 20 g/Nm^3

Figure 8. Visualizations of corona charge density and field streamlines for different particle loadings.

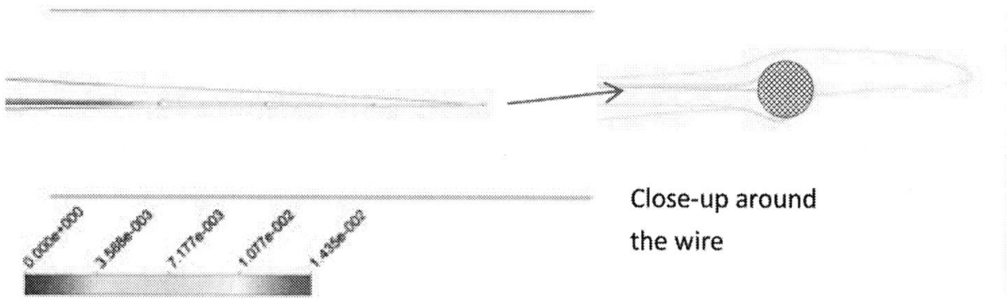

Figure 9. Trajectories of particles falling on the corona wire ($(\rho_p)_{in}$ = 20 g/Nm^3).

Concluding Remarks

CFD based ESP model has been advanced, with an enhanced coupling between electric field and particle flow. So the model can be readily applied to the early collection stage of the ESP, where dust particle loading is too high to neglect. Space charge of particles is considered, and its significance is clarified in a broad range of particle mass concentration in the gas. Dust loading significantly affects the dust collection efficiency for fine particles considered. Maximum collection efficiency exists at a critical particle loading that depends on particle size. However, as particle loading increases, collection efficiency decreases and eventually approaches asymptotically to a lower limit. The phenomenon of electrical quenching is observed, where downstream corona is extinguished or one of the emitters is polluted when the particle loading is high.

References

1. Adamiak K. (2013), "Numerical models in simulating wire-plate electrostatic precipitators: A review", *J. Electrostatics*, 71, pp.673-680.
2. Guo B.Y., Yu A.B. and Guo J. (2014), "Numerical modelling of electrostatic precipitation: effect of gas temperature", *J. Aerosol Science*, 77, pp.102-115.
3. Guo B.Y., Yu A.B. and Guo J. (2015), "Numerical modelling of ESP for design optimization", *Procedia Engineering*, 102, pp.1366-72.
4. N. Farnoosh, K. Adamiak, G.S.P. Castle (2010), "3-D numerical analysis of EHD turbulent flow and mono-disperse charged particle transport and collection in a wire-plate ESP", *J. Electrostatics*, 68, pp.513-522.
5. Choi B.S. and Fletcher C.A.J. (1998), "Turbulent particle dispersion in an electrostatic precipitator," *Applied Mathematical Modelling*, 22, pp.1009-1021.
6. Parker K.R., Applied Electrostatic Precipitation, Blackie Academic & Professional, 1997.
7. Podlinski J., Niewulis A., Mizeraczyk J. and Atten P. (2008), "ESP performance for various dust densities", *J. Electrostatics*, 66, pp.246-253.
8. Lawless P.A. (1996), "Particle charging bounds, symmetry relations, and an analytic charging rate model for the continuum regime", *J. Aerosol Sci.*, 27, pp.191-215.
9. Guo B.Y., Guo J. and Yu A.B. (2014), "Simulation of the electric field in wire-plate type electrostatic precipitators", J. Electrostatics, 72, pp.301-310
10. White H.J. (1974), "Resistivity problems in electrostatic precipitation", *J. Air Pollut. Control Assoc.* 24, pp.313–338.
11. Kasdi A. (2016), "Computation and measurement of corona current density and V–I characteristics in wires-to-plates electrostatic precipitator", *J. Electrostatics*, 81, 1-8.
12. Riebel U., Radtke R. and Loos R. (2002), "An experimental investigation on corona quenching", *J. Electrostatics*, 54, pp.159-165.

Triboelectric Charge of Spherical Glass Particles against Metal Pipeline

HoSu CHOI [1], Teruo SUZUKI [2] and Kwangseok CHOI [3,]*
[1] Christian Academy in Japan, [2] Kasuga Denki Inc.,
[3] National Institute of Occupational Safety and Health, Japan
**choiks@jniosh.go.jp*

Introduction

Insulating powders are used in many ways such as in the food, pharmacy, and chemical industries. Usually, these powders are highly charged when transported, loaded, dried, and mixed, due to repeated contact and separation between powder particles and the equipment. This results in the powders acquiring electrostatic charges. If not dissipated quickly, electrostatic charge buildup can cause inter-particle cohesion and adhesion to operating equipment, causing a production efficiency decrease. Incendiary electrostatic discharges such as spark discharges, cone discharges, and propagation brush discharges can also take place during industrial operations and may lead to fires and even explosions. Riedewald reported on an industrial fire in a pharmaceutical powder transfer operation [1]. A large number of studies involving electrostatic charge generation on powders, as well as electrostatic discharges and hazards during operation, have been reported. For example, Kaialy reviewed the main factors affecting the electrostatic charging of pharmaceuticals such as particle resistivity, size, shape, and surface roughness [2]. Gibson also reviewed industrial hazards related to electrostatic charges and discharges [3]. Our previous studies [4–6] investigated the electrostatic charging of polymer powder in mixing processes and the ignitability of aluminous, magnesium and polymer powders due to electrostatic discharges involving the effect of the ratio of N2 in the air on the ignitability.

In order to easily measure and confirm the electrostatic charge level of powders, a novel tribocharging system that utilizes the Faraday cup is presented, and the charging behavior of the sample powders is investigated in this paper.

Experimental Apparatus and method

Figure 1 shows the spiral air tribocharge apparatus presented in this study. The apparatus was divided into four main parts: the air supply part, the powder supply part, the tribocharge part, and powder collection part. The air supply part consisted of a compressor (Hitachi, PO3.7MA) that included an auto dryer (CKD, F3-217199), air controller, electronic valve, and switch. In this study, dried air was used because moisture in the air may have an effect on the electrostatic characteristics of the powder sample. The powder supply and tribocharge parts consisted of a powder port, a spiral air flow device, and a pipeline. The spiral air flow device (photo 1) was attached in front of the sample powder port in order to create spiral air inside the pipeline. The spiral air flow device had 8 holes that were 6 mm in diameter and that opened at a slant (60 degrees), which is very different from conventional models [7, 8, 9]. The straight pipeline was made up of stainless steel metal and was 1300 mm in length and 35 mm in diameter. In order to prevent external noise, the overall powder supply and tribocharge parts were shielded. The powder collection part was made up of a cyclone-type Faraday cup, unlike conventional models [7, 8, 9]. The cyclone-type Faraday cup had a separate metal container (30 mm in diameter) for easy measurement of the amount of collected powders. The other devices included two electrometers (Kethley,

6517A), a data logger (Graphtec, GL900), an electrostatic ionizer (Kasuga Denki Inc., KD-510W), an electrical balance (Shimazu, BL220H), and a timer.

The typical operating procedure for the tribocharge test is summarized as follows:
(1) Adjust the air pressure using the air pressure valve.
(2) Load 1 g of the sample powder into the powder port.
(3) Eliminate the initial charge of the sample powders using an electrostatic ionizer.
(4) Reset the electrometers (Q mode) connected to the pipeline and cyclone.
(5) Switch the air flow on, and run the sample powder through the pipeline where friction occurs.
(6) Measure the indication value when it stabilizes.
(7) Measure the amount of powder collected in the Faraday cup.
(8) Clean out the pipeline and cyclone part using an air spray in preparation for the next test.

The charge to mass ratio, q (nC/g), of the sample powder was obtained by dividing the charge, Q (nC), by mass, m (g), of the sample powder loaded into the Faraday cup:

$$q = \frac{Q}{m} \qquad (1)$$

This experiment is mainly to determine how much the charge to mass ratio, q (nC/g), of the sample powders is affected by the size of the powder and the amount of air pressure. The cyclone-type Faraday cup and the pipeline were suspended electrically with insulators. All experiments were conducted in ambient conditions (air temperature: 22 ℃; relative humidity: 55 %).

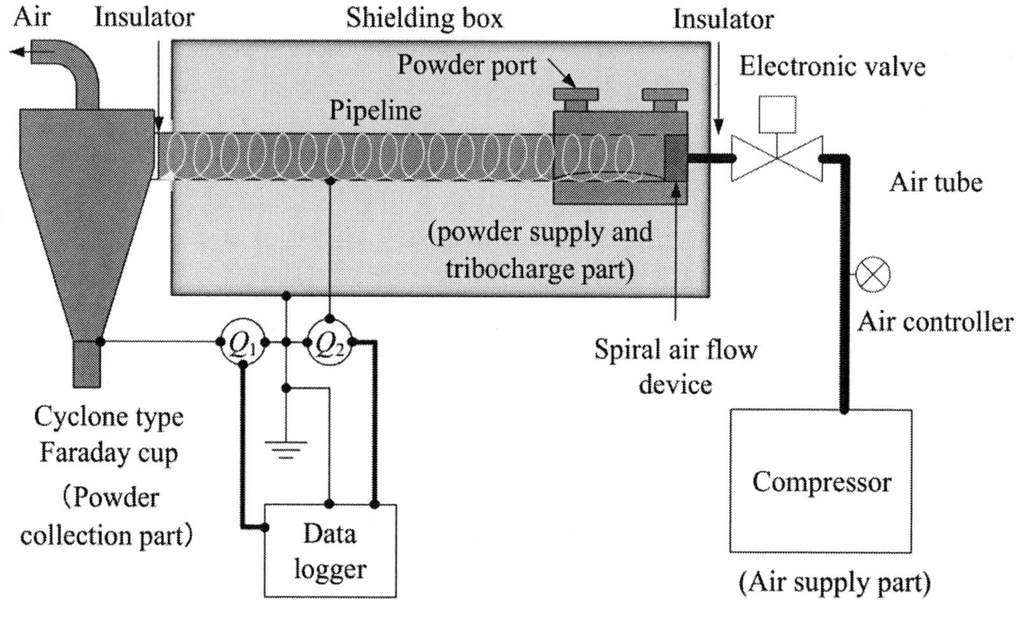

Fig. 1 Schematic of the spiral air tribocharge apparatus used in this study.

Photo 1 Structure of the powder port and spiral air flow device.

Sample powders

Glass beads are extremely good light reflectors and thus are widely used as the main component of the coating material for light-reflecting adhesive tape, background paint for traffic control sign boards, and paint for road signs. Four different grades of commercial glass beads were used in this experiment, as shown in Table 1. The particle size was determined using a laser diffraction system (LDA Win 1.21, dry type). Hereafter, they will be called by their abbreviation shown in Table 1. The analysis of the images from microscopy (Neoscope, JCM-500) revealed that the G-4 powder particles were spherical, as shown in photo 2. The other glass bead particles were the same shape as G-4.

Table 1. List of powder specimens used in experiments.

Specimen	Abbreviation	Diameter [μm]		
		D_{10}	D_{50}	D_{90}
Glass beads 1	G-1	22	62	85
Glass beads 2	G-2	93	131	175
Glass beads 3	G-3	130	180	241
Glass beads 4	G-4	162	225	294

D_X: X% volume average.

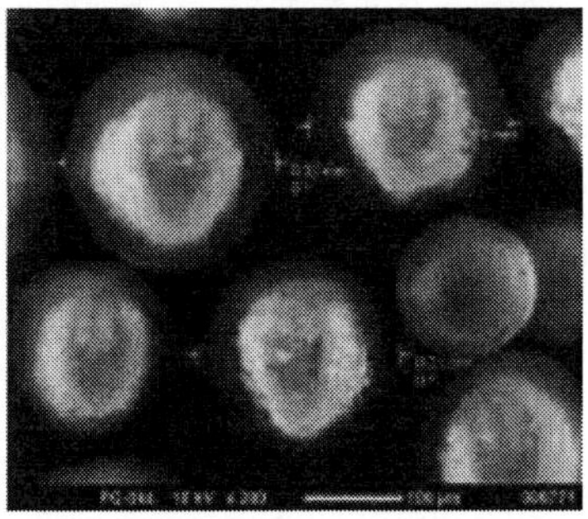

Photo 2 Shape of G-4 powder used in this experiment.

Experimental results and discussion

Figure 2 shows the typical change of charge detected over time when the G-3 particle was used and the air pressure was 0.15 MPa. It should be noted here that when no particle was used and only air was passed through the pipeline, the charge was detected as almost zero. The charge Q1 of the powder was found through the cyclone-type Faraday cup and then was confirmed through the charge Q2 obtained from the pipeline. As the powder particles passed through the pipeline for about 0.82 seconds, as seen in the Fig. 2, the charge clearly increased. Q1 and Q2 had almost the same trend and absolute values. This result confirmed the reliability of the test and showed that there was no disturbance by external noise.

In this study, all of the glass beads (G-1 to G-4) were negatively charged when stainless steel metal was used as the contact surface. The charge polarity of the materials relates to the work function (or ionization potential) of each material. The contact potential difference between two substances is defined by the difference between their fermi levels, which correlate with the work function. In general, materials with higher work functions tend to appropriate electrons from materials with lower work functions. This point will be carried out in our further study.

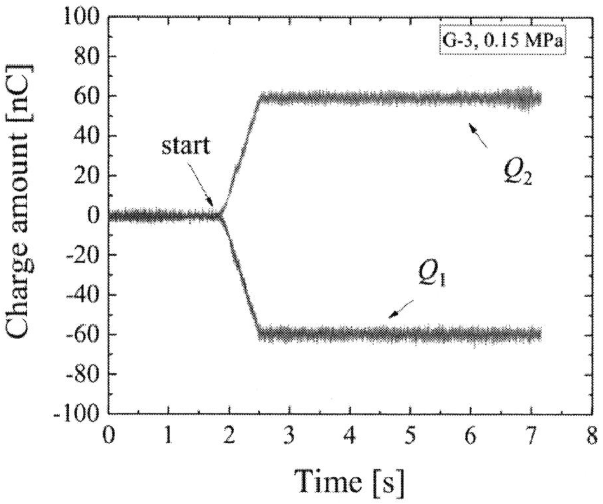

Fig. 2 Change of charge detected over time.

Figure 3 shows the charge to mass ratio of glass beads as a function of applied air. The range of applied air was between 0.1 MPa and 0.3 MPa in this study. When the applied air pressure was below 0.1 MPa, it was most likely not enough to send all of the powder through the pipeline. Each experiment was repeated five times, and the average values were reported. This graph does not contain an error bar, but these were about 10%. The results showed that the charge to mass ratio of both powders increased as the applied air increased. Although not shown in Figs. 2 and 3, the increase in air pressure also causes the slope of the graph to increase significantly. This can be explained by the fact that with increased pressure of applied air, the particles collide against the surface of the pipeline with more strength [9]. In order to investigate in more detail, the velocity of air at approximately 10 mm from the inside of the pipeline wall was measured using an anemometer (Kanomax, Anemomaster-6115). The approximate velocity of air flow was 9.5 m/s at 0.1 MPa, 12 m/s at 0.15 MPa, 15 m/s at 0.2 MPa, 20 m/s at 0.25 MPa, and 22 m/s at 0.3 MPa. It could be clearly seen that as the air pressure increased, the air velocity also increased. Another reason for this may be that because the air was applied in a spiral form, the particles may have collided more frequently against the pipeline when the air flow was increased. This exception will be investigated in detail in the future.

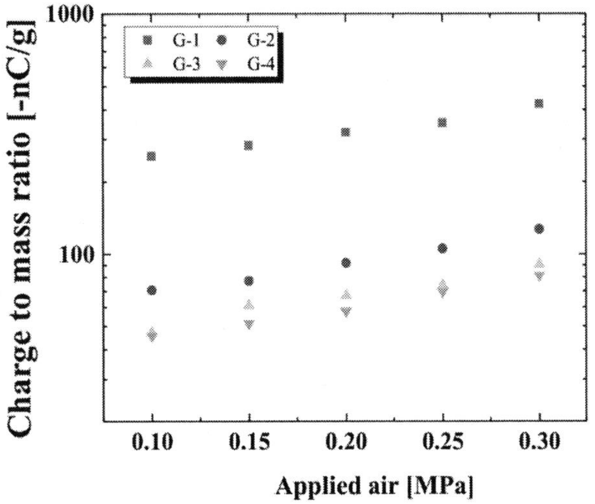

Fig. 3 Charge to mass ratio of glass beads as a function of applied air.

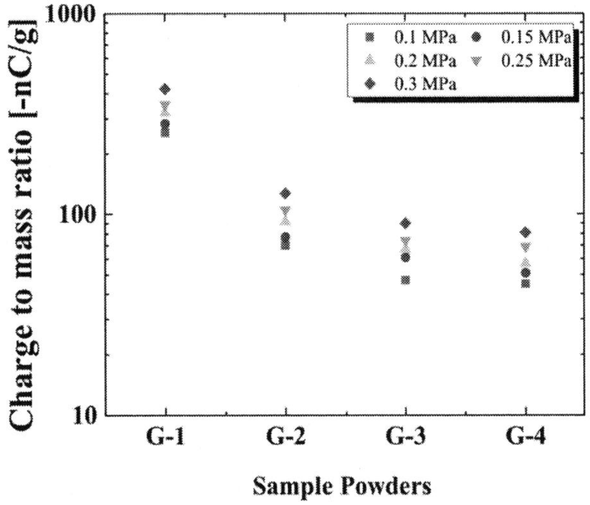

Fig. 4 Charge to mass ratio as a function of different glass bead sizes.

Figure 4 shows the charge to mass ratio as a function of different sizes of glass beads. The size distribution of each glass bead in this study was described in Table 1. The results showed that the charge to mass ratio of the glass beads increased as the powder size decreased. That is to say, smaller particles had a larger q than the larger ones. The total surface area of all of the smaller beads combined is greater than that of the larger ones, which results in a high charge to mass ratio development in the tribocharge. The value range of the q of the glass beads obtained in this study was about 40 nC/g to 100 nC/g. These values agreed with those obtained from the micronizing and pneumatic transfer process [10]. These values were high enough to generate incendiary electrostatic discharges during handling in industrial processes. This suggests that if glass beads are used with solvents in the cleaning process, they must be handled carefully, as incendiary discharges can ignite the solvents even though the glass beads themselves are nonflammable material.

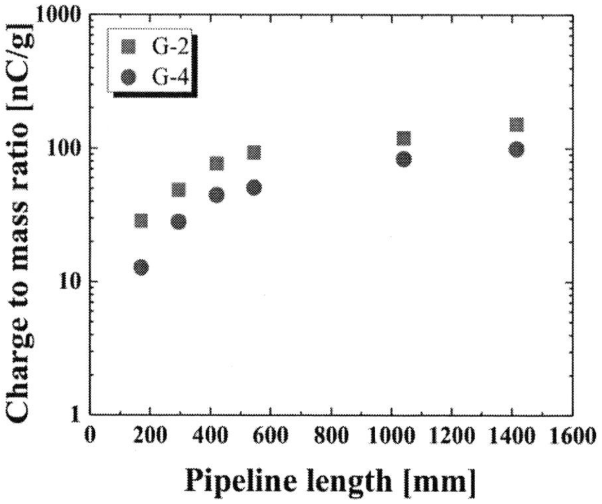

Fig. 5 Charge to mass ratio as a function of pipeline length.

Figure 5 shows the charge to mass ratio as a function of pipeline length. Six different pipeline lengths, *d*, were used: 170, 295, 420, 545, 1040, and 1415 mm. The glass beads used in this experiment were G-2 and G-4, and the air pressure was 0.35 MPa. Since the pipeline length changed with each test, it was difficult to attach the Faraday cup to the end of the pipeline. Therefore, in this study, only the value of q_2 was measured. As seen in Fig. 2, q_2 had the almost same value as q_1 but had an opposite sign. The length of the pipeline was measured starting from beginning of the powder port because the inside structure of the powder port was the same as that of the pipeline. Each experiment was repeated five times, and the average values were reported. The results showed that q_2 increased as the length of the pipeline increased for both sizes of glass beads. This can be explained by the fact that as the length increases, the time the glass beads come in contact with the pipeline also increases. The increase rate at the beginning section of the pipeline was especially large. It can also be seen through Fig. 5 that overall, G-2 had a larger q_2 than G-4 because of the total surface area. This confirms the reliability of the data shown in Fig. 4.

In addition, both beads seemed to reach saturation when the pipeline length was 1415 mm. It was compared to the maximum charge to mass ratio, q_c [nC/g] calculated from the well-known equation shown below [11]:

$$q_c = |\sigma_{max}|/(D \cdot \rho), \quad \text{--------------------------(2)}$$

where σ_{max} [kg/m^3] is the maximum surface charge density (27 μC/m^2); D [μm] is the particle diameter; and ρ is the particle density (1.5×10^3 kg/m^3).

In this study, the particle sizes of G-2 and G-4 were 131 μm and 225 μm, respectively, as shown in Table 1. The calculated maximum values q_c and experimental maximum values q_t are compared as shown in Table 2. The calculated and experimental values for both G-2 and G-4 agreed well with each other. Therefore, it can be seen that both G-2 and G-4 reached saturation.

Table 2. Calculated maximum values and experimental maximum values of the charge to mass ratio.

	q_c [nC/g]	q_t [nC/g]
G-2	137	151
G-4	80	99

Conclusion

In this paper, a novel tribocharging system that utilizes the Faraday cup was presented and the electrostatic charge behavior of glass beads was experimentally investigated. The results were as follows:
1) For all of the glass beads, the charge clearly increased as the glass bead particles passed through the pipeline. The charge amount of the glass beads and the charge amount obtained from the pipeline had almost the same trend and absolute values. This result confirmed the reliability of the test and showed that there was no disturbance by external noise.
2) All of the glass beads were negatively charged when stainless steel metal was used as the contact surface. The value range of the charge to mass ratio of the glass beads obtained in this study was about 40 nC/g to 100 nC/g.
3) The charge amount of the glass beads increased as the applied air increased. It could also be seen that as the air pressure increased, the air velocity also increased.
4) The charge amount of the glass beads increased as the powder size decreased due to the total surface area of the beads.
5) The charge amount of the glass beads (G-2, G-4) increased as the length of the pipeline. The increase rate at the beginning section of the pipeline was especially large.

Acknowledgments
The authors gratefully acknowledge Mr. Naoto Nogera for his assistance during the test.

References

1. F. Riedewald (2012): "A fire in a secondary pharmaceutical powder transfer operation," *Process safety progress*, 31, 4, pp. 390-392.
2. W. Kaialy W, "A review of factors affecting electrostatic charging of pharmaceuticals and adhesive mixtures for inhalation," *Int.J.Pharm.* Vol.503, pp.262-276.
3. N. Gibson (1997), "Static electricity — an industrial hazard under control," *J. Electrostatics,* Vol. 40-41, pp. 21-30
4. K. Choi, K. Choi, K. Nishimura (2015), "Experimental study on the influence of the nitrogen concentration in the air on the minimum ignition energies of combustible powders due to electrostatic discharges," *J. of Loss Prevention in the Process Industries.* Vol.34.pp.163-166.
5. K.S. Choi, H.Sakasai, and K. Nishimura (2015), "Experimental study on ignitability of pure aluminum powders due to electrostatic discharges and Nitrogen's effect," *J. of Loss Prevention in the Process Industries.* Vol.35.pp.232-235.

6. K.S. Choi, H. Sakasai, and K. Nishimura (2015), “Minimum ignition energies of pure magnesium powders due to electrostatic discharges and Nitrogen's effect,” *J. of Loss Prevention in the Process Industries.* Vol.41.pp.144-146.
7. N. Masui, and Y. Murata (1981), “charging tendency of the flowing powder through a metal tube (II), ” *The five Annual meeting of Institute of Electrostatic Japan,* pp. 244-247.
8. S. Watano, S. Saito, and T. Suzuki (2002), “Feedback control of powder electrostatic charge in a pneumatic conveying process,” *J. Soc. Powder,Japan*, Vol. 39, pp. 496-502.
9. P. Boschung, M. Glor (1980), “Methods for investigating the electrostatic behavior of powders, J. of elelctrostatic,” Vol. 8, pp. 205-219.
10. IEC50404(2004), “Electrostatics- Code of practice for the avoidance of hazards due to static electricity,” p. 66.
11. M.Glor (1988), “Electrostatic hazards in powder handling,” *Research studies press LTD.*, p.26.

509266 Spray-Drying of a Layered Double Hydroxide Nanosuspension

Abstract:

Over the past few years, layered double hydroxide (LDH) has attracted particular interests in both academia and industry. LDH is 2D anionic clay of brucite-like layered structure with various organic or inorganic anions located between layers. The diverse composition of LDH, and its associated physical and chemical properties open up new possibilities in various applications. LDHs are successfully applied as precursors in synthesis of supported catalysts, sorbents, carriers delivering bioactive molecules, matrices containing active pharmaceutical ingredients for control release, electrochemical biosensors, etc. LDHs are mainly synthesized in a form of nanosuspensions. The application of nano-sized sorbents or catalysts in the liquid-phase processes requires a special technique to separate them from the final product. An uncontrollable agglomeration of solid nanoparticles makes their industrial application impossible in a dry phase. Moreover, the toxic effects of nanoparticles on human health and ecosystem have been discussed in the literature.

Utilization of larger micron-size agglomerates of nanoparticles is one of the alternative ways to safely handle nanoparticles in the dry phase on the industrial scale. Spray drying is one of the well-known techniques that have been widely applied for production of microspheres from nano-sized particles. Using this technique, it is possible to control the size distributions as well as the morphology of spray-dried agglomerates. However, there are only a limited number of publications related to spray drying of LDH nanosuspensions. Moreover, spray-dried LDH agglomerates of rather wide size distribution, low surface area and irregular shape were reported in the literature.

In this study, we explore the effect of process parameters, such as spraying air flow rate, drying air temperature, slurry concentration and slurry feed rate on the size distribution, morphology, specific surface area and pore volume of LDH agglomerates fabricated by spray drying of nanosuspensions.

The round-shaped agglomerates with a diameter of more than 2 and less than 15 micrometers were obtained by spray drying of suspensions of nanoparticles with median diameter of 171 nm without addition of dispersant and binder. Our results confirmed that the spraying air flow rate has the dominant effect on the agglomerate median size and the width of agglomerate size distribution as well as on the agglomerate specific surface area. The small agglomerates with narrow size distribution were collected during drying at the high flow rate of spraying air. Our results also indicate that lowering the temperature of drying air and using the dilute nanosuspensions promote generation of small agglomerates.

The specific surface area of spray-dried LDH agglomerates is in the range of 49 to 101 m^2/g. The agglomerates of high specific surface area were obtained at low flow rate of spraying air, high slurry feed rate of concentrated slurry and using low temperature drying air. These results also confirm that drying at low drying rate yields to formation of agglomerates of high surface area. Similar trends were observed in terms of the variation of total pore volume with drying process parameters.

Thus, by adjusting the spray drying conditions, it is possible to produce agglomerates of LDH nanoparticles satisfying the product requirements.

513829 Impact of Spheroidization of UO_2 Powders on the Filling of Press Molds

Abstract:

In industry, the use of powders is very common. This underlines the interest in understanding the flow of powders in process lines (hopper, silos, transportâ€¦) and finding the link between powder physical characteristics and its rheological properties. This study deals with the flowing properties of a UO2 powder in a powder metallurgy process for nuclear fuel fabrication. A reproducible filling of the press molds is required to avoid mechanical grinding of the sintered pellets.

A powder is characterized by its particle size distribution but also by the morphology of particles. Morphology has a major impact on the cohesion and then on the flowability of powders [1]. UO2 powders are generally made up agglomerates which are composed of aggregates, themselves formed by crystallites (Figure 1). Agglomerates are primarily governing the powder flowing behavior. Their characteristics sometimes do not allow them to meet the expectations of the process, especially for the filling of the press molds.

For this reason, in a powder manufacturing process, a spheroidization operation (Figure 2) could precede the shaping of pellets to improve the filling properties.

In this study we observe the influence of particles morphology on the rheological properties of UO2 powders. Whereas there is no standard and universal method to characterize the flow of powders [2], there is direct (shear cells) and indirect methods (dynamic flowability measurements like permeability). We use both methods: the indirect methods based on qualitative measures of powder cohesion, the direct methods to determine the powder cohesion for a specific consolidation state through the Freeman shear cell [3]. While a shear cell measurement permits to determine the powder flow in quasi-static system, a permeability measurement shows the flow index in dynamic system. We identify the link between theses measurements and the quality of the filling.

References:
[1] Xiaowei Fu and others, «Effect of Particle Shape and Size on Flow Properties of Lactose Powders», Particuology, Advances in Characterization and Modeling of Particulate Processes, Apr. 2012.
[2] P. Juliano, B. Muhunthan, and G. V. Barbosa-Cánovas, « Flow and shear descriptors of preconsolidated food powders », J. Food Eng., vol. 72, no 2, p. 157-166, Jan. 2006.
[3] J. Schwedes and D. Schulze, « Measurement of flow properties of bulk solids », Powder Technol., vol. 61, no 1, p. 59-68, Apr. 1990.

508203 Modelling a Twin Screw Granulator Using the Discrete Element Method

Abstract:

Wet granulation is a process used to create larger stable agglomerates (granules) from fine powders. This has many desirable outcomes such as improving flowability, compactibility, and homogeneity. Granulation is commonly employed in the food, pharmaceutical, detergent, and fertilizer industries, but despite its wide adoption it is often inefficiently operated, with high recycle ratios in continuous processes and high rejection rates in batch processes. Wet granulation is the most common type of granulation and, in pharmaceutical applications, it is a critical step in tablet manufacturing that affects the uniformity and compactibility of the final dosage form.

Traditionally batch granulation was the favoured granulation process method in the pharmaceutical industry due to the challenges faced from adopting a continuous approach, such as cost or inability to monitor product quality reliably. Since the introduction of Quality by Design (QbD) and Process Analytical Technology (PAT) by the FDA in 2003 there has been a move towards new manufacturing techniques such as continuous processing.

Although tremendous efforts have been made to gain scientific insight into the granulating process, a fundamental understanding of wet granulation is still lacking due to the complexity of the mechanisms involved. With twin screw granulation becoming a popularly employed method of wet granulation, an in-depth understanding of particle enlargement in the granulating process is necessary in order to improve the quality of the final product without the need for large-scale Design-of-Experiment studies.

Despite the extensive experimental research carried out for twin screw granulators in recent years, there has been little computational work carried out. This paper employs the Discrete Element Method (DEM) to study a 25 mm diameter, GEA ConsiGmaâ„¢ 1 twin screw granulator with a typical 60Â° forward configuration for kneading elements. The DEM simulations were conducted using the commercial code EDEM with a DEM contact model developed for cohesive solids implemented through an API. The contact model is based on an elasto-plastic contact with adhesion and uses hysteretic loading and unloading paths to model the elastic-plastic contact deformation. In these simulations, the adhesion is used to capture the effect of the binder liquid without the complication of modelling the liquid directly. The adhesion parameter is a function of the plastic contact overlap. The model has previously been shown to be able to predict the stress-history-dependent behaviour depicted by a flow function of the material. In this study the effect of particle size, gap tolerance and cohesion on the residence time distribution and local solid fraction at different elements in the granulator is assessed. The level of cohesion used is related back to uniaxial test results which can be used to measure the flowability of a cohesive granular solid.

508216 Hybrid Multiscale Modelling of a Twin Screw Granulator

Abstract:

Wet granulation is a process used to create larger stable agglomerates (granules) from fine powders. This has many desirable outcomes such as improving flowability, compactibility, and homogeneity. Granulation is commonly employed in the food, pharmaceutical, detergent, and fertilizer industries, but despite its wide adoption it is often inefficiently operated, with high recycle ratios in continuous processes and high rejection rates in batch processes. Wet granulation is the most common type of granulation and, in pharmaceutical applications, it is a critical step in tablet manufacturing that affects the uniformity and compactibility of the final dosage form.

Modelling processes such as wet granulation is not straightforward, but may be achieved using techniques such as Population Balance Modelling (PBM) and Discrete Element Method (DEM). PBM groups particles into different classes based on their properties and uses empirically derived differential equations to describe key processes such as aggregation or breakage. However, although PBM tracks the size and number of particles over time, it cannot account for the distribution of the particles and the effect of this local variability.

By contrast, DEM simulations track individual particles, and as such, phenomena such as segregation or liquid migration can be studied. DEM simulations also consider the effect that the geometry and its dynamics have on the particles' behaviour, which is something that PBM cannot account for directly. However, DEM simulations of fine materials can be computationally expensive and this often results in reduced numbers of particles being used to make the simulation computationally tractable. This is usually achieved through increased, unrealistic particle sizes or the simulation of isolated local zones without considering adjacent zones which may affect the behaviour.

Traditionally, large-scale Design-of-Experiment (DoE) studies would be needed to estimate key rate parameters for use in PBM modelling of granulation and this can be both costly and time-consuming. DEM simulations provide an alternative method to estimate these parameters. By coupling the two simulation methods, one can utilise the key benefits of each method to better simulate complex processes.

This paper presents a hybrid approach to modelling twin screw granulation via DEM–PBM coupled simulations to study a 25 mm diameter, GEA ConsiGma™ 1 twin screw granulator with a typical 60° forward configuration for kneading elements. The DEM simulation was conducted using the commercial code EDEM with a DEM contact model developed for cohesive solids implemented through an API. The PBM simulation is carried out with the commercial code gFORMULATE which is built on top on the advanced modelling platform gPROMS. A bi-directional coupling is implemented with information shared between the two simulation tools at specified intervals. Preliminary results for this coupled approach are presented and discussed.

509577 Blocking Rules for Discharging Granular Materials from a Flat Bottom Hopper

Abstract:

Hoppers and silos are widely used in storing granular materials in various industries, such as agricultural, chemical, food and pharmaceutical industries. Three key aspects need to be considered when hoppers are used to store powders and grains: 1) stress analysis; 2) hopper blockage ; 3) mass flow rate. A hopper must be designed strong enough to sustain the stress generated from the bulk materials during storage and discharge. On the other hand, a hopper must be designed properly so that the stored bulk materials can be discharged, i.e. without blockage and formation of ratholes and arches. Furthermore, the mass flow rate needs to be predicted and controlled. In this study, a combined experimental and numerical study was performed to determine the blocking rule for discharging granular materials from a flat bottom hopper. Granular materials of various sizes were considered and the minimal orifice size through which the bulk materials can be discharged was determined experimentally using a Flodex tester and numerically using DEM. The dependency of the minimal

orifice size on particle size was explored. An semi-empirical model was also developed. Using this model, the critical orifice size at which the flow will stop, i.e. blockage takes place, can be determined from the mass flow rate data. It was found that excellent agreement between the numerical results and experimental measurements was obtained. Both experimental and numerical results showed that, even for spherical particles, layered dead zones could be formed during the discharge from the flat bottom hopper. In addition, mono-layered dead zones were developed when large particles were used. Furthermore, the developed blocking rules were also verified experimentally and numerically.

508310 Characterization of Mesoscopic Structure in Cohesive Powder, Neat or Blended, By X-Ray Computed Tomography and Prediction By the Discrete Element Method

Abstract:

Understanding and predicting structural characteristics of cohesive powder systems is a technical challenge in many powder technology processes and applications. The discrete element method (DEM) [1] is a promising approach to model the structure and flow behavior of cohesive granular systems. Advances in experimental techniques such as X-ray computed tomography (XRCT) have now enabled characterization of fine cohesive powder systems with sub-micron resolution. Combined with standard bulk powder characterization methods (e.g. density), this technique allows for mesoscopic structural characteristics of cohesive powder systems to be compared with DEM simulation. In this study, DEM and XRCT are used to characterize the mesoscale structure, under controlled consolidation conditions, of two moderately cohesive 'model' (roughly spherical) powders with distinct material properties. The DEM simulations are based on a constitutive elasto-plastic model [2] that makes use of the JKR theory [3] to account for cohesive interparticle forces. Density and void structure obtained from DEM simulation are compared to characterization results obtained from XRCT image analysis. XRCT and DEM simulation approach is applied to examine structures of binary mixtures of the two model powders. The ability of DEM simulation and the contact model to accurately predict structural characteristics of single component and binary granular systems is discussed.

References

[1] P. A. Cundall and Otto D.L. Strack. A discrete numerical model for granular assemblies. Geotechnique, 29(1):47–65, 1979.

[2] C. Thornton and Z. Ning. A theoretical model for the stick/bounce behavior of adhesive, elastic-plastic spheres. Powder technology, 99(2):154– 162, 1998.

[3] K.L. Johnson, K. Kendall, and A.D. Roberts. Surface energy and the contact of elastic solids. In Proceedings of the Royal Society of London A: Mathematical, Physical and Engineering Sciences, volume 324, pages 301–313. The Royal Society, 1971.

511638 Numerical Research of Hydrodynamics in Gas-Solid Micro Fluidized Beds

Abstract:

Due to the advantages of less energy consumption, more safety and good operability, the miniaturization of fluidized bed, or named micro fluidized bed, has been extensively studied by a lot of researchers. The Micro Fluidized Bed Reaction Analyzer (MFBRA) was adopted by the Institute of Process Engineering (IPE), Chinese Academy of Sciences (CAS), to study the gas-solid reaction kinetics under isothermal differential condition[1, 2]. In the MFBRA, the hydrodynamics of micro fluidized beds play a vital role on the heat and mass transfer and which is a key to the accurate measurement of gas-solid reaction kinetics. In order to optimize and design the MFBRA, numerous experiments have been conducted to study the hydrodynamics of micro fluidized beds[3, 4]. Unfortunately, those works mostly focused on the bed pressure drop and minimum fluidization velocity, nevertheless, the detail information, such as local solid volume fraction and particle velocity profile in micro fluidized beds are reported rarely. It is because the existing experimental techniques, such as, optical fiber and capacitance probes are hard to measure those local parameters without disturbing the flow field.

The objective of this work is to study the hydrodynamics of micro fluidized beds and to investigate the isothermal characteristics during kinetics testing of graphite powder combustion in MFBRA. In this study, Discrete Particle Model (DPM) and Two-Fluid Model (TFM) are selected to investigate micro fluidized beds, respectively. The DPM is carried out by the commercial software EDEM-Fluent Coupling simulations. And for TFM, we used the Eulerian multiphase model in the ANSYS Fluent. From our numerical simulations, the detailed hydrodynamic characteristics of micro fluidized beds are explored systematically by comparing different column sizes and superficial gas velocity. Meanwhile, it is found that both DPM and TFM have the ability to model the MFBRA under practical operation condition with high superficial gas velocity. In the term of isothermal characteristics in micro fluidized beds, the temperature field distribution as well as the heat transfer between heat-carrier and graphite powder are explored by the combination of simulations and experiments.

Reference:

1. Xu, G., *Micro fluidized bed reaction kinetics analyzer. Supported by instru-ment development project of Chinese Academy of Sciences, China, Project numberY2005014. (in Chinese).* 2005.
2. Yu, J., et al., *Isothermal differential characteristics of gasâ€"solid reaction in micro-fluidized bed reactor.* Fuel, 2013. **103**: p. 29-36.
3. Liu, X., G. Xu, and S. Gao, *Micro fluidized beds: Wall effect and operability.* Chemical Engineering Journal, 2008. **137**(2): p. 302-307.
4. Guo, Q.j., Y. Xu, and X. Yue, *Fluidization Characteristics in Micro-Fluidized Beds of Various Inner Diameters.* Chemical Engineering & Technology, 2009. **32**(12): p. 1992-1999.

507312 An Improved Bubble-Based Drag ModelÂ for Accurate Coarse-Grid Two-Fluid Modeling of Geldart a Powder Bubbling Fluidization

Abstract:

An improved bubble-based energy minimization multi-scale (EMMS) drag model was formulated in which the original bubble-based EMMS drag model at macroscale bed level was extended to microscale sub-grid level for accurate coarse-grid two-fluid modeling (TFM) of Geldart A powder bubbling fluidization. At both the macroscale and microscale levels, the physically heterogeneous structures were modeled by three pseudo phases, i.e., bubble phase, emulsion phase, and inter-phase, where conservation and stability condition equations were applied. Through such scale and phase decompositions, the improved bubble-based EMMS drag model was formulated through a two-step macro-to-micro scheme, where the obtained drag coefficient is related to phase volume fractions, velocities, and accelerations within each coarse grid. Modeling results show that TFM combined with the improved bubble-based EMMS drag model can accurately reproduce the heterogeneous structures for a laboratory-scale Geldart A powder bubbling fluidized bed. At relatively coarse grid, the predicted axial and radial distributions of solid concentration agree very well with experimental measurements. This indicates that our proposed improved bubble-based EMMS drag model incorporated TFM is promising for accurate and economical simulation of engineering-scale Geldart A powder bubbling fluidization.

507860 Numerical Simulation on Fine Particle Transport Behaviour in Electrostatic Precipitators

Abstract:

Electrostatic precipitators (ESPs) are a major air pollution control device to remove particles from treated gas. Extensive research has been done on several aspects such as electrostatics, fluid dynamics, charging mechanism, and particle dynamics. The overall particle-collection efficiency of a modern industrial wire-plate ESP exceeds 99%. However, the removal efficiency of submicron particles by conventional ESPs is as low as 70â€"80% owing to the low charge carried by fine particles and the increase in mobility with decreasing size. Submicron particles possess a high superficial area, thus favouring toxic heavy metals and other pollutants. Therefore, fine particles are collected with much lower efficiency and further research is required to improve this aspect.

In this work, a numerical study on the charging and transport behaviour of fine particles has been carried out based on wire-plate electrostatic precipitators (ESPs) with multiple wire electrodes. The effect of the applied wire voltage, inlet height, and wire distance on particle charging and transport behaviour, and the influence of the precipitator structure on particle trapping are analysed in detail. The 2-D numerical model involves electric field, space-charge density, gas flow, and particle trajectories. The fluid-flow velocity is higher than 1 m/s in this study and the EHD flow has a negligible. A Lagrangian approach for particle transport is utilized with the trajectories of individual particles tracked by solving their equation of motion. The electric field and gas flow are calculated using the finite-volume method. The flow simulation is done using the commercial software FLUENT with other parts coded as user-defined functions. The following main results can be drawn:

(1) It was observed that the Y-displacement of the particles exhibited a universal increase for all particle sizes, especially for larger particles, at low injection positions (Y < 25 mm). The larger the particle size, the higher the Y-displacement of particles because larger particle sizes improve both the particle charging level and electric field strength. No significant improvements were found for submicron particles.

(2) Increasing the spacing of the discharge electrode within a certain range and therefore increasing the electric field, it was observed that the particle-trapping effect improved and the outlet level exhibited a universal increase for all particle sizes, especially for large particles.

(3) The ESP models M1 and M2 were used to investigate their influence on the outlet height. Except for particles with a diameter of 2 µm, all other particle sizes showed an increased outlet height in M2 compared to that in M0. In addition, the different discharge-electrode arrangements, especially in M2, exhibited significant effects on the charging and transport behaviour of particles with a diameter of 1 µm.

(4) Particles with an inlet height of 5 and 10 mm were studied in M1 and M2 precipitators. The changed discharge-electrode arrangement improves particle trapping, especially in M2. This is because of the electron trajectories being closer to the discharge electrodes (M2), giving the particles more charge and leading to a greater electric field.

507879 Lattice Boltzmann Simulations of Porous Particulate Flows

Abstract:

The comprehensive understanding of the particle-fluid interaction is very important for fluid-particles system, and thus many studies have been conducted for the particle-fluid system[1-3] of solid impermeable particles. In the real industrial fields such as the catalyst process in chemical engineering[4], the particles usually possess the porous structure which will have effects on the fluid-particles interaction[5-6]. However, there has been little open literature discussing the particle-fluid system of porous permeable particles. Therefore, the objective of the present study is to investigate the effect of the porosity and permeability on the porous particulate flows which are formulated by the volume-averaged governing equations[7] in terms of intrinsic phase average velocity. The lattice Boltzmann equation model (LBE)[8] is used to solve the governing equations. Firstly, we validate the LBE model by the numerical results existing in the literature for the flow around and through one porous particle. Then, the porous particles settling against gravity in a quiescent fluid, i.e., the DKT process[9] for porous particles, is investigated. The effects of the porosity and Darcy number on the porous particulate flows are studied numerically in details. Also, we present the simulation results of the fluidization of 512 porous particles. Our results demonstrate the significance of the porous structure of particles for the particles dynamics behaviors.

Reference

[1] Ladd, A.J.C., 1994. Numerical simulations of particulate suspensions via a discretized Boltzmann equation Part I. Theoretical foundation. J. Fluid Mech., 271, 285-310.

[2] Hill, R.J., Koch, D.L., Ladd, A.J.C., 2001. The first effects of fluid inertia on flows in ordered and random arrays of spheres. J. Fluid Mech. 448, 213-241.

[3] Van der Hoef, M.A., Beetstra, R., Kuipers, J.A.M., 2005. Lattice Boltzmann simulations of low-Reynolds-number flow past mono- and bidisperse arrays of spheres: results for the permeability and drag force. J. Fluid Mech. 528, 233-254.

[4] Zhao, Y.F., Li, H., Ye, M., Liu, Z.M., 2013. 3D numerical simulation of a large scale MTO fluidized bed reactor. Ind. Eng. Chem. Res. 52, 11354-11364.

[5] Noymer, P.D., Glicksman, L.R., Devendran, A., 1998. Drag on a permeable cylinder in steady flow at moderate Reynolds numbers. Chem. Eng. Sci. 53, 2859-2869.

[6] Zhu, Q.Y., Chen, Y.Q., Yu, H.Z., 2014. Numerical simulation of the flow around and through a hygroscopic porous circular cylinder. Comput. Fluids 92, 188-198.

[7] Wang, L., Wang, L.-P., Guo, Z.L., Mi, J.C., 2015. Volume-averaged macroscopic equation for fluid flow in moving porous media. Int. J. Heat Mass Trans. 82, 357-368.

[8] Chen, S.Y., Doolen, G.D., 1998. Lattice Boltzmann method for fluid flows. Annu. Rev. Fluid Mech. 30, 329-364.

[9] Fortes, A.F., Joseph, D.D., Lundgren, T.S., 1987. Nonlinear mechanics of fluidization of beds of spherical particles. J. Fluid Mech. 177, 467-483.

508370 Modeling of a Novel Multi-Particle Collision Model for Gas-Solid FlowsÂ

Abstract:

Fluidized beds are widely applied in the chemical, petrochemical, metallurgical, environmental and energy industries due to their favourable mass and heat transfer characteristics. In past few decades computational fluid dynamics has proven to predict hydrodynamics of fluidized beds. DPM (Euler-Lagrange) [1] and TFM (Euler-Euler) [2] are two most common modelling approaches for gas-solid flow in fluidized beds. In TFM the solid phase itself is treated as a continuum phase interpenetrating with the gas phase. Since the concept of individual particles has disappeared, the effects of particle-particle interactions in the TFM has to be incorporated in a statistically averaged way, via the solid phase pressure and viscosity. TFM is preferred model to study large scale fluidized beds but model suffer from high numerical diffusion. And the KTGF theory in TFM assume elastic collisions and is limited only to spherical particles [2]. In DPM, the gas phase is considered as the continuous phase whereas the particle trajectories are obtained by integrating Newtonâ€™s equations of motion. Particle-particle and particle-wall interactions are explicitly taken into account using various physical models such as the soft sphere model. DPM proves to be accurate modelling of gas-particle flow. However, lagrangian models quickly become computationally too expensive because of the large number of particle interactions. Therefore coarse graining of discrete particles is an emerging model that can handle large number of particles effectively. The new coarse-grained model developments are of vital importance for the advancement of large-scale modelling of gas-particle flows that are flexible to extended for polydispersity, and non-spherical particles. In this study a novel lagrangian multi-particle collision (MPC) method is developed for predicting dense particulate-fluid flow.

The MPC solver is developed in OpenFoam [5], where the fluid phase is treated as a continuum and the solids are modelled by lagrangian particles. The MPC model grew out of the multiphase particle-in-cell (MP-PIC) model [3], that employs a fixed Eulerian grid, and particles are traced in a lagrangian fashion, including a parcel approach where each simulated particle represents a large collection of real particles. Each parcel follows Newtonâ€™s equations of motion, where inter-phase momentum transfer is accounted through mapping and interpolating quantities back and forth between parcel locations and the eulerian grid. Particle collisions in MP-PIC are not fully resolved but are derived from particles stresses that are theoretically calculated from Harris and Crighton [4] stress closure. In the MPC model, we implemented a functional form for the particle stress tensor where the coefficients are obtained computationally by means of DPM simulations. The idea is to extend the model from spherical particles to non-spherical particles, for which theoretical particles closures are currently not available in the literature. The viscous stress and random forces are typically ignored in MP-PIC model due to their insignificant effect [3]. Importantly, we incorporated viscous stress and random forces representing scattering of particles by collisions, which are obtained from DPM simulations by measuring the actual variance. Details of model development and validation is presented in this study.

Results from MPC model are compared with MPPICFoam and, with DPM simulations from DEM-LIGGGHT [6], and MFiX [7] code respectively. Pseuduo-2D fluidized beds of dimension 0.01x0.15x0.45 m is considered for comparison with experiments [8]. Comparison is made on the basic of pressure drop, void fraction, bubble dynamics and solids motion. We observed that MPC is in good agreement with DPM, and experiments, when compared to MPPICFoam. Unrealistic over-packing of solids is predicted in MPPIC due to limitation of Harris and Crighton [4] model and the velocity field updates, whereas in MPC these limitations overcomes and no over packing of particles is predicted. Solid circulations are in good agreement with DPM results, where solid moving above in the center and downward near to the wall. Time-average porosity plots revels the path of bubble motion and is in close agreement with results of DPM. In conclusion a novel multi-particle collision model is developed and validated to study gas-solid fluidized beds. The model is efficient to handle large number of particles. In future study, we plan to extend the model for dense flow of non-spherical particles.

Acknowledgments

The authors thank the European Research Council for its financial support under its consolidator grant scheme, contract no. 615906 (NonSphereFlow).

References

1. N.A. Patankar, D.D. Joseph, 2001. Modeling and numerical simulation of particulate flows by the Eulerian Lagrangian approach, International Journal of Multiphase Flow, Volume 27, Issue 10, October 2001, Pages 1659-1684.
2. Gidaspow, D., 1994. Multiphase flow and fluidization continuum and kinetic theory description, Academic Press, Boston.
3. Andrews, M.J., O'Rourke, P.J., 1996. The multiphase particle-in-cell (MP-PIC) method for dense particulate ows, Int. J. Multiphase Flow 22, 379-402.
4. Harris, S.E., Crighton, D.G., 1988. Solutions, solitary waves and voidage disturbances in gas-fluidized beds, J. Fluid Mech. 195, 437.
5. OpenFOAM, 2015. The Open Source CFD Toolbox, Ver. 3.0.0. ESI Group, Paris, France.
6. https://www.cfdem.com/liggghtsr-open-source-discrete-element-method-particle-simulation-code
7. https://mfix.netl.doe.gov/mfix/
8. M. Goldschmidt, 2001. PhD thesis, Hydrodynamic Modelling of Fluidised Bed Spray Granulation.

508490 Dpm Analysis of Large Fluidized Catalytic Cracking (FCC) Reactors

Abstract:

DPM Analysis of Large Fluidized Catalytic Cracking (FCC) Reactors

Abstract:

Computational Fluid Dynamic (CFD) modeling is a powerful tool for equipment design and development and is widely used at UOP, particularly for the reactor and regenerator design of FCC process unit. Through CFD modeling, a technical solution was developed for a large capacity FCC process unit purposing an elevated amount of propylene production.

A key challenge in FCC process design is how to optimize the size of the reactor vessel, which is a function of the number and size of reactor cyclones (catalyst separationdisengagers). UOP developed a new cyclone cluster configuration via CFD study of the reactor internals, including UOP's VSS™ riser termination device and reactor cyclones, which reduce the reactor size significantly compare to the conventional design, while maintaining effective catalyst / vapor separation and mechanical reliability. The objective of the current CFD study is to evaluate the catalyst separation efficiency and performance of the new riser termination device technology, which accommodate more cyclones in a smaller reactor vessel.

To better understand the flow pattern, flow distribution and the gas residence time inside the UOP VSS™ , the analysis was broken down into two sections. Part 1 focused on single gas phase only and in part 2 the particles behavior is studied using the Discrete Phase Model (DPM) in Fluent.

The single gas flow analysis results showed that the flow in the VSS chamber section has the same rotational pattern as in conventional FCC/VSS units, where the catalyst will separate from the gas products. Also the gas flow distribution at each chamber arm is uniform, which is critical to catalyst separation efficiency.

The DPM analysis confirmed an equal catalyst particle distribution through all the cyclones and a high efficiency separation in the riser termination device. Having this information from CFD allows for the new design to move forward.

509623 Catalytic Reactor Design Using Multiphase CFD

Abstract:

Catalytic Reactor Design Usng Multiphase Cfddimitri Gidaspow the 2011 Book , "Computational Gas-Solids Flows and Reacting Systems ", Edited By S. Pannala, M. Syamlal and Tom O'Brien ,Engineering Science Reference, Shows the Progress We Have Made in Using Multiphase CFD in Modeling Fluidized Bed Chemical Reactors and Other Multiphase Systems, Such As the Pyroplastic Flow Emitted from Volcanoes. in This Presentation the Design of Two Types of Catalytic Reactors Will be Described: A Fischer_ Tropsch Slurry Bubble Column Reactora Polymerization Reactor the CFD Design of a New Slurry Bubble Column Reactor Was Already Described By Gidaspow ,Et Al in Chemical Engineering Science 134,784-799 (2015).in This Presentation the Design Will be Extended to Obtain 100, 000 Barrels of Diesel per Day, Twice That of the Largest Commercial Reactors. the Production of Synthesis Gas Using Oxygen or Carbon Dioxide in Catalytic Fluidized Beds Needed for the Fischer-Tropsch Reactor Will Also be Discussed Briefly. for the Polymerization Reactor Our CFD Code Described in " Computational Techniques " By D. Gidaspow and V. Jiradilok , Nova Science, 2009 Was Modified for Polymerization Reactors. Such Reactors and Their CFD Modeling Are Described in a Chapter in Pannala's Et Al Book from Professor Rodney Fox's Group. They Point out That the Fine Particle (80 microns) Catalyst Radial Distribution, Is a Remaining Issue the Computations to be Presented Will Show That the Catalyst Will Have a High Concentration at the Reactor Walls for Various Operating Conditions. This Leads to a High Rate of Reaction at the Reactor Walls and May Result in Melting of the Polymers and Sheet Formation , Forcing Reactor Shut-Down. in Some of Our Simulations There Is Large Bubble and Slug Formation Similar to Those Mentioned in Fox's Chapter. Acknowledgement : I Thank Dr. Mayank Kashyap of SABIC for a Valuable Discussion and His Papers " Application of Multiphase Flow CFD in the Gas Phase Polymerization Process " Fluidization XV. May 24,2016 and a Related Paper Presented at Doe NETL, July 2016

509627 Simulation of Large Particle Turbulent Fluidization in Riser Reactors By Coarse Grain DEM-CFD

Abstract:

The complex two-phase flow in the riser is an important element of many industrial circulating fluidized bed reactors. In some cases, e.g. gas-solid polymerization, relatively large particles are transported upward in a turbulent regime, in which the high gas velocity results mostly from heat removal requirements. Strong solids acceleration, radial and axial velocity profiles for both the gas and solid, intense clustering and inhomogeneities make it impossible to reliably predict the riser head or local bed density profile. Detailed numerical simulations proved capable of showing some of the peculiarities of the flow, but limitations in the scale, lack of grid independence, need for special drag corrections and sub-grid models still hinder confident use of such tools in design and scale-up, particularly in industrial applications characterized by particle size and Reynolds numbers that are not well covered by the available scientific literature.

In the present contribution, coupled Discrete Element Method-Computational Fluid Dynamics (DEM-CFD) simulations are carried out using a coarse graining technique for the solid particles. According to this method, small number of particles are grouped into a computational particle (*parcel*) and represented by it as a spherical "lump," to reduce the computational cost of the simulations and allow lab-scale systems to be represented in full detail. The investigation focuses on the influence of the coars-graining degree, i.e. the number of real particles represented by a *parcel*, on the simulated flow hydrodynamics both in terms of its advantages (a quantification of the computational savings) and disadvantage, primarily the real capability of the method to resolve the behaviour observed and produce results in line with experimental findings. The gas and solid motion are studied, with the aim to characterize the clustering phenomena observed during the up-flow and its effect on the pressure drop. Assuming prescribed gas and solid inlet velocity fields, the potential tendency to show core-annulus flow, the intensity of solids clustering and the local bed density within the riser are examined at different number of particles per *parcel*.

509640 Numerical Study of Particles Shape Effects on Solid-Liquid Fluidizations

Abstract:

Numerical study of particles shape effects on solid-liquid fluidizations

Esmaeil Abbaszadeh Molaei, Zongyan Zhou*

Laboratory for Simulation and Modelling of Particulate Systems, Department of Chemical Engineering, Monash University, Clayton, VIC 3800, Australia

Solid-liquid fluidizations have been observed in many industries due to their high heat and mass transfer rate, high operation flexibilities, and reduced back mixing of fluid and solid phases. In most of these multiphase operations, particle properties such as size, density, and shape may experience significant changes during the process because of attrition, coalescence, comminution or chemical reactions. The difference in particle properties can result in significant segregation problem in liquid fluidized beds. To gain an understanding of solid-liquid fluidisation behaviour, most existing studies, either experimentally or numerically, have focused on solid flow behaviour with mono-sized particles or binary and ternary mixtures with different particle sizes or densities. In spite of substantial studies on the hydrodynamics of liquid fluidized beds, the effects of particles shape on the solid-liquid flow behaviour are still less reported. Except for two physical experiments, the previous studies considered particles as spheres, and few studies have been remarked on liquid fluidization of non-spherical particles. However, particles are generally irregular in most of the processes such as ore beneficiations using liquid-solid fluidized beds separator. Particle shape can cause different flow behaviour in liquid fluidizations rather than that observed for ideal spherical particles. Nonetheless, the answers to some fundamental questions such as how and why particle shape affects flow phenomena are not clear, and hence still poorly understood.

Therefore, in this work, a combination of computational fluid dynamics (CFD) and discrete element model (DEM) [CFD-DEM] is used to examine the effects of particle shape on the flow characteristics in liquid fluidizations. In the simulations, ellipsoidal particles are used as they can represent a wide range of particle shapes from oblate to prolate particles. Different aspect ratios are employed, and results are analysed mainly in terms of particle flow patterns, pressure drop, and bed expansion. In addition, force and particle orientation analysis is also conducted to examine the effect of particle shape and explain the occurrence of entrainment phenomenon.

The results show that ellipsoidal particles intend to enter to freeboard region and entrainment phenomenon occurs, and consequently, the interface between the bed and freeboard cannot be observed clearly. In addition, the pressure gradient variation with bed height is uniform for spheres; however, this uniformity is replaced with â€œdistorted parentheses-shapeâ€ profile for disc-shape and elongated particles. Finally, bed expansion analyses show that spherical particles have the lowest bed expansion. On the other hand, disc-shape particles intend to expand more significantly than spherical and elongated particles. This causes the entrainment phenomenon which has been found to be due to the effects of particles shape on particles terminal velocity.

* Corresponding author. Tel: 61 3 9905 0846; email: zongyan.zhou@monash.edu.

514318 CFD Investigation on Gas-Solids Flow and Heat Transfer in Two Fluidized Catalyst Cooler

Abstract:

Our previously experimental work showed that the internal circulation of solids in an annular catalyst cooler (ACC) had an ability to enhance the heat transfer between bed and wall, comparing with the base catalyst cooler (BCC). The current work simulated in detailed hydrodynamics of gas-solids flow and heat transfer to advance our understanding of the ACC by using the two-fluid model. The effective particle diameter $d_p^{âŽ}$ was determined to 170 Î¼m and the specularity coefficient Ï† was 0.3 in the 3D simulations. The catalyst cooler can be divided to four distinct regions including draft tube region, annulus region, bottom region and particle diffluence region according to the distributions of solids volume fraction and solids velocity. The effect of gas-solids flow on bed-to-wall heat transfer was compared in the different regions of the BCC and ACC. And the effects of operational gas velocity and bed structure were discussed. The results showed that the proper selection of the ratio of gas velocities and the position of the heat transfer tube were needed to form a stable internal solids circulation in the ACC. The ACC had a combined hydrodynamic feature of up-flow and down-flow catalyst coolers with bigger solids volume fraction and smaller particle resident time, which are beneficial for improving the heat transfer between solids and wall.

Hydrodynamics and mixing characteristics of a new-type particle mixer

Mengxi Liu, Chunxi Lu, Zhenliang Meng

State Key Laboratory of Heavy Oil Processing, China University of Petroleum, Beijing 102249, PR China

E-mail: mengxiliu@sina.com (M. Liu)

Abstract:

In order to further enhance the particle mixing in the internally circulating fluidized bed, especially along the radial direction, a new-type particle mixer with slots on the draft tube was proposed in the present work. The hydrodynamics of the mixer was investigated by using a multi-scale computational fluid dynamics with the EMMS drag. The prediction in terms of cross-sectionally averaged solid holdup and solid velocity were compared to experimental data for validation. The results show that in draft tube a low pressure region can be observed near the slots and in annulus the pressure in the vicinity of the slots seems larger than that in the region between the adjacent slots, which induces horizontal particle immigration and mixing. Nearly 25.4% to 29.7% of gas in the draft tube bypasses and flows through the slots into the annulus. The ratio of the mass flux of the tracer gas bypassing into the annulus to that of total tracer gas is approximately equal to 0. With increase of the superficial gas velocity in the draft tube from 0.3 m/s to 0.45 m/s, the total solid mass flux increases from 4.87 kg/s to 10.86 kg/s, while the mass flux through slots decreases from 2.69 kg/s to 0.52 kg/s.

1. Introduction

The characteristics of particle flow and mixing have a considerable influence on the mass and heat transfer, as well as the product quality. According to the previous study, the mixing degree of particles along the axial and radial direction is quite different [1-5]. Mostoufi and Chaouki [4] discovered that the diffusivities varied from 3.3×10^{-3} m^2/s to 5.6×10^{-2} m^2/s along the axial direction and 2.6×10^{-4} m^2/s to 1.5×10^{-3} m^2/s along the radial direction. Berruti et al. [6] also revealed that the particle mixing rate for radial direction was an order of magnitude smaller than for the axial direction. Therefore, the weak particle flow along the radial direction greatly limits the mixing ability of a fluidized bed. Internally circulating fluidized bed (ICFB) has been widely recognized as an efficient fluidized bed reactor, in which orderly and macroscopic particle circulation occurs. Compared with the conventional fluidized bed, ICFB offers multiple particle circulation between the draft tube and the annulus, which leads to relatively strong particle convection, diffusion and shear, and thereby improves the particle mixing performance along the radial direction [7]. In order to further enhance the particle mixing for some processes in which a uniform particle mixing is required, a new-type particle mixer with slots on the draft tube was proposed in the present work.

With increase of the computational power, the computational fluid dynamics (CFD) provides an additional tool to predict particle flow and mixing in the local region, which is difficult with the experimental methods. In particular, the gas-solid flow in a gas-solid fluidized bed can be accurately described by the energy minimization multi-scale (EMMS) model [8]. In this paper, the hydrodynamics of the mixer was investigated by using a multi-scale CFD based on the EMMS drag model [9, 10].

2. Experimental apparatus, Materials and Operating conditions

The structure of the particle mixer is shown in Fig.1. The mixer has a height of 5 m and an inner diameter of 286 mm. A draft tube with a height of 1 m and an inner diameter of 215 mm is coaxially mounted into the column. The four rectangular slots of 80×40 mm are symmetrically opened on the draft tube at a height of h=860 mm, in order to strengthen the particle radial flow. A gas distributor plate and a ring distributor are installed below the draft tube and the annulus respectively. The height of the gap between the gas distributor plate and the draft tube is 60 mm. For the sake of the convenience, the mixer is divided into six zones: the pre-mixing zone, feed influence zone, draft tube zone, gas-solid separation zone, annulus zone and the bottom zone. A perforated plate is mounted at the bottom of the pre-mixing zone, which is connected with a central downcomer, extending into the draft tube. Particles are first mixed in the pre-mixing zone and then the pre-mixed particles flow downward through the central downcomer and into the feed influence zone. In the feed influence zone, the pre-mixed particles are mixed with the particles moving upward in the draft tube, flow upward and circulate between the draft tube and the annulus. In the annulus, particles move downward to the bottom. Then a part of particles flow out of the mixer and the other part flow horizontally into the draft tube and recirculate.

The solid particles used in the experiment were FCC catalyst (ρ_p=1498 kg/m^3), with an average diameter of 76 μm. Air (ρ_g=1.225 kg/m^3) was introduced into the mixer to fluidize the particles. The superficial gas velocity ranged from 0.3 to 0.45 m/s in the draft tube (based on the cross-sectional area of the draft tube) and remained at 0.05m/s in the annulus (based on the cross-sectional area of the annulus). The mass flux of cold particles remained at 13.8 kg/m^{2}s. As we only investigated the influence of slots on the flow and mixing of particles in this paper, no hot particles were induced into the mixer.

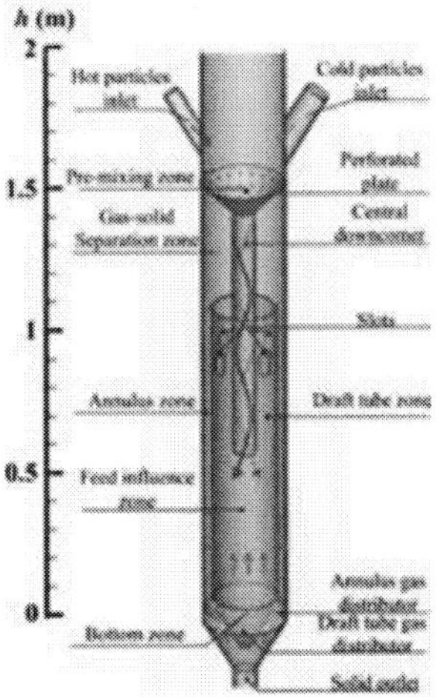

Fig. 1 Schematic diagram of the new-type particle mixer

3. CFD model and simulation method

The hydrodynamics of the mixer was performed using the Eulerian- Eulerian of Ansys Fluent

®13.0. The governing equations [11] include a set of mass and momentum equations for each phase, as shown in Table 1.

A 3-D mesh containing 580,000 cells was generated by using Gambit®2.4. The no-slip boundary was assigned for the gas phase and the partial-slip boundary condition developed by Johnson and Jackson [12] was used for the solid phase. In order to ensure a steady-state operation of the mixer, simulations were first performed for 30 s, and then the time-averaged data was analyzed for another period of 10 s. Other parameters settings in the model are shown in Table 2.

Table 1 Governing equations

Gas-phase Continuity equation $\frac{\partial}{\partial t}(\varepsilon_g\rho_g)+\nabla\cdot(\varepsilon_g\rho_g\boldsymbol{u}_g)=0$

Solid-phase Continuity equation $\frac{\partial}{\partial t}(\varepsilon_s\rho_s)+\nabla\cdot(\varepsilon_s\rho_s\boldsymbol{u}_s)=0$

Gas-phase momentum equation $\frac{\partial}{\partial t}(\varepsilon_g\rho_g\boldsymbol{u}_g)+\nabla\cdot(\varepsilon_g\rho_g\boldsymbol{u}_g\boldsymbol{u}_g)=-\varepsilon_g\nabla p+\nabla\cdot\boldsymbol{\tau}_g+\varepsilon_g\rho_g\boldsymbol{g}-\beta(\boldsymbol{u}_g-\boldsymbol{u}_s)$

Solid-phase momentum equation $\frac{\partial}{\partial t}(\varepsilon_s\rho_s\boldsymbol{u}_s)+\nabla\cdot(\varepsilon_s\rho_s\boldsymbol{u}_s\boldsymbol{u}_s)=-\varepsilon_s\nabla p-\nabla p_s+\nabla\cdot\boldsymbol{\tau}_s+\varepsilon_s\rho_s\boldsymbol{g}+\beta(\boldsymbol{u}_g-\boldsymbol{u}_s)$

Granular temperature equation

$$\frac{3}{2}[\frac{\partial}{\partial t}(\varepsilon_s\rho_s\Theta_s)+\nabla\cdot(\varepsilon_s\rho_s\Theta_s\boldsymbol{u}_s)]=(-p_s\boldsymbol{I}+\boldsymbol{\tau}_s):\nabla\boldsymbol{u}_s+\nabla\cdot(\kappa_s\nabla\Theta_s)-\gamma_s-3\beta\Theta_s$$

Gas-phase stress $\boldsymbol{\tau}_g=\varepsilon_g\mu_g[\nabla\boldsymbol{u}_g+(\nabla\boldsymbol{u}_g)^T]-\frac{2}{3}\varepsilon_g\mu_g(\nabla\cdot\boldsymbol{u}_g)\boldsymbol{I}$

Solid-phase stress $\boldsymbol{\tau}_s=\varepsilon_s\mu_s[\nabla\boldsymbol{u}_s+(\nabla\boldsymbol{u}_s)^T]+\varepsilon_s(\lambda_s-\frac{2}{3}\mu_s)(\nabla\cdot\boldsymbol{u}_k)\boldsymbol{I}$

Solid-phase pressure $p_s=\varepsilon_s\rho_s\Theta_s[1+2(1+e_s)\varepsilon_s g_0]$

Solid-phase shear viscosity $\mu_s=\mu_{s,col}+\mu_{s,kin}+\mu_{s,fr}$

Where $\mu_{s,col}=\frac{4}{5}\varepsilon_s\rho_s d_p g_0(1+e_s)\sqrt{\Theta_s/\pi}$, $\mu_{s,kin}=\frac{10\rho_s d_p\sqrt{\Theta_s/\pi}}{96\varepsilon_s(1+e_s)g_0}[1+\frac{4}{5}g_0\varepsilon_s(1+e_s)]^2$, $\mu_{s,fr}=\frac{p_s\sin\phi}{2\sqrt{I_{2D}}}$

Radial distribution function $g_0=[1-(\varepsilon_s/\varepsilon_{s,max})^{1/3}]^{-1}$

Diffusion coefficient $\kappa_s=\frac{150\rho_s d_p\sqrt{\Theta_s/\pi}}{384(1+e_s)g_0}[1+\frac{6}{5}g_0\varepsilon_s(1+e_s)]^2+2\rho_s\varepsilon_s^2 d_p(1+e_s)g_0\sqrt{\Theta_s/\pi}$

Collisional energy dissipation $\gamma_s=\frac{12(1-e_s^2)\varepsilon_s^2\rho_s g_0\Theta_s^{3/2}}{d_p\sqrt{\pi}}$

Heterogeneity index $H_d=\frac{\beta_e}{\beta}$, $\beta_e=\beta H_d$

Drag based homogeneous structure

$$\beta=\begin{cases}\frac{3}{4}C_d\frac{\varepsilon_s\varepsilon_g\rho_g|\boldsymbol{u}_g-\boldsymbol{u}_s|}{d_p}\varepsilon_g^{-2.65}, \varepsilon_g>0.8\\ 150\frac{\varepsilon_s^2}{\varepsilon_g}\frac{\mu_g}{d_p^2}+1.75\varepsilon_s\frac{\rho_p}{d_p}|\boldsymbol{u}_g-\boldsymbol{u}_s|, \varepsilon_g\le 0.8\end{cases}$$

Where $C_d=\begin{cases}0.44 & \mathrm{Re}_s\ge 1000\\ \frac{24}{\mathrm{Re}_s}(1+0.15\mathrm{Re}_s^{0.687}), & \mathrm{Re}_s<1000\end{cases}$, $\mathrm{Re}_s=\frac{\varepsilon_g\rho_g d_p|\boldsymbol{u}_g-\boldsymbol{u}_s|}{\mu_g}$

Table 2 Simulation Parameters

Gas-solid model	Eulerian-Eulerian
Solids Pressure	Lun et al.
Particle-wall specularity coefficient	0.5
Gas-wall friction coefficient	No-slip
Pressure-Velocity Coupling	Simple
Gradient discretization	Least squares cell based
Momentum discretization	Second-order upwind
Volume fraction discretization	Quick
Max. number of iterations per time step	40
Pressure outlet (atm)	1
Inventory (kg)	56
Incipient solid holdup	0.6

4. Results and discussion

4.1 Model validation

Fig. 2 gives the distribution of cross-sectionally averaged solid holdup and particle velocity in the mixer. As shown in Fig. 2(a), the solid holdup is larger in the bottom zone (h=0.04 m) and due to the accelerated upward motion of particles, the solid holdup decreases with increase of the axial height (when $h \leq 0.16$ m). The cross-sectionally averaged solid holdup is nearly constant in the draft tube and gas-solid separation zone of $0.4 \leq h \leq 1.25$ m. The cross-sectionally averaged particle velocity is shown in Fig. 2(b). The particle velocity is smaller in the bottom zone and keeps almost invariable in the draft tube and the gas-solid separation zone. Compared with the experimental data, the particle velocity seems slightly higher, while the solid holdup in the simulation agrees reasonably with the experimental results.

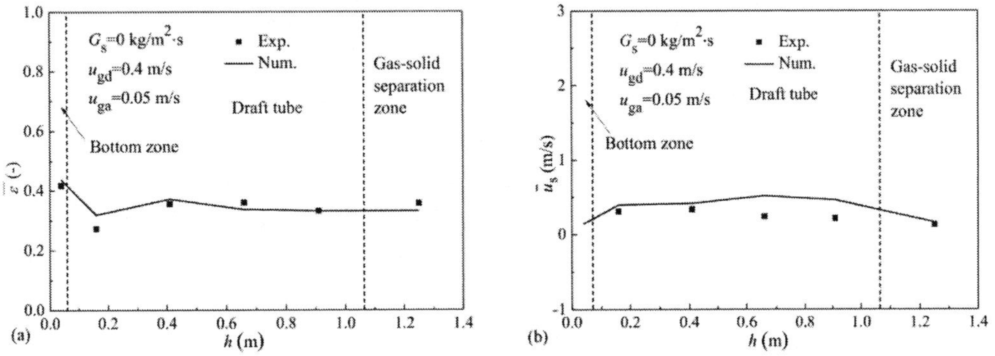

Fig. 2 Distribution of cross-sectionally averaged solid holdup and particle velocity as a function of axial heights

4.2 Pressure distribution near the slots

Fig. 3 shows the time-averaged contours of static pressure on the cross-section of h=0.9 m, the cross-section on the middle of the slots. It is seen that in the draft tube a low pressure region can be observed near the slots, smaller than the pressure in the region between the adjacent slots. This pressure drop forced particles inside the draft tube to move towards the slots and pass through them, which significantly facilitates the mixing of the particles along the radial direction. In the annulus the pressure in the vicinity of the slots seems larger than that in the region between the adjacent slots, inducing horizontal particle immigration and mixing.

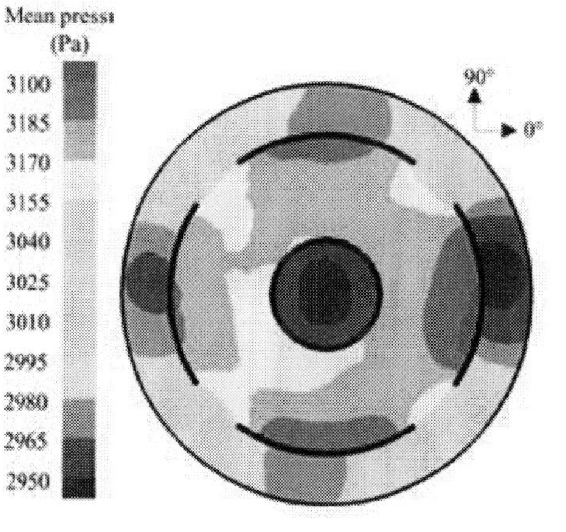

Fig. 3 Time-averaged contours of static pressure at the cross-section of the slots (u_{gd} =0.4 m/s, G_s=13.8 kg/m^{2}s, h=0.9 m)

4.3 Gas bypassing

In the present work, the gas bypassing behavior was investigated by CFD simulation, and a tracer gas having identical physical properties with air was defined. After the mixer running stably, the tracer gas was introduced into the mixer from the gas distributor plate to replace the original gas. A cross section (h=0.06 m) on the exit of the annulus was selected and the mass flux of the tracer gas through the cross section was registered. The results are shown in Fig. 4. f_B refers to the ratio of the mass flux of the tracer gas through the cross section (h=0.06 m) to that of total tracer gas. It is seen that f_B varies between -0.05 and 0.05 with a frequency of 10 Hz, indicating an unstable gas bypassing from the draft tube to annulus. This suggests that the gas bypassing into the annulus was finally brought out of the annulus by the circulating particles. The average f_B is approximately equal to 0, much smaller than that in a conventional ICFB (usually within 15%) [13], because massive particle circulation in the mixer makes it difficult to bypass from the draft tube to annulus.

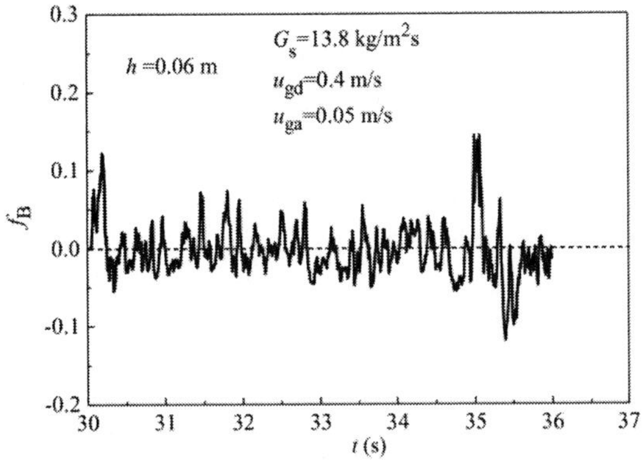

Fig. 4 Gas bypassing from draft tube to annulus in the bottom zone

As discussed previously, in the mixer there exists a pressure gradient from the draft tube to the vicinity of the slots, and then the annulus. Such a pressure difference leads to horizontal flows of particles and gas from the draft tube to the slots and then to the annulus. In the mixer the upside and downside of the slots is located at h=0.94 m and 0.86 m, respectively. Mass flux of the gas bypassing through the slots was obtained by monitoring the gas flux through the lower cross section (h=0.7 m) and the upper cross section (h=1 m) in the draft tube. Fig. 5 shows the gas mass flux under different operating conditions. The total gas mass flux in the draft tube increases with increasing gas velocity in the draft tube. With increase of the gas velocity from 0.3 to 0.45 m/s, the mass flux of the gas bypassing through the slots increases from 0.157 to 0.221 kg/m^{2}s. But the mass flux of bypassing gas does not increase linearly with the gas velocity. It is also seen that the mass ratio of the gas through the slots to that in the draft tube drops from 29.7% to 25.4%.

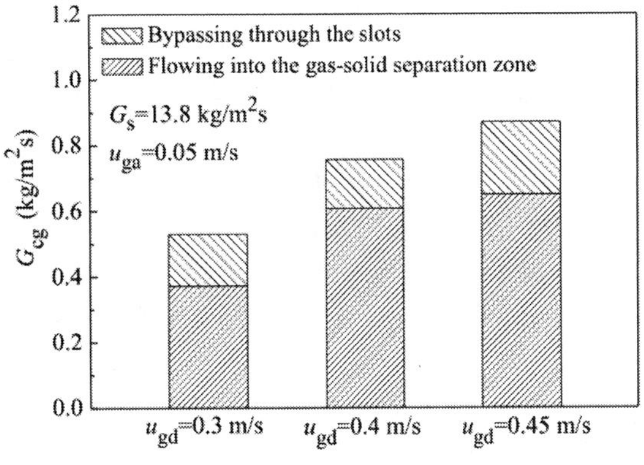

Fig. 5 Gas mass flux through the slots and the gas-solid separation zone as a function of superficial gas velocity in the draft tube

4.4 Particle mass flux

Fig. 6 shows the particle mass flux in the mixer. It is seen that the mass flux increases with increase of the superficial gas velocity in the draft tube. The mass flux through the slots accounts for about 50% of the total mass flux when u_{gd}=0.3 m/s. Because the axial particle velocity rose with increase of the superficial gas velocity from 0.3 m/s to 0.45 m/s in the draft tube, making it difficult to change the direction of the particles. Meanwhile, the radial particle velocity above the draft tube increases. Consequently, with the increase of the superficial gas velocity in the draft tube , the mass flux through the slots considerably decreased, from about 55.3% of the total mass flux (about 2.69 kg/s) at u_{gd}=0.3 m/s to only 4.8% (about 0.52 kg/s) at u_{gd}=0.45 m/s. Meanwhile, the total particle mass flux in the mixer increased from 4.87 to 10.86 kg/s as u_{gd} increased from 0.3 to 0.45 m/s.

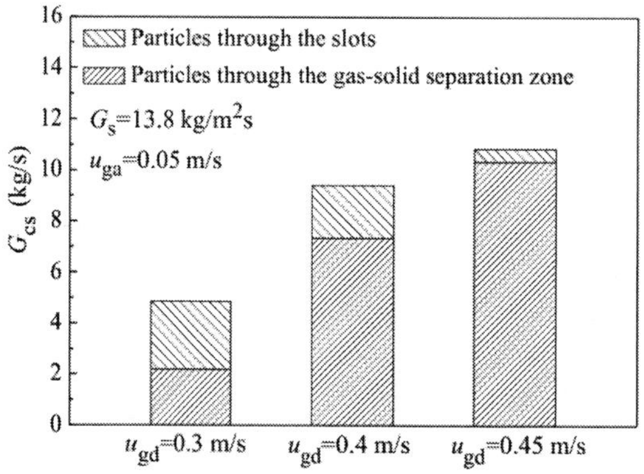

Fig. 6 Particle mass flux through the slots and the gas-solid separation zone as a function of superficial gas velocity in the draft tube

5. Conclusions

The characteristics of particle flow and mixing have a considerable influence on the mass and heat transfer, as well as the product quality. In order to further enhance the particle mixing in the ICFB, especially along the radial direction, a new-type particle mixer with slots on the draft tube was proposed in the present work. The hydrodynamics of the mixer was investigated by using a multi-scale computational fluid dynamics with the EMMS drag. The prediction in terms of cross-sectionally averaged solid holdup and solid velocity were compared to experimental data for validation. The results show that in draft tube a low pressure region can be observed near the slots and in annulus the pressure in the vicinity of the slots seems larger than that in the region between the adjacent slots, which induces horizontal particle immigration and mixing. Nearly 25.4% to 29.7% of gas in the draft tube bypasses and flows through the slots into the annulus. The ratio of the mass flux of the tracer gas bypassing into the annulus to that of total tracer gas is approximately equal to 0. With increasing the superficial gas velocity in the draft tube from 0.3 m/s to 0.45 m/s, the total solid mass flux increases from 4.87 kg/s to 10.86 kg/s, while the mass flux through slots decreases from 2.69 kg/s to 0.52 kg/s.

Notation

G_s	solid mass flux (kg/m^2 s)	G_{cg}	gas mass flux in the mixer (kg/m^2 s)
G_{cs}	solid mass flux in the mixer (kg/s)	h	height (m)

p	pressure (Pa)	R	radius of the reactor (mm)
r	radial position of measurement points (mm)	t	time (s)
u_{ga}	superficial gas velocity in the annulus (m/s)	u_{gd}	superficial gas velocity in the draft tube (m/s)
$\bar{u}$	cross-sectionally averaged axial particle velocity (m/s)		

Greek

ε_g	voidage	ε_s	solid holdup
$\bar{\varepsilon}_s$	cross-sectionally averaged solid holdup	ρ	density (kg/m^3)

References

[1] F. Niklasson, H. Thunman, F. Johnsson, B. Leckner, Estimation of solids mixing in a fluidized-bed combustor, Ind Eng Chem Res, 41 (2002) 4663-4673.

[2] L.H. Shen, M.Y. Zhang, Effect of particle size on solids mixing in bubbling fluidized beds, Powder Technol, 97 (1998) 170-177.

[3] L.H. Shen, M.Y. Zhang, Y.Q. Xu, Solids Mixing in Fluidized-Beds, Powder Technol, 84 (1995) 207-212.

[4] N. Mostoufi, J. Chaouki, Local solid mixing in gas-solid fluidized beds, Powder Technol, 114 (2001) 23-31.

[5] D.Y. Liu, X.P. Chen, Lateral solids dispersion coefficient in large-scale fluidized beds, Combust Flame, 157 (2010) 2116-2124.

[6] F. Berruti, D.S. Scott, E. Rhodes, Measuring and Modeling Lateral Solid Mixing in a 3-Dimensional Batch Gas Solid Fluidized-Bed Reactor, Can J Chem Eng, 64 (1986) 48-56.

[7] M.M. Fang, K. Luo, S.L. Yang, K. Zhang, J.R. Fan, Computational Fluid Dynamics-Discrete Element Method Investigation of Solid Mixing Characteristics in an Internally Circulating Fluidized Bed, Ind Eng Chem Res, 52 (2013) 7556-7568.

[8] J. Li, M. Kwauk, Particle-fluid Two-phase Flow: the Energy-minimization Multi-scale Method, Metallurgical Industry Press, Beijing, 1994.

[9] Z.S. Shi, W. Wang, J.H. Li, A bubble-based EMMS model for gas-solid bubbling fluidization, Chem Eng Sci, 66 (2011) 5541-5555.

[10] K. Hong, Z.S. Shi, W. Wang, J.H. Li, A structure-dependent multi-fluid model (SFM) for heterogeneous gas-solid flow, Chem Eng Sci, 99 (2013) 191-202.

[11] D. Gidaspow, Multiphase flow and fluidization: Continuum and Kinetic Theory Descriptions, Academic Press, Boston, 1994.

[12] P.C. Johnson, R. Jackson, Frictional–collisional constitutive relations for granular materials, with application to plane shearing, J. Fluid Mech. 176 (1987) 67–93.

[13] B.H. Song, Y.T. Kim, S.D. Kim, Circulation of solids and gas bypassing in an internally circulating fluidized bed with a draft tube, Chem Eng J, 68 (1997) 115-122.

514498 Power Spectral Density Analysis of Pressure Signal in 18 m Circulation Fluidized Bed Riser

Abstract:

Studies on the gas-solids flow characteristics using Power Spectral Density based on the pressure data from different axial locations in a riser of diameter 80 mm and height 18 m. Results show that the pressure signal fluctuation is mainly affected by gas-solids interaction, which included gas-solids friction, particle-particle collision and particle-particle aggregation. With the increasing solids circulating rate, frequency of particle interactions including particle-particle collision and particle-particle arrogation become strong and the pressure fluctuation also increased. It can be seen that the pressure signal fluctuation intensity decreased gradually with the increasing axial position of the riser. The low frequency but high power phenomenon nearly disappears in the fully developed zone near the top of the riser. Based on the results, it can be concluded that the gas-solids interaction is stronger at the bottom of riser. On the other hand, the gas-solid interaction is more homogeneous, and the pressure fluctuation is relatively weak in the fully developed zone.

514514 Characteristics of Pressure Fluctuations in Particle-Transport Inclined Pipe of a Circulating Fluidized Bed

Abstract:

In a large circulating fluidized bed device, the characteristics of pressure fluctuation in a particle-transport inclined pipe under different solid circulating flux were investigated based on measuring the dynamic pressure of gas-solid two-phase flow by using FCC equilibrium catalyst as the particles. The experimental results showed that with increase of particle mass flux, the flow patterns of FCC particles above and below the butterfly valve in the inclined pipe both successively changed, and the fluctuation amplitude of dynamic pressure increased. The standard deviation of the dynamic pressure below the butterfly valve possessed a linear relation with the particle mass flux, which could be used to characterize the real-time dynamic pressure in the inclined pipe. By wavelet analysis, it is indicated that there exists a main frequency caused by inclined pipe flow structures in the pressure fluctuations above and below the butterfly valve, besides, there is a secondary frequency in the pressure fluctuation below the butterfly valve caused by gas turbulence and gas-solid interactions. With the increase of particle mass flux, the wavelet energy ratio of the secondary frequency gradually decreased. The fluctuation pressure caused by discharge in the inclined pipe is a low frequency pressure and the excitation source of inclined pipe vibration.

508323 Dynamic Modeling of Attrition and Reactions in Circulating Fluidized Bed Reactors

Abstract:

Fluidized bed reactors are widely applied in chemical processes due to the excellent gas-solid contact. A problem that occurs in fluidized beds is the mechanical stress that particles are exposed to during the fluidization, which causes particle attrition. Additionally, fluidized beds are often operated in batch mode, which will cause ever-changing operating conditions with time. This makes fluidized bed modeling especially challenging, because the distributed particle properties, like size, wetness or reaction state, have to be tracked over the whole course of the simulation.

At the Institute of Solids Process Engineering at Hamburg University of Technology, novel process simulation software named DYSSOL is developed, particularly tailored for solids processes. The software is based on the flowsheeting concept. This means that chemical unit operations, like for example chemical reactors or separators, are put into process modules, which are interconnected into a process flowsheet and then mathematically solved. To make dynamic simulations available, new dynamic models had to be proposed for the involved equipment. The models themselves are based on empirical and semi-empirical correlations for hydrodynamics and kinetic particle reaction models are integrated.

The dynamic flowsheeting software was then used to simulate a system of interconnected fluidized bed reactors, which are used for Chemical Looping Combustion of solid fuels. This process is especially interesting, because heterogeneous reactions take place. Further, the process is a reactor-regenerator system causing the reactive solid to be oxidized and reduced and, hence changing its conversion state. Attrition plays a major role for the cost effective operation of the process.

In a small lab scale bubbling bed reactor, the jet and bubbling attrition rate constants for the solid oxygen carrier were experimentally determined under ambient conditions. In a small stand-alone cyclone the cyclone attrition was investigated. Furthermore, the reaction kinetics of the solid material with fuel gases was experimentally determined. All this information was included into the flowsheet simulation of a pilot scale circulating fluidized bed reactor, which is actually operated in our institute to deliver validation data.

The dynamic flowsheet simulation of the pilot plant, which was enhanced with kinetic and attrition information of smaller scale facilities, was conducted over 1000 seconds. A fuel load change was simulated and the chemical conversion was tracked. Over the whole simulation, the attrition and small particles leaving the reactor system was also covered. When comparing the simulation with experiments it could be shown that the gas conversion was sufficiently calculated and the conversion state of the oxygen carrier tracked correctly. The attrition modeling, however, with its parameters measured had to be adapted, so that it fitted the outlet particle flows measured under reaction conditions.

The financial support of DFG (Deutsche Forschungsgemeinschaft) within the priority program SPP 1679: â€œDynamic simulation of interconnected solids processesâ€ is gratefully acknowledged; grant number HA 6935/2-1.

514599 Analysis of FCC Cyclone Fault Diagnosis Technology Based on Particles Information

Abstract:

Cyclones are key equipment in fluid catalytic cracking unit (FCCU), which are used for separating the catalyst particles from oil gas or flue gas. Hence, it is very necessary to carry out condition monitoring and fault diagnosis for the running cyclone. The properties of particles vary at different positions in the particle circulation loop, such as size distribution, morphology and element composition, etc. In consequence, the particles contain much information about operation. Once the cyclone operates abnormally, the information of run-off catalyst particles appear abnormal too, and there exists a corresponding relationship between them. So, taking the catalyst particles as the carrier of fault information, this paper put forward cyclone fault diagnosis technology by extracting catalyst particle information, such as particle packing density, particle size distribution, particle micrographs etc, to build the relationship model between the cyclone fault and catalyst particle information.

514600 The Multi-Hole Throttling Effect and the Erosion Characteristics of the High Pressure Natural Gas

Abstract:

Based on Euler-Lagrange multi-phase flow numerical simulation method, the single hole and multi-hole throttling process of the high pressure natural gas was studied by using the RNG *k*-*Îµ* turbulent model, the Sutherland Viscosity Law compressible fluid viscosity model, and the advanced erosion model numerical simulation. The calculated mass flow variation law in the throttle choke is comparable with the theoretical calculations, which verified the adopted models and calculating methods. And then the throttling effect and the erosion characteristic of the high pressure natural gas was studied to check its specific law with different throttle holes by using the advanced erosion model. The results showed that, high pressure natural gas can reduce pressure and expansion after flowing through the single hole and multi-hole throttle choke, in which lead a high-speed jet region came up near its outlet. But the maximum jet velocity of multi-hole throttle choke was significantly lower than that of the single hole. At the same time the multi-hole uniform jet may reduce the vortex entrainment and the wall erosion damage, in which can be easier to mix the cold core and achieve stable pressure recovery at the role of the multi-jet mixing effect. Therefore, the multi-hole throttling can effectively restrain the erosion damage and the gas hydrate formation probability.

514601 Numerical Simulation of Flow Field In a Gas Pipe Distributor of the FCCU Regenerator

Abstract:

The flow field in the pipe distributor used in regenerator of FCCU (Fluid Catalytic Cracking Unit), was analyzed based on the numerical simulation. Firstly, the distributing gas performance of the pipe distributor was investigated. The result shows that the pressure increases along the branch pipe of the pipe distributor and the axial flow rate decreases in the branch pipe, which leads to the nozzle flow rate increasing gradually. Furthermore, an obvious bias flow was found in the inlet of branch pipe, which brings some nozzle flow rate less than the designed flow rate and causes an uneven gas distribution in the fluidized bed as well. Meanwhile, the momentum equation with variable mass flow was applied to analyze the flow in the pipe distributor. The flow type in the branch pipe belongs to the "momentum exchange controlling model". The pressure in the branch pipe always increases from the inlet to the end, and the pressure distribution is not consequentially uniform. The uneven gas distribution of the pipe distributor results in fluidized gas non-uniformity in the bed and erosion in the branch pipe and nozzle. Finally, according to the pressure characteristics in fluidized bed system, relationship between the pressure drop of branch pipe and pressure drop of nozzles, as well as of the critical nozzle pressure drop and the designed nozzle pressure drop, were discussed. The gas distributor performance could be optimized by means of the structural improvement of branch pipe.

514602 Full-Loop Simulation of Gas-solids Flow in Circulating Fluidized Bed of FCC System

Abstract:

Circulating fluidized beds are widely used in petrochemical, coal, metallurgy, and other industrial fields. A typical circulating fluidized bed (CFB) includes the riser, gas-solid separator, downer and solids flow control device. To obtain detailed information and get better understand of gas-solids flow in CFB, computational fluid dynamics (CFD) had been applied extensively in many systems such as circulating fluidized bed combustion (CFBC) systems and fluid catalytic cracking (FCC) systems. In most studies, the riser, the core part in CFB system, was isolated as a single simulation object by applying proper boundary conditions. However, there are some limitations of riser-only simulation, without consideration of dynamics in solids circulation rate and effects of other part of CFB system. Therefore, the full-loop simulation is a more promising way to describe complex gas-solids interaction in CFB by treating the whole CFB system as a simulation object. Nowadays, most of current full-loop simulations were focused on CFBC systems, there was little work about FCC system, with different type of particles and recycle mechanism from CFBC system.

In this work, both full-loop and riser-only simulations were carried out by TFM approach to investigate a pilot-scale gas-solid circulating fluidized bed of FCC system which has been set up in our group. The simulation condition was chosen according to experiments with different superficial gas velocity, valve opening and fluidizing air flow rate. Detailed comparison between full-loop and riser-only numerical simulations has been conducted with respect to the hydrodynamics inside the riser, which showed full-loop simulation had advantages on predicting dynamics of solids circulation rate and the gas-solids flow behavior. Distribution of pressure, solids holdup and velocity in system was investigated which was helpful to understand the hydrodynamics in CFB. The effects of operating condition were then studied, which revealed the system stability mechanism.

506758 Immobilization of Sulfur-Oxidizing Bacterium, *Thioalkalivibrio Sp.* D301 on Magnetic Nanoparticles and Biodesulfurization

Abstract:

There is an urgent need to develop desulfurization methods for halo-alkaline wastewater. *Thialkalivibrio sp.* D301, an obligatory chemolithoautotrophic, haloalkaliphilic, sulfur-oxidizing bacterium, was used in the removal of sulfide in wastewater. We developed a process to immobilize D301 cells by using superparamagnetic Fe_3O_4 nanoparticles (NPs) under halo-alkaline conditions. Fe_3O_4 NPs were synthesized by coprecipitation of ferric and ferrous salts in the presence of N_2 gas. TEM image shows the spherical shape of Fe_3O_4 NPs with an average diameter of 10 to 15 nm. The particles had superparamagnetic properties. The saturation magnetization (Î´$_s$) of the Fe_3O_4 NPs was 73.985 emu g$^{âˆ'1}$ with zero remanence and coercivity. Analysis of zeta potentials and hydraulic diameters showed that the Fe_3O_4 NPs and bacterial cells interact mainly through the nano-size effect. The NPs-coated cells have the same sulfur oxidization activity as that of free cells. Experiment results showed that the immobilized cells still retained >90% of their sulfur oxidization activity after six times of reuse. The cells coated with Fe_3O_4 NPs did not interact through electrostatic attraction. The mechanism of NPs adsorption mainly involved the nano-size effect. Magnetic immobilization is a convenient technique for HA-SOB immobilization under halo-alkaline conditions and is a promising technology for large-scale industrial applications.

507944 Simulation of Bulk Solids and Granular Systems By Using Combined Discrete Element Models

Abstract:

Granular materials can be found in many industrial processes from chemical to energy, agriculture, mining, pharmacy and many other fields. They are almost always opaque, making the acquisition of experimental data below the surface difficult. DEM (Discrete Element Method) simulations are an effective way to study the physics of granular materials because they can simulate the micro-dynamic behavior of granular material on an individual particle scale and the motion of every particle can be traced. Since DEM was first considered, many modified discrete element models have been developed. However, most existing DEM simulation are just for spherical particles or a specific type of non-spherical particles. In this work, a combination of several types of discrete element models including spherical models, super-ellipsoids, multi-sphere models (with no interaction and deformation between sub-spheres), and bonded-spheres models (with interaction and deformation between sub-spheres) were used to simulate the flow of granular materials with many different shapes. Several simulation cases including the flow of particles in tumblers, blenders, and other devices were performed to validate the modeling approaches. The simulation results show that it is practical and efficient to combine different discrete element models together for simulating both spherical and non-spherical particles in bulk solids and granular systems.

515703 Material Optimization of Perovskite Films using High Throughput Synthesis and Multi-Dimensional Analysis

Abstract:

Over the past years, Perovskites received special attention in the field of renewable energy as they are one of the prime candidates for energy storage and conversion materials. Perovskites act as the active light harvesting layer of a solar cell and provide excellent performance.[1] A common synthesis strategy is through the liquid phase under inert atmosphere.[2,3] However, this comes with the challenge of a huge parameter space that in principle would have to be first synthesized, then characterized and finally optimized based on the therefrom deduced knowledge.

To overcome these limitations, we follow a fully automated approach, which allows automated high throughput (HT) synthesis and HT characterization of perovskites with high accuracy. As an efficient route to explore the parameter space, we use design of experiments (DoE) to find optimized settings for process parameters such as synthesis and annealing time and temperature, but also material aspects like the molar ratio of the reactants.

For all experiments, a Chemspeed© technologies (CST) swing XL platform with a liquid handling tool and a heatable shaking rack was used. The robotic platform is enclosed within an Mbraun© glovebox with inert atmosphere (Nitrogen). After precursor preparation, deposition was realized by a liquid handling tool on glass substrates inserted in a customized aluminum 24 well plate, which is then transferred onto a hot plate for film formation.

In a first step, after annealing at different temperatures, a characterization cascade was developed and applied that contains the subsequent elimination of samples that exhibit undesired properties and side products. Fast absorption analysis, photoluminescence (PL) measurements and finally X-ray diffraction (XRD) were performed to identify optimum chemistry and process parameters. Although this is a very reliable procedure, it is still comparatively demanding in terms of the time effort. Therefore a more efficient, purely optical approach by which the experimental parameter space is only explored in terms of rapid emission analysis was developed in a second step. Noteworthy, both strategies led to the same optimum conditions, however, at significantly reduced time effort.

In conclusion, our work shows how by the combination of automation, DoE and efficient characterization, optimum conditions for new materials like perovskites are rapidly identified. This opens the door towards faster material development which is important whenever new perovskite structures, e.g. lead free materials for industrial applications, are strived for.

Literature

1. Green, M.A.; Ho-Bailie, A.; Snaith, H.J., "The emergence of perovskite solar cells", Nature Photonics 2014, 8, (7), 506-514.
2. Baikie, T.; Fang, Y.; Kadro, J.M.; Schreyer, M.; Wei, F.; Mhaisalkar, S.G.; Graetzel, M.; White, T.J., "Synthesis and crystal chemistry of the hybrid perovskite (CH 3NH3)PbI3 for solid-state sensitised solar cell applications", J. Mater. Chem. A 2013,1 , 5628-5641.
3. Yan, K.; Long, M.; Zhang, T.; Wei, Z.; Chen, H.; Yang, S.; Xu, J., "Hybrid Halide Perovskite Solar Cell Precursors: Colloidal Chemistry and Coordination Engineering behind Device Processing for High Efficiency", J. Am. Chem. Soc. 2015, 137, (13), 4460–4468.

509509 Solid Liquid Separation Via Particle Flow Instability

Abstract:

A new separation strategy using novel particle-flow-instability physics is successfully developed for a *difficult-to-separate suspension* in which there is some combination of a small density difference between solid and liquid, high viscosity, and small-sized particles. The method we propose here requires no dilution for medium, high or extremely high viscosity slurries and can produce effective solidâ€"liquid separation with low capital and operational inputs. The results have relevance for many biotechnological and/or chemical processes.

509543 Milling and Grindability Assessment of Pharmaceutical Materials

Abstract:

Milling is one of the most important unit operations in the pharmaceutical industries. Most of the Active Pharmaceutical Ingredients (APIs) and excipients are produced by crystallisation, and grown to a sufficiently large particle size suitable for isolation and drying. Therefore most of the materials have to be milled to desirable particle sizes. Milling is influenced by a combined effect of milling dynamics (input power, mill geometry, etc) and material properties. Considerable research has been carried out on analysing the particle behaviour in various mills. Nevertheless, the prediction of the milling rate and milled product properties (e.g. product size distribution) is still very challenging.

In this work, the milling behaviour of several pharmaceutical materials in the PicoPlex mill (pin mill unit of the Picoline, Hosokawa Alpine) has been studied. A combination of experiments and Distinct Element Method (DEM) simulations is used to explore the dynamic behaviours of the materials in the mill. The results are then used to validate a novel approach of grindability assessment of pharmaceutical materials using the Scirocco disperser of the Malvern Mastersizer 2000 (Bonakdar et al. 2016). Based on this method the ease of grindability of the materials in the mill can be predicted.

Bonakdar, T., Ali, M., Dogbe, S., Ghadiri, M., and Tinke, A. (2016). A method for grindability testing using the scirocco disperser. *International Journal of Pharmaceutics,* 501: 65-74

507857 Bi-Directional Thermal Control of Twin Screw Granulation Process Via a Specialised Annular Heat Pipe

Abstract:

The continuous process of twin screw granulation forms a hotbed in pharmaceutical research as it is expected to succeed the current batch production of granules. In essence, the scale up of the material output is a function of production time, instead of equipment size (for batch granulators). This way, a â€œjust-in-time productionâ€ of the desired amount of material can be performed according to specific needs. In this work, a twin screw granulator was utilised in the granulation of both the placebo and active pharmaceutical formulations. Subsequently, the same equipment was used in granulating ceramic raw material powders for tile pressing. During processing, the spatial constraint in the barrel and screw compartments caused friction, resulting in excess heat generation. For the thermally-sensitive pharmaceutical formulations, this must be controlled to avoid denaturation of the granule products. A common solution is by using flowing cool water, but a good control has been proven to be difficult to achieve as different granulation zones generate different heat levels (kneading blocks > conveyer blocks). Heat pipe is a passive and highly effective thermal conductor and stabiliser. A carefully-designed annular heat pipe, filled with a small amount of methanol was successfully employed to stabilise the temperature in the barrel space, cooling the granulating material before it reached its critical temperature. For the ceramic powders, the methanol filling of the heat pipe was replaced by water and it was then heated up to provide an in-situ drying operation as the powders granulating. The in-situ drying in this instance, reduces energy consumption for the downstream drying processing as the product granules contain less moisture to be removed. The annular heat pipe, with its bi-directional operation of heating and cooling is a technically viable temperature controller for twin screw granulators. Compared to the established active cooling methods, the heat pipe provides a passive alternative, with only a discreet amount of fluid required to last its operational lifetime, thereby providing an economic thermal control solution.

Mixing Grains with Different Elongation in a Rotating Drums

Claudia Piacenza[1,2], *Marco Marconati*[1], *Colin Hare*[1], *Andrea C. Santomaso*[2], *Marco Ramaioli*[1,*]

[1]*Department of Chemical and Process Engineering, Faculty of Engineering and Physical Sciences, University of Surrey, Guildford GU2 7XH, UK*
[2] *APT Lab, Department of Industrial Engineering, University of Padova, Padova, Italy*

[*] *Email: m.ramaioli@surrey.ac.uk*

Abstract

Mixing grains with different properties is a remarkably challenging process, relevant to many industrial applications. Rotating drums have been used extensively as model systems to study granular media flow and mixing and segregation. Numerous studies considered the mixing of grains with different sizes, but only few studies considered shape and elongation, which have been already identified as important characteristics affecting packing and inducing segregation. In this contribution, the mixing of binary mixtures of grains having the same volume, but different elongations was studied experimentally. The mixing dynamics of a layered granular medium was characterized in a rotating drum, highlighting the impact of the grain shape, of the drum angular speed and of the drum filling ratio. A mixed or segregated state is reached very rapidly, but experiments are continued to verify that the state reached is a steady state. The experiments demonstrated that below a critical difference of elongation, the grains can be mixed effectively. Conversely, when grain elongations are very different, a central core is formed, rich in the more elongated grains. In this case only limited mixing can be achieved. These results can guide the formulation of mixtures of grains with different elongations.

1. Introduction

Mixing is a process operation aimed at obtaining a homogeneous system starting from two or more components. On the contrary, when there is a lack of homogeneity within the mixture segregation occurs. Segregation phenomena represent a relevant problem in the case of mixing solid particles having different physical properties such as particle size, density, shape or surface roughness. Differently from liquids and gases, solid particles are not thermal activated system [1] and so in order to obtain a certain degree of mixing they need to be mechanically activated. Typically three mixing mechanisms are largely recognized:

- dispersion, analogous to diffusion in fluids, it involves single particles that randomly move due to collision between each other;
- convection, it involves a portion of the bulk material, with a given direction and velocity;
- shear mixing: it is peculiar to granular solids and it is a combination of the two previous mechanisms.

Powder mixers are relatively simple machines of low capital cost. They are classified into two types: tumbling mixers and agitated mixers. Tumbling mixers are the simplest. A totally enclosed vessel is rotated about an axis causing the particles within the mixer to tumble over each other on the mixture surface. Tumbling vessels provide low-shear environments. Convective mixers instead are characterized by the action of impellers or paddles to move the powders around and generate a well-mixed product. Convective mixers provide more shear into the mixture and can be used with cohesive materials.

Rotating drums belong to the tumbling mixers. They are extensively used in the chemical and process industries. With respect to mixing processes, rotating drums are often used to study and understand mixing and segregation phenomena; as a result, many experimental and numerical studies were developed on granular behavior in rotating drums [1-3]. Different granular flow regimes can occur in the drum, each one with its specific flow behavior. In particular, different flow regimes can be assessed by using the dimensionless Froude number, $Fr = \omega^2 R/g$ [4], where ω [s^{-1}] is the drum rotation speed, R [m] the drum radius and g [m/s^2] the gravitational acceleration. With $10^{-4}<Fr<10^{-2}$ and filling level $\varphi>0.1$ rolling regime takes place. This motion is characterized by a uniform, steady flow of a particle layer on the surface (cascading or active layer) while the larger part of the bed (plug flow region) is moved upwards by solid rotation with the rotational speed of the wall. A uniform and good mixing is achievable with this type of motion, but when particle size distribution is very broad segregation can appear. The rolling bed is preferred, since is the simplest regime providing favorable conditions for heat transfer and high quality of the product even when mass flow rates are large.

When particles to be mixed have different properties such as different size, different density or different shape segregation typically occurs and particles tend to separate. [5, 6]. Difference in size is by far the most important cause of particle segregation. Particle size distribution influences also strongly bulk properties such as packing fraction and bulk modulus [7].

Different size segregation mechanisms can be identified:

- trajectory mechanism: according to the Stokes' law the velocity of fine particles is lower than that of big particles. This can cause segregation when small particles move through air;
- percolation mechanism: if a bulk material is disturbed, a rearrangement in the particle packing occurs and for fine particles it is easier to fall down in the inter-particle voids;
- inertial mechanism: in pouring and filling operations bigger particles tend to flow selectively to the bottom of the pile since they have a higher momentum;
- elutriation mechanism: in presence of air the very fine particles can remain in suspension while larger ones tends to settled. This can occur in charging storage vessels or hoppers.

A homogeneous binary granular mixture, with differences in size and density, in a two-dimensional rotating drum typically gives radial segregation. Considering a rotating drum operating in the rolling regime filled with particles of different sizes, bigger particles mainly tend to the edge of the system, while smaller particles accumulate to the centre of the drum. The phenomenon is fast and the rotational speed of the drum seems not to affect the phenomenon; the size ratio of the particles is, instead, the parameter that mainly affects size segregation phenomena. Most of the experimental investigations about segregation in the rotating drum involved spherical or nearly-spherical particles; the experimental studies of the flow of non-spherical particles are, in fact, rather limited [6,8]. The shape of the particles, however, strongly influences the motion in the system. In particular, the static and dynamic angles of repose of non-spherical particles are found to be very different from the ones measured for spherical particles suggesting that differences in particle mobility can affect the segregation process [9].

2. Materials and methods

2.1 Particles

Both spherical and non-spherical particles were used for the experiments. Their size (diameter d for spheres, major and minor axis for ellipsoids a and b respectively) , aspect ratio AR, projected area A_p, volume V, and true density ρ are reported in Table 1.

Table 1. Physical properties of the particles

Type	d [mm]	a [mm]	b [mm]	AR [-]	A_p [mm^2]	V [mm^3]	ρ [g/cm^3]
Spheres S	8	-	-	1	50.3	268	1.09
Ellipsoids E1	-	10.48	6.98	1.5	57.5	267	0.96
Ellipsoids E2	-	18.44	5.26	3.5	76.2	267	0.98

Spherical particles with size 8 mm were used for the experiments. The ellipsoids were 3D printed with *AR* equal to 1.5 and 3.5. They were printed with the Ultimaker 3D printer and made of polylactic acid (PLA). The software used to realize the 3D design of the particles used in this work is FreeCAD, which is an open-source parametric 3D modeller allowing to obtain STL files.

The true density of all the particles used for the experiments was measured by gas pycnometry (Multivolume Pycnometer 1305, Micromeritics). The measured density of the spheres was very close to the typical density of the nylon 66 (1.14 g/cm^3). The PLA ellipsoids had a density that was lower than the typical density of the PLA (1.24 g/cm^3). This is because the particles were printed with a set infill density of the 20%, so they where partially empty inside.

2.2 Tumbling mixer

The quasi-2D drum was made of two transparent polycarbonate disks with a diameter of 30 cm and a thickness of 6 mm; between them a wooden spacer was placed, with the same external diameter, an internal diameter of 25 cm and a thickness of about 9.1 mm. The spacer allowed to create an internal volume of finite thickness separating the two polycarbonate disks.

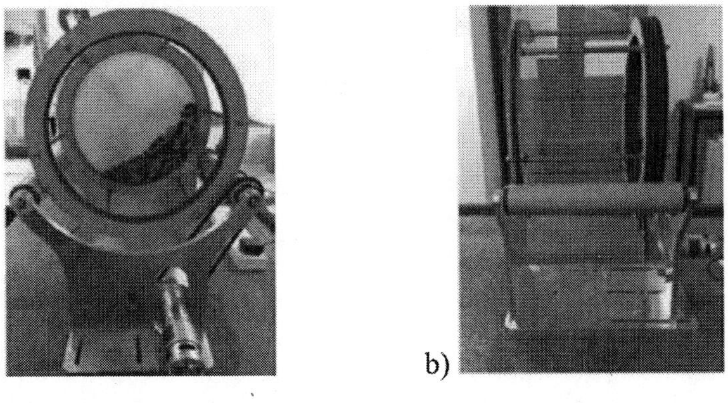

Figure 1. a) Front view and b) side view of the rotating drum (the drum is on the left side of the rigid structure).

The internal lateral wall of the volume was covered with sandpaper in order to provide friction and avoid undesired slip of the particles. To keep the drum in vertical position and allow its rotation, three additional wooden spacers were used, at a distance of about 16.3 cm. This rigid structure could rotate at variable rotational speed thanks to a motor connected to a voltage supplier. Two rollers mounted on a rigid support made the rotation of the drum possible; a pulley transmitted the motion to the driving roller, while the second roller was passive. The rotational speed could be changed by varying the motor voltage. The structure could be removed from the rollers and by unscrewing the bolts, the drum could be easily opened so that to load/unload the particles.

2.3 Image acquisition

Frontal pictures of the drum were taken by a color camera and analyzed by the image processing software ImageJ [10]. To avoid the reflection of the camera and other objects on the front transparent disk, a squared LED panel was used which allowed to have a dominant diffused light illuminating the drum. The LED had a power of 8 W and a color temperature of 4000 K; it was placed behind the drum and opposite to the camera. A circular sheet of paper, which was attached to the drum between it and the LED, diffused the light properly. The camera was a Basler ace camera – model acA2000-165uc, with a maximum resolution of 2040 × 1086 pixels. The focus and the diaphragm of the camera could be adjusted manually.

2.4 Experimental plan

The attention was focused on mixtures of particles with different shapes, since segregation phenomena due to size had been widely investigated in the past. Three mixtures were considered, with two possible initial top-bottom segregated configurations. Filling level and rotational speed where changed also as detailed in Table 2.

Table 2. Experimental tests performed with differently shaped particles.

AR 1 [-]	*AR* 2 [-]	ω [rpm]	φ [-]	Init. Config.
1	1.5	4.22	25%	E1/S
1	1.5	4.22	25%	S/E1
1	1.5	4.22	47%	E1/S
1	1.5	4.22	47%	S/E1
1	1.5	11.51	25%	E1/S
1	1.5	11.51	47%	E1/S
1	3.5	4.22	25%	E2/S
1	3.5	4.22	25%	S/E2
1	3.5	4.22	47%	E2/S
1	3.5	11.51	25%	E2/S
1	3.5	11.51	25%	S/E2
1	3.5	11.51	47%	E2/S
1.5	3.5	4.22	25%	E2/E1
1.5	3.5	4.22	25%	E1/E2
1.5	3.5	4.22	47%	E2/E1
1.5	3.5	4.22	47%	E1/E2

*AR*1 and *AR*2 indicate the aspect ratios of the particles in the binary mixture. It is reminded that the spheres (*AR*=1) and the two types of ellipsoids (*AR*=1.5 and *AR*=3.5) had the same volume. The E1/S initial configuration indicates a segregated configuration with short ellipsoids above spheres (this was assumed as basis configuration); the S/E1 configuration is inverted with spheres above short ellipsoids. A similar convention is applied to the experiments considering mixtures of ellipsoids. All the experiments were performed with binary mixtures with a composition of 50% and 50% on a number (and hence also volume) basis. The filling level φ of the 25% was obtained with a total number of particles equal to 200; with 320 total particles the filling level of the drum was equal to 47%.

2.5 Image analysis procedure

Image analysis techniques were used to identify the particles on the basis of their different shape. The procedure was automated using customized ImageJ macros. Contrast and brightness of the images were first adjusted. Color images were then set to gray scale and segmented by applying a threshold. The operation was carried out twice to separate light and dark particles with different thresholds. Watershed segmentation was then used to separate touching objects. This last operation was more difficult for the particles at the border of the drum; in some case, in fact, the ellipsoids were erroneously identified as spheres. Also, the presence of the border of the drum complicated the watershed operation.

The centre of mass of each particle ($x_{m,i}$ and $y_{m,i}$) was estimated and the radial distance (d_i) from the centre of the drum (x_c, y_c) calculated as:

$$d_i=[(x_c-x_{m,i})^2+(y_c-y_{m,i})^2]^{0.5}$$

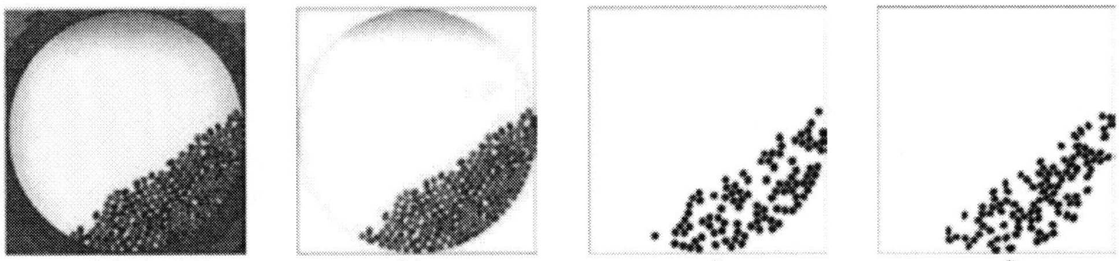

a) b) c) d)

Figure 2. Example of image processing steps. a) Original colour image; b) grayscale conversion and circle cut from the original picture; segmentation applied to c) light particles and d) dark particles.

The mean value of distance (d_m) for the spheres and for the ellipsoids was then calculated as:

$$d_m = \Sigma\, d_i/N$$

Only for very elongated ellipsoid (AR=3.5) the watershed operation failed resulting in over-segmentation errors. Single particles could not be identified unequivocally in this case; only one center of mass for all the elongated ellipsoid was estimated and the average distance (d_m) of the particles from the drum center calculated.

3. Results and discussion

Figure 3 shows two typical experiments, carried out at the lowest rotational speed and filling ratio, with ellipsoids (AR=1.5 at left and AR=3.5 at right) on top and spheres on bottom in the initial and final configuration. It is evident even by visual inspection that the final state presented some level of radial segregation and that a different final segregation intensity was reached in the two systems.

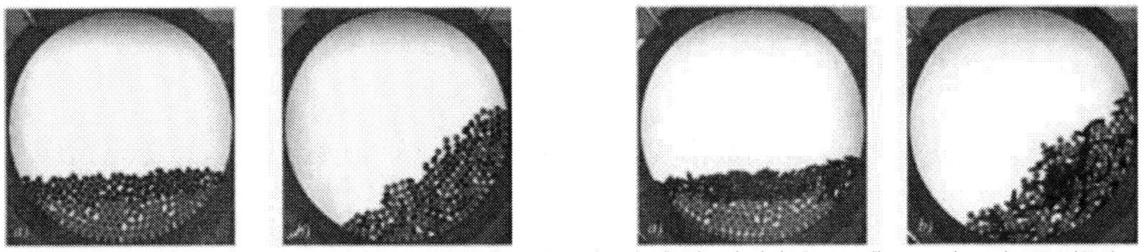

Figure 3. Drum at a) t=0 and b) t=30 min. On the left the initial configuration is E1/S (using ellipsoids with AR=1.5) and on the right the initial configuration is E2/S (AR=3.5). In both experiments ω=4.22 rpm and φ= 25%.

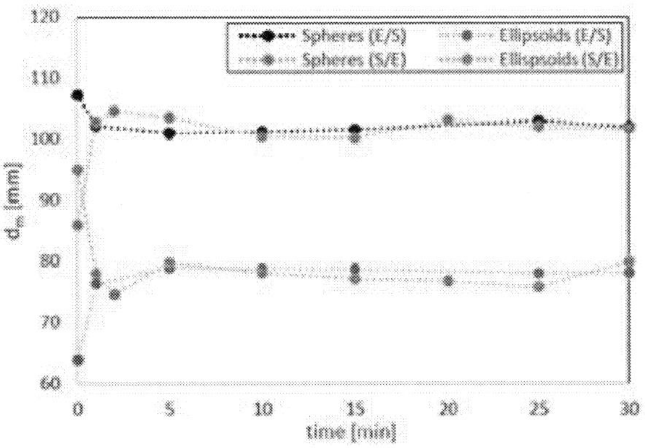

Figure 4. Evolution of the mean radial distance vs. time for the mixture of spheres and ellipsoids with AR=3.5. The initial configuration E2/S is compared to the initial configuration S/E2.

A more quantitative analysis was obtained by plotting the trend of the mean radial distance (d_m) of the two species at different time instants. As a typical example of the system evolution, the case of E2/S

is shown in Figure 4. The d_m reached a steady value after a short time. Dedicated experiments on a shorter time scale (2 minutes, not shown here) revealed that steady state value for d_m was reached after 30s for mixtures with ellipsoids E1 (AR=1.5) and after 60s for mixtures with E2 (AR=3.5).
At steady state, the radial distance was in both cases smaller for the ellipsoids, meaning that they mainly remained concentrated on top or at the center of the granular bed while the spheres mainly concentrated at the periphery of the drum (close to the drum wall). This tendency was much stronger for ellipsoids E2. So a slower segregation dynamics compensated by a larger segregation intensity was observed with the more elongated particles.
As Figure 4 shows, inverting the initial configuration from E/S to S/E did not affect the dynamics and the final result in all the experiments performed with these particles. Initial configuration resulted to have no effect on the segregation propensity of the mixtures analyzed.

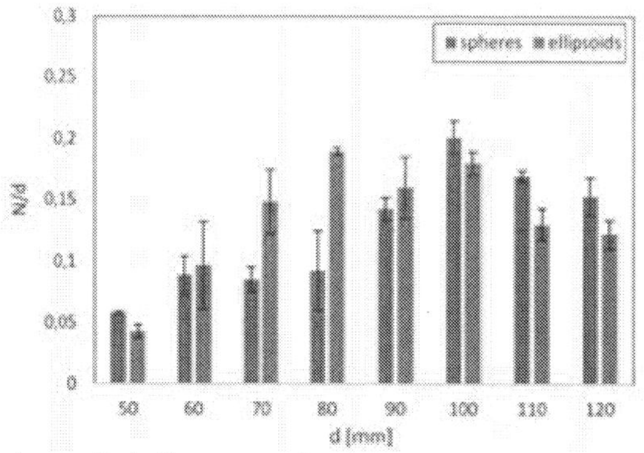

Figure 5. Distribution of the radial distance of spheres and ellipsoids with AR=1.5, obtained by averaging the results of three repetitions. Other conditions are t=30 min, ω=4.22 rpm, φ= 25%.

A more in-depth analysis can be carried out considering the distribution of the radial distance of the spheres and ellipsoids separately. The radial distance from the centre of the drum (d) was divided in classes of 10 mm and the number of each type of particles (N) in each class was computed. To obtain the distribution, N was normalized on the radial distance as N/d and plotted versus d. In the case of completely homogeneous mixtures and assuming that the medium takes the shape of a circular sector, N/d should be constant and equal for the two types of particles in each class. The peak of the distribution of the ellipsoids was found at lower distance than the peak of the distribution of the spheres. Moreover, at lower distance, the number of ellipsoids was higher than the number of spheres, especially for distances between 70 and 90 mm; at higher distance (closer to the drum wall), instead, the number of spheres was higher than the number of ellipsoids, thus confirming the previous analysis based on d_m.

Also, binary mixtures made of the two types of ellipsoids E1 and E2 (AR=1.5 and AR=3.5 respectively) were taken into account at the same filling levels of the previous experimental tests and at the lower rotational speed. Stable segregation was again observed with intermediate results with respect to the two previous cases (S+E1 and S+E2). So as a general comment on these experiments, stable segregation was always observed when mixing particles having same volume but different AR. The intensity of segregation increased with increasing the difference in AR between the particles. In particular particles with larger AR tended to accumulate at the center of the granular bed, while particle with lower AR accumulated at the periphery. Segregation occurred in the surface active layer where the more elongated particles, because of their reduced propensity to flow (particle rolling was prevented by their elongated shape), were systematically overtaken by the less elongated (and faster) particles. In this way a spontaneous separation of particles occurred in the active layer, with those having a lower AR arriving first at the drum wall, while those having an higher AR remaining trapped in the core.

3.1 Effect of filling level

The comparison of the behavior of the mixtures at the two filling levels can be carried out by normalizing the mean radial distance d_m of the two types of particle on the mean distance of all the particles d_c, calculated as $(d_{m,\ spheres}+d_{m,\ ellips.})/2$. The trends of d_m/d_c are shown in Figure 6

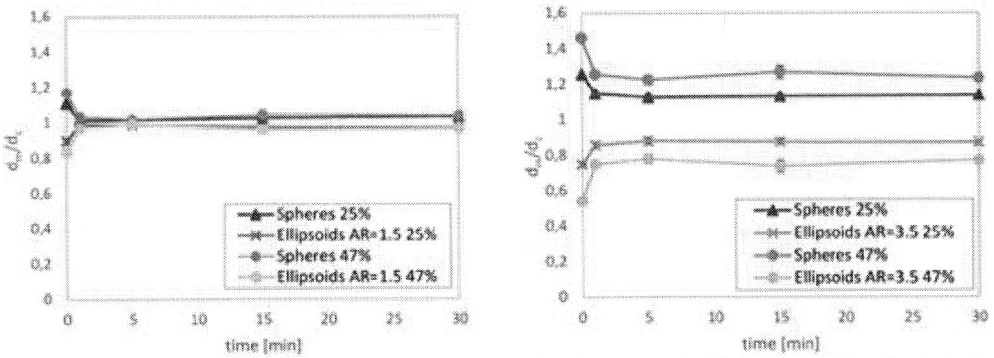

Figure 6. Trend of d_m/d_c vs. time for the mixture of spheres and ellipsoids with AR=1.5 (left) and AR=3.5 (right) at two different filling levels.

In the case of mixture of spheres and ellipsoids with AR=1.5, the curves at different filling levels overlap, showing that the behaviour of the mixture was not affected by the filling level of the drum. In the case of the other two mixtures (S+E2 and E1+E2, not shown), the curves did not overlap , especially for the case S+E2 shown in Figure 6 (right). An explanation can be found in terms of bed geometry. When increasing the filling level (below 50%) the length of the surface active layer increased as well. This gave more chances for particles to separate, in particular when the difference in AR was large enough (S+E2 and E1+E2). However for the system S+E1 where the difference in AR was the smallest, the difference in active layer length in the two cases was not sufficient to create observable differences in the segregation pattern.

3.2 Effect of rotational speed

In addition to experiments carried out at 4.2 rpm, tests were repeated at 11.5 rpm with spheres and ellipsoids (25% filling level). Figure 7 shows that rotational speed did not affect the segregation patterns, at least in the range of the analyzed speeds.

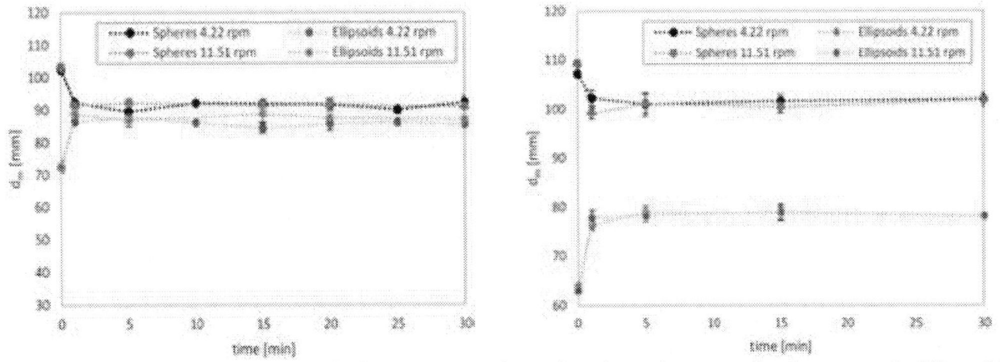

Figure 7. Trend of the mean radial distance vs. time for the mixture of spheres and ellipsoids with AR=1.5 (left) and AR=3.5 (right) at two different rotational speeds.

Conclusions

From the experimental tests described in the previous paragraphs, it was found that binary mixtures of particles with different shape but with the same volume do not mix uniformly and segregation was always observed. In particular, the particles with lower aspect ratio tended to stay closer to the drum walls, while the more elongated particles were found preferentially at the centre of the granular bed. The most significant case was represented by the mixture of spheres and ellipsoids with AR=3.5; for this mixture, in fact, the segregation was more evident and the non-spherical particles clearly accumulated in the centre of the granular bed. The same final state was obtained regardless of whether the longer particles were initially located to the top or the the bottom of the mixture. A proportionality between AR difference and segregation intensity was observed. Some variables such as rotational speed and initial bed configuration were found to not affect the segregation dynamics and intensity. Only filling level (in addition to AR) was found to affect segregation intensity. A mechanistic explanation was given in terms of differences in particle flowability occurring in the surface active layer (confirmed also by an analysis of the angles of repose in the drum). In this specific case therefore percolation and inertial effects (typical of size segregation) were not the main driving mechanisms for the observed segregation (the particle indeed had the same volume and similar mass so that sterical and inertial phenomena were minimized). Segregation mainly occurred because of differences in particle velocity in the active layer due to a reduced mobility of particles with larger AR.

References

1. GDR MiDi (2004), "On dense granular flows", The European Physical Journal E, 14, 4, 341–365.
2. Jarray A., Magnanimo V., Ramaioli M., Luding S. (2017) Scaling of wet granular flows in a rotating drum, EPJ Web of Conferences, 140,03078.
3. Volpato S., Canu, P., Santomaso A.C. (2017), "Simulation of free surface granular flows in tumblers", Advanced Powder Technology 28(3), 1028-1037.
4. Mellmann J. (2001), "The transverse motion of solids in rotating cylinders – forms of motion and transition behaviour", Powder Technology, 118, pp 251-270.
5. Windows-Yule C.R.K., Scheper B.J., van der Horn A.J., Hainsworth N., Saunders J., Parker D.J., Thornton A.R. (2016), "Understanding and exploiting competing segregation mechanisms in horizontally rotated granular media", New Journal of Physics, 18, 2, 023013, 2016.
6. Santomaso, A.C., Olivi, M., Canu, P. (2004), "Mechanisms of mixing of granular materials in drum mixers under rolling regime", Chemical Engineering Science, 59 (16), 3269-3280.
7. Kumar N., Magnanimo V., Ramaioli M., Luding S. (2015), "Tuning the bulk properties of bidisperse granular mixtures by small amount of fines", Powder Technology, 293, 94-112.
8. Ramaioli M., Pournin L., Liebling Th. M., (2005), "Numerical and experimental investigation of alignment and segregation of vibrated granular media composed of rods and spheres", Proceedings of the 5th International Conference on Micromechanics of Granular Media, 2, 1359-1363.
9. Santomaso, A.C., Canu, P. (2007), "Single particles properties vs. bulk flowability", Proceedings of 5th International Conference for Conveying and Handling of Particulate Solids.
10. Schneider, C. A., Rasband, W. S., Eliceiri, K. W. (2012), "NIH Image to ImageJ: 25 years of image analysis", Nature methods 9(7): 671-675.

507441 Generation of Particles with a Special Morphology By Desublimation of Copper Phthalocyanine

Abstract:

The morphology of submicron particles becomes more and more important concerning new functional materials. For example the special electric properties of particles can be used in electronic devices [1]. Moreover the electric, optical or catalytic properties of particles can be optimised by adjusting the morphology of the particles. This may imply a risk for the health but makes it possible to realize specific product characteristics. The morphology of the particles should therefore be considered e.g. for filter tests and for the developing and testing of sensor systems.

To explore the characterization of the particle morphology the DFG-AiF cluster project "multi-parametric characterization of particle-based functional materials using innovative online measurement systems" (MPaC) includes the development of online sensor systems for the size and structure of the particles as well as the development of generation methods for particles with a special morphology. The production of spherical particles can be achieved with different commercial aerosol generators. Several generation techniques for special particles are described in literature [2] [3] [4]. To test the sensor systems with particles of different shapes a new type of aerosol generator was developed. The aim for the new generator is to produce the particles in the gaseous phase and provide the particles directly for sensor or filter testing.

The developed generator uses the desublimation of copper phthalocyanine (organic pigment) to structure the surface of titanium dioxide. The bulk material copper phthalocyanine is dispersed into a nitrogen gas stream and transported to a tube furnace. Here the bulk material sublimates and the gas stream carries the material to a laval nozzle where the adiabatic expansion takes place. This leads to a rapid decrease of the temperature behind the laval nozzle and an oversaturation in the gaseous phase. In addition to the decrease of the temperature core particles of titanium dioxide were initiated behind the laval nozzle. The copper phthalocyanine desublimates on the core particles and causes a structured surface. The morphology of the generated particles is characterized by SEM. The particles show a punctual coating of the titanium dioxide particles with copper phthalocyanine. The size of the titanium dioxide particles is about 500 nm and the punctual coating of copper phthalocyanine measures a few nanometres.

The generation method and the design of the generator will be presented and the morphology of the generated particles will be described.

[1] K. Deppert, M. H. Magnusson, S. B. Carlsson, B. Gustafson, T. Junno, . L. Montelius, B. J. Ohlsson, L. Samuelson und L. E. Wernersson, „Future electronic devices by aerosol techniques," *Journal of Aerosol Science,* Nr. 31, 2000.

[2] J. Raula, A. Kuivanen, A. Lähde, H. Jiang, M. Antopolsky, J. Kansikas und E. . I. Kauppinen, „Synthesis of image-leucine nanoparticles via physical vapor deposition at varying saturation conditions," Journal of Aerosol Science, Nr. 38, 2007.

[3] M. Heurlin, M. H. Martin, D. Lindgren, M. Ek, R. L. Wallenberg, K. Deppert und L. Samuelson, „Continuous gas-phase synthesis of nanowires with tunable properties," Nature, Nr. 492, 2012.

[4] K. Nielsch, K. Deppert, M. H. Magnusson, G. Dumpich, E. Kruis und H. Fissan, „Formation of ultrafine particles from powders of compounds with incongruent sublimation behavior," Journal of Aerosol Science, Nr. 28, 1997.

516184 Rubbery Milling of Seed Endosperms for Improved Sustainability by Natural Functionality Preservation

Abstract:

Increasing demand for functionalised goods addressing individual needs requires a profound understanding of process-induced alterations. Development of models and process-structure-property relationships are essential to control the resulting quality and properties of processed food. A wide range of food systems contain solid or semi-solid particulates as major structure building blocks. Generation and tailoring of such particulates strongly depends on their size distribution, shape, morphology, phase behavior (amorphous/crystalline), interfacial properties and related inter-particle force interactions.

A novel processing technology was developed widening the frame of possible alterations, aiming towards tailoring such particle properties generating novel food structures or preserving techno-functional characteristics of the raw material. The technology employs targeted adjustments of the initial moisture content in combination with milling temperature. The resulting transition from the glassy to the rubbery state generates different material properties and processing behaviour. The technology is highly versatile, owing to the universality of moisture induced state transitions.

Seeds from locus bean gum or guar gum were wetted with excess water in order to adjust a rubbery state of the seed endosperm material. In this state milling within a novel type of impact mill with superimposed drying and sifting steps causes more gentle material disintegration compared to conventional dry milling. This preserves the high molecular weight fraction of e.g. galactomannans (micro-scale structure) from degrada-tion and allows at the same time the adjustment of a micro-porous morphology (macro-scale structure) improving wetting, water uptake and dispersing characteristics of the resulting thickener powders. Such structure preservation /modification increased the thickening characteristics extraordinarily as will be demonstrated.

From this prominent rubbery milling example we have entered in related systematic work addressing a wider range of seed endosperms with various â€œnature-synthesizedâ€ func-tionnalities to be further addressed.

/1/ Pollard, M., Eder, B., Fischer, P., and Windhab, E. J. (2010) "Characterization of galacto-mannans isolated from legume endosperms of Caesalpinioideae and Faboideae subfamilies by multidetection aqueous SEC." *Carbohydrate Polymers*, 79(1), 70-84.

/2/ Illmann, S., Stoob, M., Sieber, C., Konz, A.K. and Windhab, E.J. (2012) "New Thickening Quality of Galactomannan Polysaccharides by Tailored Impact Milling" *XVIth Int. Congress on Rheology*, Lisbon, Portugal, August 5-10

503422 Particle Technology Education at Purdue University

Abstract:

Degree programs and coursework at Purdue University focused on particle technology will be presented. Additional information may be found at: https://engineering.purdue.edu/CP3

506453 Teaching Particle Technology in Portugal â€“ University of Coimbra

Abstract:

Education in Particle Technology started in Portugal in the 80â€™s (last years of the decade). The Chemical Engineering degree of the University of Coimbra was the first engineering degree, in the country, to include this topic in the curriculum. Particle Technology was recognized by us, at that time, essential for the education of chemical engineers, in parallel to education in the traditional areas of Chemical Engineering which were still focused on monophasic systems.

Since then, Particle Technology has always been present in the Chemical Engineering curriculum of the University of Coimbra, in different configurations. During the first years particle characterization was combined with design of operations involving particulate systems (grinding and pneumatic and hydraulic transport). From 2005 to 2010, with the curriculum reformulation in relation with the Bologna Process, the two topics started to be addressed in different courses: one course on Materials Characterization and the process topic in a course entitled Structuring and Stabilization Processes, dealing with colloids and emulsions, grinding, agglomeration and spray drying and, for the first time, an integrated modelling strategy for particulate processes based on the population balance formulation was introduced in this course. The two courses were compulsory, at this time.

From the school year 2011/2012 till present, the configuration of this area in the Chemical Engineering degree of the University of Coimbra changed again, even if not deeply (see https://apps.uc.pt/courses/EN/course/361). The course on materials characterization, which includes general notions on particle characterization is kept as an autonomous course included in the 1st cycle of the Chemical Engineering degree of the University of Coimbra. A second course on design of processes involving particulate systems now exists in the 4th year of the MSc degree (1st year of the 2nd cycle), entitled Disperse Systems Technologies, which is an optional course, and which deals with all types of disperse systems, i.e gas, liquid or solids as disperse phase. This course starts with a review of the mechanisms/interactions occurring at the interface with the disperse phase, goes on, again, with a chapter on population balances modelling, and proceeds with strategies for disperse systems stabilization, linking this subject with formulation concepts, ending with the design of some processes involving particulate systems, such as colloids and emulsions, crystallization, grinding, agglomeration and spray drying. This course is also offered to other engineering degrees, such as Materials Engineering and Biomedical Engineering.

Also at the PhD level a course on Disperse Systems Technologies can also be offered during the formal lectures period of the PhD program on Chemical Engineering (or others), depending on the students background and interests.

Since the introduction of the Particle Technology topic in the curriculum of the Chemical Engineering degree of the University of Coimbra, the feedback from employers has been very positive, and the competences in this field acquired by our students have been quite valued by the job market. Till today, the Chemical Engineering degree of the University of Coimbra is still the one, in Portugal, with a stronger emphasis on this topic in the curriculum. In our curriculum, the topic of Particle Technology interacts directly with another course on Product Design, also in the 4th year of the MSc degree and, of course, also with the Project Design course (5th year of the MSc degree), besides other more traditional courses in a Chemical Engineering degree such as in Reaction Engineering and with courses related to environmental technologies.

At present, we have started developing multimedia tools for the teaching/learning of this topic, especially in the field of particle characterization, to be made available on-line, as is the case of the interactive video on nanoparticles characterization (http://labvirtual.eq.uc.pt/siteJoomla/index.php?option=com_content&task=view&id=277&Itemid=470, in Portuguese). Also, the European Federation of Chemical Engineering (EFCE) is considering investing in this type of tools for the field of particle technology, as the result of a collaboration between the Working Groups of Characterization of Particulate Systems and Education.

Acknowledgement

The authors would like to acknowledge the financial support of Pest/C/EQB/UI0102/2013 from FCT/MCTES (PIDDAC), co-financed by the European Regional Development Fund (ERDF) through the program COMPETE (POFC).

506629 Highlights of Particle Technology Teaching in Singapore

Abstract:

Our approach to education in particle technology is to incorporate elements of our research in this field, relevant industrial case studies and apply a Technology Enhanced Learning (TEL) approach for implementing pedagogical interventions to enhance the learning experiences of students.

For supercritical CO_2 processing, case studies in extraction and drying are used to teach different chapters in the curriculum such as:

- Particle characterization
  - The bulk density, size, shape of real industry samples are given to students for analysis using software such as ImageJ
- Flow through packed bed
  - The pressure drop and scale-up of packed bed design for supercritical CO_2 flow through packed bed are used as a worked example and assignment problem to help students appreciate the role of density and viscosity in the pressure drop in a packed bed
- Drying
  - Supercritical CO_2 as an alternative drying media as compared to hot air or superheated steam. Case studies and examples provided

For gasification and chemical looping combustion applications, case studies in the treatment of wastes and carbon capture are similarly used as part of the syllabus:

- Particle characterization
  - The effect of polydispersity on various fluidization characteristics (entrainment, transport dis-engaging height, clusters, segregation, etc)
  - The effect of agglomeration
- Reliability of correlations and simulations
  - Correlations in literature give predictions orders-of-magnitude apart
- Scale-up
  - Rule-of-thumbs in scaling up the various regimes

A design software has been developed within the paradigm of Technology Enhanced Learning (TEL) to facilitate learning via a repeated practice approach. The software generates detailed solution steps to typical engineering design problems encountered within this module. Students can compare the complete solutions generated with their own hand calculations, and thereby apply a repeated practice approach towards their learning of engineering design calculations. The software has been successful in enhancing the performance of students in the final examination.

507116 Particle Technology Course at the University of Salerno

Abstract:

The Course of Particle technology at the University of Salerno is a 6 EC, that is a 60 hours course including lectures, as well as classroom and laboratory practical. The course dedicated to Master student in Chemical Engineering. It a compulsory course within the â€œinnovative process and nanotechnologies curriculumâ€ can be optioned by any mechanical or chemical engineering student at the master level.

The course includes basic particle size and shape characterization concepts, fundamentals on interparticle forces, fundamentals of powder mechanics and silo discharge, fundamentals on-fluid particle interactions and fluidization.

The main learning outcome is the ability of the student to understand the phenomena occurring in processes involving particulate solids and to carry out basic design operation on storage and fluidized bed units.

507382 Professional Master of Engineering Degree in Particle Technology

Abstract:

Particles are ubiquitous in the process industries yet nearly absent in the undergraduate education of chemical engineers. This poster descibes a professional Master of Engineering degree program at University of Delaware (the MEPT) that closes that gap by exclusively focusing on particulate matter, its processing, and its incorporation into products. The program can be completed in a single year although we welcome already-employed students who wish to take it part time. Now in our fourth year, we will describe our program and our experiences including guiding distance learners successfully through the program.

In its 1-year realization, a core curriculum of 18 credits is augmented by 6 credits of electives which the student may use to tailor the program to their specific interests. The program is capped by a 10-week, 6-credit summer internship. The 12 core credits taken in the fall teach the fundamentals of particle transport, particle properties & characterization, particle formation and transformation kinetics and the mathematics of particulate systems. The 6 core credits taken in the spring teach the student to apply the fundamentals to both (a) equipment design (in a unit operations course) and (b) product and process design (in a product engineering course).

The presenters of this poster are the program directors, both with long industrial careers preceding this venture into education. In addition, we have developed a strong network of industrial partners who share a wealth of additional experience and expertise with our students both as adjunct professors and as internship mentors. We have successfully placed international students in internships at US companies via the curriculum practical training program (CPT) who have gone on to practice in industry after graduation under the optional practical training program (OPT).

507780 Recent Developments in Particle Technology at the Universidad Tecnica Federico Santa Maria

Abstract:

The Department of Mechanical Engineering of the Universidad Técnica Federico Santa María formed in 2003 a Research Center for Transporting Granular Materials (CITRAM) to support educational activities and experimental research, as an extension for those ME students taking the regular course on Bulk Solids Handling and those developing a thesis in this area.

Over the years, many ME students have studied, investigated and modeled different topics including: pneumatic conveying of solid particles of different nature, size and density; identification of flow patterns developed in the pipeline, minimum conveying velocity and pressure drop; abrasive wear using a novel rubber-wheel tester; segregation of ores in stockpiles and silos; flow of granular materials in hoppers and chutes; heat transfer of granular materials; among others. The aim of this poster is to present and share the results of this work.

508520 Using Perusall to Enhance Student Learning of Particle Technology at Graz University of Technology

Abstract:

Perusall is a digital platform that allows students to collaboratively annotate textbooks, papers, and other similar classroom reading assignments. In order to enhance student learning, Perusall was introduced to help teach Particle Technology at Graz University of Technology (TU Graz, Austria). At Graz University of Technology, two Particle Technology courses exist in sequence; the first course is focused on fundamentals topics, whereas the second course is focused more on applications and separations topics. Approximately 60-75 students are registered for each course. For both courses, Perusall was used to assign reading materials, which were primarily journal articles that complemented the course topics. Although these assignments were non-compulsory due to national requirements, students were asked to use Perusall to read and annotate the course readings. Student participation and input regarding this digital approach to learning was recorded. This poster will highlight how Perusall was implemented in both Particle Technology courses as a tool to enhance student learning.

508524 A Graduate Course in Fluidization and Gas-Solid Flow Systems

Abstract:

The objective of this course is: a) to provide students with a fundamental understanding and design of Fluidization and Fluid- Particle flow systems, b) to provide students with an understanding of Computational Fluid Dynamics (CFD) and the kinetic theory approach to fluid-particle flows c) to provide students with the knowledge required to solve flow, heat and mass transfer problems in processes which involve particles and d) to provide students with sufficient knowledge to read and understand technical articles in areas related to fluid-particle processes. The course is designed for Master and PhD students in Chemical, Mechanical and aerospace engineering. The course includes: Fluidization concepts and regimes including bubbling, turbulent and circulating beds; Particle characterization, diagnostics and size reduction and agglomeration/granulation; Concept and design of Standpipes, non- mechanical valves, gas distributors, cyclones and hoppers; CFD approach to fluid-particle flow systems and fluidization including kinetic theory and dispersed phase modeling; Computational transport phenomena of fluid-particle systems including method of moments; and examples from energy, chemical and pharmaceutical industries. In addition the students will visit PSRI to see large-scale fluidized beds in operation.

The students are required to finish a team project. The teams consist of 3-4 students working on a project which will count as their final exam. The projects will primarily involve background research (about 80%) but some of the investigation (about 20%) should involve taking an additional step from what is currently known. This additional step can take the form of a research proposal, new theory, new insight based on current information, new calculations, new simulation and/or new data. The results of the projects will be communicated to the class and to the Course Instructors via a written report of 10-20 pages, and an oral presentation involving the professors and all of the group members.

508742 Education of Fluid-Solid Multiphase Flow at Department of Mechanical Engineering, Osaka University

Abstract:

In mechanical engineering field, education curriculum in undergraduate course concentrates on the basic subjects of mechanical engineering such as mathematics, mechanics, material mechanics, fluid mechanics, thermodynamics, control engineering, design, etc. Concerning the course relating to powder technology, we have a class of "Fluid-Solid Multiphase Flow" in the curriculum of graduate school of engineering.

The contents of "fluid-solid two-phase flow" is as follows;

1 Introduction: examples of multiphase flow

2 Definitions and fundamental concepts

3 Particle response time and Stokes number

4 Fluid force acting on particle

5 Fluid force acting on particle in dense system

6 Particle-wall and particle-particle interaction

7 Heat and mass transfer

8 Continuous phase flow equations

9 Modeling of fluid-solid multiphase flow

10 Discrete particle modeling and simulation

11 Numerical simulation method for continuous phase flow

This course includes basic concepts of powder technology, dynamics of particles in fluid flow, derivation of basic equations for locally-volume-averaged fluid flow, discrete particle modeling including DEM (Discrete Element Method), and basics of numerical analysis and modeling.

509420 Bulk Solids Handling Education at the KSU Bulk Solids Innovation Center

Abstract:

Kansas State University Bulk Solids Innovation Center (BSIC) was started in 2015 with university and industry partnership and is located in Salina, Kansas. This center is dedicated to bulk solids handling research, education and outreach and equipped with world class pneumatic conveying system (both dense and dilute) full scale test lab and other material testing equipment. The vision of this center is to become valued resources for companies who produce or handle bulk solids during their operations. Currently, this center offers short courses in various subject lines related to bulk solids handling such as storage and flow, pneumatic conveying systems, discrete element method, dust explosion and control and mixing and segregation. Duration for these short courses varies from two to three and half days and offered as live course with class room teaching and practical training. The fundamental course on bulk solids handling is offered as distance education online format. This center is also in the process of starting two year online (distance education) master degree program named professional master of technology specialized in bulk solids handling and one year graduate certificate program (online) in bulk solids handling.

509587 Structure of Spherical Silica Colloid Film Prepared By Electrophoretic Deposition in Pulsed Direct Current Electric Field

Abstract:

Colloidal crystal is of an ordered structure comprising regularly arrayed (in 2D) or deposited (in 3D) spherical colloidal particles. Owing to its unique optical properties resulting from Bragg reflection, it is expected to have it applied to optical devices, such as tunable laser and color materials not suffering from color fade-out. Several methods have been proposed and attempted to fabricate 3D colloidal crystals, including the dip coating method, the sedimentation method, and the convective self-assembly method. The 3D colloidal crystal fabrication, however, requires precise operations as well as prolonged time. Electrophoretic deposition is the phenomenon that can attain migration and deposition of colloidal particles toward the electrode upon application of an electric field; it is possible to fabricate a thin film of particles on the electrode with relative ease in comparatively short time. Furthermore, there is additional possibility of controlling the structure of particle thin film because it involves many adjustable operation parameters.

In this study we tried to assemble a 3D ordered structure of particle thin film with spherical silica colloid in ethanol, and found that the hexagonal close-packed structure could be realized by applying a pulsed direct current electric field with 5 to 50 Hz frequency. Based on the calculation results, it is revealed that the deposited particles led to essentially one-particle layer during ON time of the pulsed field and were relaxed through Brownian motion during OFF time of the pulsed field over the given frequency range. It is suggested then that the hcp structure was formed by deposition and re-deposition behavior caused by the pulsed direct current electric field.

DYNAMIC WETTING OF MULTICOMPONENT PARTICLE SYSTEMS

J. Kammerhofer[1], L. Fries[2], J. Dupas[3], L. Forny[2], S. Heinrich[1], S. Palzer[4]

[1]*Hamburg University of Technology, Germany*
[2]*Nestlé Research Center, Switzerland*
[3]*Nestlé Product Technology Center, Switzerland*
[4]*Nestlé S.A., Switzerland*

Introduction

The performance of a food powder can be characterized by its reconstitution behavior which can be generally separated into four steps, wetting and capillarity, sinking, dispersion and dissolution. Wetting as first step is often the rate limiting one and, therefore, of special interest [1]. The dynamic wetting process is driven by penetration of liquid into the powder due to capillary forces, which strongly depends on the surface composition of the material [2]. Thus, the complexity of the composition of food powders is a challenging aspect for capillary wetting [3].

Many authors focused on the relation of heterogeneous surfaces or bulk compositions of food powders and their wetting behavior [2,4–7]. Different components in food powders lead to heterogeneous wetting properties. The interactions between the liquid and the solid phase are described by the contact angle θ. A chemically heterogeneous powder surface cannot be characterized by one contact angle but by a mixed contact angle, which is expressed by the Cassie-Baxter equation (Eq. 1), where θ_{CB} is the apparent contact angle, θ_1 and θ_2 are the contact angles of component 1 and 2 and f_1 and f_2 are the area fractions of component 1 and 2 [8]:

$$\cos\theta_{CB} = f_1 \cdot \cos\theta_1 + f_2 \cdot \cos\theta_2. \qquad (1)$$

Wettability studies get even more challenging if well soluble materials such as carbohydrates are involved [9]. Dissolution of sugars results in an increase of viscosity, which influences the wetting process. This aspect was studied by Forny et al. [10] who found a decreasing wetting rate into dextrose agglomerates with increasing viscosity. Another consequence of dissolution is the alteration of the particle size and, thus, of the pore volume [11].

A general aim in the food industry is the introduction of reliable methods to predict the behavior of food powders during rehydration. This can be achieved by modeling the process, either with empirical relations or with physically based models. A well-known approach for capillary imbibition into single pores is the Washburn equation (Eq. 2) [12], where $h(t)$ is the time dependent height of the liquid in a pore with a radius r_P. The liquid properties are described by the surface tension σ and the viscosity η. The interactions between the liquid and the solid phase are expressed by the contact angle θ:

$$h(t) = \sqrt{\frac{r_P \cdot \sigma \cdot \cos\theta}{2 \cdot \eta} \cdot t}. \qquad (2)$$

Eq. (2) forms the basis for several capillary rise models which were already applied in food science [13]. An approach of Benavente et al. [14] developed for capillary imbibition of porous rocks based on the Washburn equation was used for modelling the rehydration of freeze-dried carrots [15]. The occurrence of discrepancies between theoretical and experimental penetration rates can be explained by the neglection of food related phenomena such as swelling or dissolution.

Our study should provide a step towards enhanced predictions of capillary wetting of heterogeneous, soluble food materials. We introduce a model for liquid penetration into a model food powder containing hydrophilic and hydrophobic surfaces and consider changing liquid properties due to dissolution. Experimental investigations were performed and compared to theoretical penetration rates.

Materials and methods

Materials

Glass beads (SiLibeads, Sigmund Lindner, Germany) with an average size $d_{50,3}$ of 287 µm and a sphericity Ψ of 0.92 made of soda lime glass were used as inert model powder system. The material has a solid density of 2469 kg/m³. The characteristic size values are summarized in Table 1. "Hydrophobic" glass material was obtained by using a cleaning step with peroxymonosulfuric acid which consists of sulfuric acid and hydrogen peroxide in a volume ratio of 3:1, respectively, for 15 minutes. Afterwards, the glass was rinsed in distilled water until the waste water shows a pH-value of 7. Secondly, a silanization step was performed by modification of the clean surface with triethoxyoctylsilane in toluene (10 mM) for 30 minutes. After the silanization reaction the glass beads were rinsed with toluene and acetone and dried under the fume hood. Sulfuric acid (96%), hydrogen peroxide (35%), triethoxyoctylsilane, toluene and acetone were purchased from Carl Roth (Germany). For characterization of the hydrophilic and hydrophobic glass material in terms of contact angle, glass slides (Marienfeld superior, Germany) were treated equally as the glass beads.

Table 1. Characteristic size values of glass powder and sucrose.

Characteristic size values	Hydrophobic glass	Sucrose
$d_{10,3}$ [µm]	240	231
$d_{50,3}$ [µm]	281	287
$d_{90,3}$ [µm]	318	355
Ψ	0.92	0.84
ε_{tap}	0.40	0.42
d_{32} [µm]	278	272
S_v [1/mm]	234	263

Sucrose is a water soluble, common ingredient in several food products. Therefore, we used crystalline sucrose powder (Unser Feinster, Nordzucker, Germany) as our model food powder with a lower sphericity of 0.84. A fraction of 200-370 µm was sieved in order to get a size fraction which is comparable to the glass bead fraction for experimental investigations. The characteristic size values are given in Table 1. The solid density of sucrose is 1553 kg/m³. Distilled water and hexane (Carl Roth, Germany) were used as wetting liquids. Their properties are summarized in Table 2.

Table 2. Liquid properties at 20 °C [16,17].

	Density ρ [kg/m³]	Surface tension σ [mN/m]	Viscosity η [mPa·s]
distilled water	998	72.8	1.00
n-hexane	661	18.4	0.33

Methods

The particle size distribution of glass beads and sucrose powder was determined by dynamic image analysis (CAMSIZER X2, Retsch Technology, Germany). The material density was measured by means of a helium pycnometer (Multivolume Pycnometer 1305, Micrometrics, USA). Furthermore, the tapping porosity ε_{tap} was determined for pure glass beads, for pure sucrose and for all mixtures in a cylinder with a defined volume after tapping the sample for 10 times.

Experimental penetration rates for liquid into the different powder mixtures were measured by using the Washburn setup of the K100 tensiometer (Krüss, Germany). This equipment includes a cylindrical glass tube (radius r_c of 5 mm), which is connected to a balance. Below the tube a moveable and tempered container is installed wherein the wetting liquid is stored. 3.5 g of powder is filled into the cylinder and tapped 10 times in order to achieve a uniform packing. The bottom of the tube is closed with a filter paper. The penetration of liquid into the powder bed starts when the liquid gets in contact with the cylinder. The mass gain vs. time is recorded as well as the liquid temperature. The measurement is controlled with LabDesk (Krüss, Germany). For our experimental study, different mixtures of hydrophobic glass powder and sucrose powder were tested. Water was used as wetting liquid. For each powder mixture the measurement was repeated at least 3 times. All penetration experiments were performed at a water temperature of 20°C. Additionally, the penetration experiments were repeated with hexane as wetting liquid to determine the capillary constant K_{hex} of the powder packings (see Eq. 26).

Modelling

For predicting penetration rates into a food system containing sucrose particles, it is required to consider the effect of solubility by developing a model. Dissolution of sucrose during the penetration process leads to changing liquid properties, such as density, surface tension and viscosity [11]. Furthermore, due to the shrinkage of the particles, the pore volume increases and the surface area of sucrose decreases. The two latter influencing factors caused by particle shrinkage are neglected in our model since we focused on the first few seconds of the penetration step. It is assumed that both effects, the decrease of surface area and the increase of pore volume, are marginal compared to the changing liquid properties within the first seconds, which is discussed in detail in the chapter results and discussion. Secondly, the effect of heterogeneity in terms of contact angles has to be considered. This is realized by using Eq. (1) to calculate an apparent mixed contact angle of the system depending on the content of hydrophilic and hydrophobic surface.

Our model, which was developed in a former study [18], was slightly modified in this work. It is based on two differential equations, the differential mass transfer equation (Eq. 3) and the differential Washburn equation (Eq. 4). We implemented a coupled system of both differential equations, which is solved by means of the solver ode45 in MATLAB. The dissolution of sucrose is described by Eq. (3), where M_s is the mass of sucrose, β is the mass transfer coefficient and ρ_{sat} the density of a saturated sucrose solution of 1332.7 kg/m^3 (20°C), which is taken from literature [19]:

$$\frac{dM_s}{dt} = \beta \cdot a \cdot (\rho_{sat} - \rho_l), \qquad (3)$$

$$\frac{dh}{dt} = \frac{r_p \cdot \sigma \cdot \cos\theta_{CB}}{4 \cdot \eta \cdot h}, \qquad (4)$$

$$r_{p,eff} = \frac{2 \cdot K_{hex}}{A^2 \cdot \varepsilon^2}. \qquad (5)$$

The pore radius r_p of the system can be approximated by an effective pore radius $r_{p,eff}$. The $r_{p,eff}$-value of the pore network can be calculated by using the capillary constant K_{hex}, which is experimentally determined with the Washburn setup and hexane as wetting liquid, the cross-sectional area and the porosity (Eq. 5). $r_{p,eff}$ is inserted in Eq. (4) for the pore radius r_p.

Figure 1 provides an overview about the modelling approach. A detailed explanation about the operating principle follows.

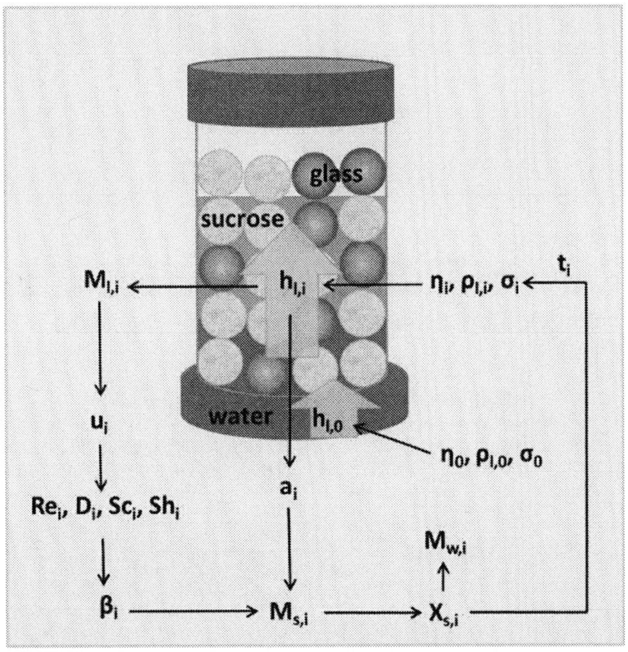

Figure 1. Schematic model of liquid penetration into sucrose-glass system.

At $t = 0$ the model starts to run with the liquid properties of water, which are given in Table 2. Eq. (4) determines for each time step a penetration height h. This parameter is used to calculate the penetration mass of liquid M_l (water and dissolved sucrose) (Eq. 6) and the velocity of the rise u (Eq. 7):

$$\frac{dM_l}{dt} = \rho_l \cdot \varepsilon \cdot A \cdot \frac{dh}{dt}, \tag{6}$$

$$u = \frac{dh}{dt}. \tag{7}$$

The velocity u is necessary to determine the Reynolds number Re (Eq. 8) where d_s represents the average diameter of the particles in the powder bed:

$$Re = \frac{u \cdot d_s \cdot \rho_l}{\eta}. \tag{8}$$

The determination of Reynolds number, Sherwood number Sh and Schmidt number Sc is necessary in order to define the mass transfer coefficient β for each step (Eq. 10). Furthermore, β depends on the diffusion coefficient D, which is calculated with the Stokes-Einstein equation (Eq. 9), where k_B is the Boltzmann constant with a value of $1.38 \cdot 10^{-23}$ J/K [16], T is the ambient temperature of 20°C and R_0 is the hydrodynamic radius of sucrose with a value of $0.49 \cdot 10^{-6}$ mm [20]:

$$D = \frac{k_B \cdot T}{6 \cdot \pi \cdot \eta \cdot R_0}, \tag{9}$$

$$\beta = \frac{Sh \cdot D}{d_s}. \tag{10}$$

The Sherwood number is estimated by the correlation for flow through packed beds of solids (Eq. 11), where Sh_{sphere} is the Sherwood number for a flow over single spheres and f_a is a form factor [21]. Eq. (12) shows the correlation for the form factor if the bed consists of spheres of the same size [21]:

$$Sh_{bulk} = f_a \cdot Sh_{sphere}, \tag{11}$$

$$f_a = 1 + 1.5(1 - \varepsilon). \tag{12}$$

The Sherwood number Sh_{sphere} results from Eq. (13), where Sh_{lam} and Sh_{turb} are the Sherwood numbers for a laminar and a turbulent flow, respectively. With Eq. (14) and Eq. (15), Sh_{lam} and Sh_{turb} are calculated by inserting Re and Sc:

$$Sh_{sphere} = 2 + (Sh_{lam}^2 + Sh_{turb}^2)^{1/2}, \tag{13}$$

$$Sh_{lam} = 0.664 \cdot Re^{1/2} \cdot Sc^{1/3}, \tag{14}$$

$$Sh_{turb} = \frac{0.037 \cdot Re^{0.8} \cdot Sc}{1 + 2.443 \cdot Re^{-0.1} \cdot (Sc^{2/3} - 1)}. \tag{15}$$

The penetration height in the packed bed is also used to determine the mass transfer area a (Eq. 16), which depends on the height-independent mass transfer area a_h (Eq. 17). This parameter is derived by the total surface area of the particles divided by the height of the packed bed. f_1 gives the proportion of sucrose surface in the system:

$$a = a_h \cdot h \cdot f_1, \tag{16}$$

$$a_h = \pi \cdot (1 - \varepsilon) \cdot \frac{3}{2} \cdot \frac{d_c^2}{d_s}. \tag{17}$$

Eq. (3) provides us the mass of sucrose M_s, Eq. (6) the mass of liquid M_l and Eq. (18) calculates the sucrose load for each time step. With Eq. (19) the mass of water can be determined:

$$X_s = \frac{\frac{M_s}{M_l}}{1 - \frac{M_s}{M_l}}, \tag{18}$$

$$M_w(t) = \frac{M_s}{X_s}. \tag{19}$$

The sucrose load of the liquid is used to determine the liquid properties. The correlation between the sucrose load and the solution properties are taken from literature [22].

Results and discussion

Experimental penetration rates

We investigated the dynamic penetration of liquid into a powder system using the Washburn tool of the K100 tensiometer. The device measures the increasing water mass in the powder by recording the mass^2-time-plot. The powder systems consist of mixtures of sucrose as hydrophilic, soluble component and silanized glass powder as hydrophobic, inert component with different surface area proportions reaching from 100% sucrose to 50% sucrose and 50% glass beads. A summary about the mixtures is provided in Table 3 where f_1 and f_2 present the surface area fractions of sucrose and hydrophobic glass beads, respectively. By means of the volume specific surface areas of sucrose $S_{v,s}$ and glass $S_{v,g}$ and the material densities of sucrose ρ_s and glass ρ_g, the fractions of surface area can be converted into the fractions of mass, which were used for preparation of the samples. In order to characterize the powder mixtures in terms of contact angle, the Cassie-Baxter contact angle (Eq. 1) and its cosine were calculated as shown in Table 3. For the calculations the dynamic contact angle of sucrose of 31.3° and of hydrophobic glass of 83.8° were taken from former investigations [18] and were inserted in Eq. (1) as θ_1 and θ_2, respectively. The mixed contact angles range from 31.3° for the 100% hydrophilic surface to 61.2° for the 50:50-mixture.

Table 3. Composition of mixtures containing sucrose particles and hydrophobic glass beads, Cassie-Baxter contact angle of mixtures and the cosine of this contact angle.

Surface area Sucrose f_1 [%]	Surface area hydrophobic glass f_2 [%]	Cassie-Baxter CA θ_{CB} [°]	$\cos\theta_{CB}$ [-]
100	0	31.3	0.85
95	5	35.2	0.82
80	20	45.2	0.71
50	50	61.2	0.48

In order to compare the penetration of water into the different mixtures of sucrose and glass beads, we determined experimental penetration rates. They are obtained by the slope of the linear increase of the mass²-time-plot, which is divided by the cross-sectional area of the cylinder. A coefficient of determination R^2 of at least 0.98 is set for selection of the linear part and its slope. The time interval was chosen between 0.2 s and 2.3 s. Each experiment was repeated 3 times and a mean penetration rate and its standard deviation were calculated. The penetration rates of the mixture containing 100%, 95%, 80% and 50% sucrose are presented in Figure 2. It can be observed that an addition of hydrophobic surface area of 5% and 20% leads to an increase of the penetration rate from 3.78 $kg/s^{0.5}/m^2$ (100%) to 4.12 $kg/s^{0.5}/m^2$ (95%) and 4.06 $kg/s^{0.5}/m^2$ (80%). Due to the high standard deviations the penetration rates of 100%, 95% and 80% do not differ significantly, even if an increasing trend is visible. A further addition of hydrophobic surface area of 50% results in a significant decrease in penetration rate to 3.21 $kg/s^{0.5}/m^2$. In order to understand these trends, we have to distinguish between two main effects, which are linked to the wetting of hydrophobic surfaces and the wetting of soluble material. An increasing amount of hydrophobic surface decreases the capillary penetration into a powder bed. This effect was studied and described in a former study [23]. Secondly, the contact of water and sucrose results in the dissolution of sucrose leading to an increasing viscosity of the liquid. The increased viscosity has also a decelerating influence on the water penetration [18]. Thus, we have to consider two factors slowing down the capillary wetting, but both in a different way. The penetration of water into a sucrose system can be improved by the addition of hydrophobic surface of 5% and 20%. It seems that the hydrophobic surface (in small or intermediate quantities) reduces the dissolution and the consequent viscosity increase since there is less contact area between sucrose and water. This effect changes in the 50:50-mixture. Even if the contact area between sucrose and water is further decreased, the high amount of hydrophobic surface area in this sample leads to a decrease in penetration rate. Hence, we distinguish between an accelerating effect of hydrophobic surface at small and intermediate quantities and a decelerating effect at high quantities during the capillary wetting of sucrose.

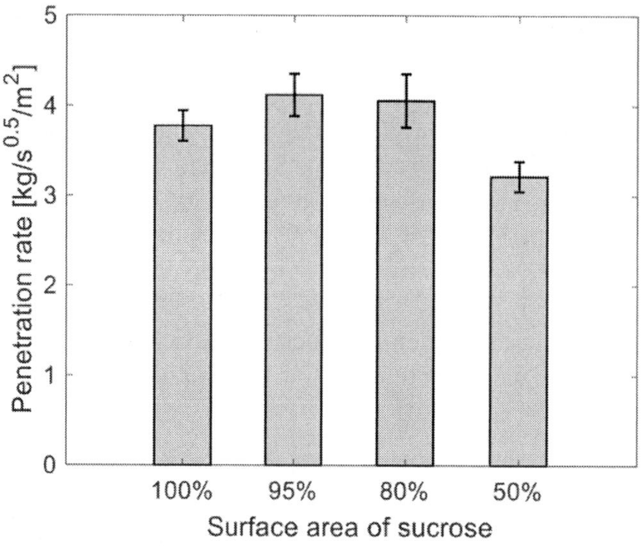

Figure 2. Penetration rates of water into different mixtures of sucrose and hydrophobic glass beads.

The wetting experiment was repeated with hexane as ideally wetting liquid, in order to get more information about the pore network. By assuming a spreading of hexane on hydrophilic sucrose and hydrophobic glass material ($\theta = 0°$) the capillary constant can be determined by inserting the slope of the linear increase of the mass^2-time-plot and the liquid properties from Table **2** into Eq. (20):

$$K_{hex} = \frac{{M_w}^2}{t} \cdot \frac{\eta}{{\rho_l}^2 \cdot \sigma}. \tag{20}$$

Capillary constants of all combinations of sucrose and glass beads were measured and are presented in Table 4. The capillary constant represents a geometric description of the packing. By mixing two powders, sucrose and glass beads, with a different particle size distribution and a different sphericity we receive different capillary constants for each mixture. Additionally, the porosity of the mixed powders was determined and is also presented in Table 4. It can be seen that the capillary constants slightly increase with an increasing amount of glass beads in the sucrose sample, while the porosity is only higher for the samples containing 5% and 20% of glass beads. Eq. (5) allows us to calculate effective pore radii of the packings with a cross-sectional area of the Washburn cylinder of $7.85 \cdot 10^{-5}$ m^2 and the porosities which are also shown in Table 4. This equation assumes a bundle of straight, parallel, cylindrical pores and does not consider the real pore shape, orientation and waviness. Nonetheless, the equation provides us an effective pore radius for each mixture, which can be used for describing the penetration into these systems. We obtained a pore radius of 6.78 µm for the sucrose bed. An increased amount of glass beads in the sample (5% and 20%) decreases the pore radius, even if the porosity is increased. This means that the system has a larger number of small pores in the samples containing 5% and 20% of glass surface, while the powder beds no and 50% of glass surface consist of a smaller number of big pores. Furthermore, the higher capillary constant of the 50:50-mixture compared to the pure sucrose indicates a more geometrically homogenous system with a high number of spherical glass beads in the sample. Higher amounts of smooth surface due to more glass beads and less edges due to less sucrose particles support a homogeneous and, thus, a faster uptake of hexane into the system.

Table 4. Capillary constant, porosity and effective pore radius of different mixtures of sucrose and hydrophobic glass beads.

Surface area sucrose f_1 [%]	Surface area hydrophobic glass f_2 [%]	Capillary constant K_{hex} [mm^{-5}]	Porosity ε [-]	Effective pore radius $r_{p,eff}$ [µm]
100	0	3.69	0.42	6.78
95	5	3.81	0.45	6.10
80	20	3.83	0.44	6.41
50	50	3.89	0.42	7.14

Comparing the penetration rates in Figure 2 and the capillary constants in Table 4, it is obvious that the faster penetration of water into the mixtures containing 95% and 80% sucrose cannot be explained by geometric aspects of the powder beds. While the fastest rise of hexane can be observed for the 50:50-mixture, the water penetration is the slowest into this mixture. This finding emphasizes that the two differently decelerating effects of viscosity increase due to dissolution and the presence of hydrophobic surface are responsible for the water penetration behavior.

Modelling of penetration behavior

Besides experimental investigation we developed a model to simulate the penetration of water into a heterogeneous system containing soluble sucrose particles and hydrophobic glass beads. For this purpose, the porosities and the effective pore radii were taken from Table 4, the water properties from Table 2 and the dynamic contact angles of 31.3° for water on sucrose and 83.8° for water on hydrophobic glass as mentioned already above. For the determination of the Reynolds number we inserted the $d_{50,3}$-value of sucrose of 287 µm (Table 1). All parameters calculated by the model are evaluated within a range of 0 to 6 s. Since the utilized Sherwood correlation is only applicable in the range of Reynolds number from 0.1 to 10^4, the model loses its validity after approximate 5.6 s.

Since we focused on the constant wetting rates at the beginning of the capillary rise into a powder system, we only included the change of liquid properties, which are viscosity and density, in a first step. Due to a relatively small increase in surface tension of only 8% by increasing the sucrose load from 0 to 1.5 g_s/g_w, the value of surface tension was kept constant in our model [24]. With the same increase in sucrose load the viscosity rises by 60 times and density increases by about 20% [22]. Neglecting the geometric changes due to dissolution was determined to be tolerable when comparing penetration during the first two seconds [18]. A reduction in particle volume of about 15% leads to a viscosity increase of about 160% [18].

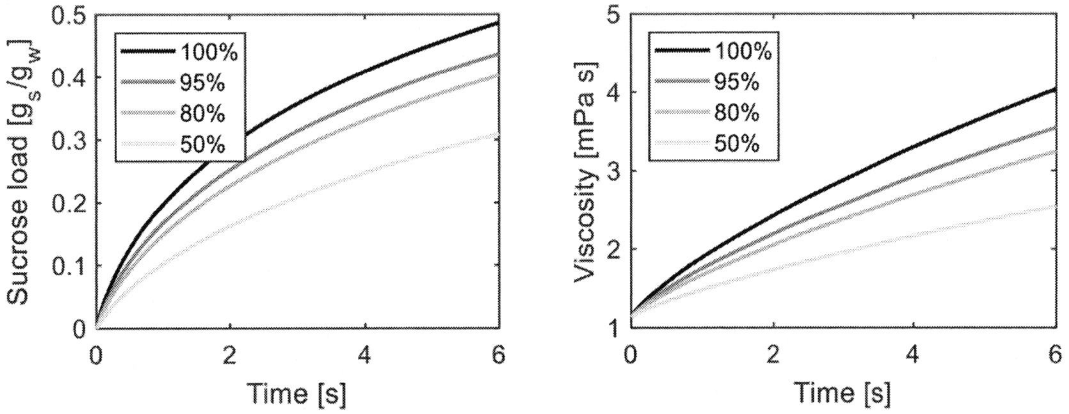

Figure 3. Sucrose load (left) and viscosity (right) of water over time for different mixtures of sucrose and hydrophobic glass beads; legend presents surface area of sucrose.

By applying the model, we calculate the parameters sucrose load and viscosity of the water over the initial penetration time of 6 s which are presented in Figure 3, left and right, respectively. Both parameters were modeled for the same four powder mixtures as in the experimental investigations. Figure 3, left provides the solution of Eq. (3) and shows a significant increase of sucrose load over time and a decreasing trend with increasing hydrophobic surface in the sample. Due to less contact area between sucrose and water with increasing glass surface in the mixture, the dissolution rate decreases resulting in less dissolved sucrose in water. After 6 s the sucrose load in the 100% sample rises to 0.49 g_s/g_w, while in the 50:50-mixture the sucrose load is only 0.31 g_s/g_w. The model assumes an ideal mixing of sucrose in water which results in a constant local concentration of sucrose in the sample.

The same trends can be observed for the viscosity since the viscosity of the liquid directly depends on the sucrose load. We can observe an increase of viscosity from 1.0 to 4.1 mPa·s for pure sucrose and an increase to 2.6 mPa·s for the 50:50-mixture within 6 s. Since the viscosity increase plays a major role during liquid penetration in our model [18], the differences in viscosity of the four mixtures influence the capillary rise.

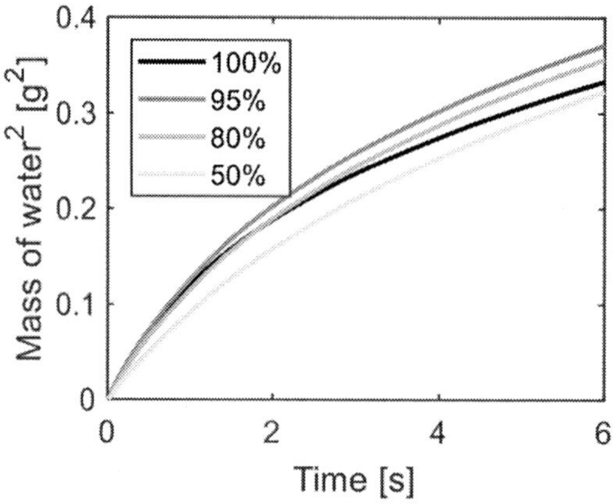

Figure 4. Washburn plot (mass^2 over time) for liquid penetration into different mixtures of sucrose and hydrophobic glass beads; legend presents surface area of sucrose.

In order to evaluate the influence of both, the solubility and the hydrophobic surface on the liquid penetration into the sample, Figure 4 provides the mass of water^2-time-plot for the four different cases. It is immediately apparent that the trend has changed in this diagram. During the first seconds the slope is similar for the sample 100%, 95% and 80%, while the slope of the 50% is lower. This means that in the initial period the hydrophobic surface has the major reducing impact on the penetration velocity. After about 2 s, the slope of the 100% sucrose-curve starts to decrease significantly faster compared to the samples 95% and 80% and approximates the slope of the 50%-curve. It is obvious that during the second period starting at approximate 2 s, the impact of the viscosity increase due to dissolution gains importance. The fast increase of viscosity in the 100% sample results in a decelerating effect on the rise, while the addition of 5% and 20% of hydrophobic surface to the samples seems to have a stabilizing impact on the liquid flow. Comparing all samples the mixture containing 5% hydrophobic surface provides the best result in terms of fast liquid penetration. The increase of the apparent mixed contact angle from 31.3° to 35.2° with the addition of 5% of hydrophobic surface appears to be marginal compared to the slower increase of viscosity. Furthermore, the experimental penetration rates in Figure 2 show the same trend which can be observed in Figure 4 when analyzing the slope during the first 2 s seconds. The highest experimental penetration rate was also obtained for the mixture containing 5% hydrophobic surface.

Conclusion

In this work, the dynamic water penetration into heterogeneous and soluble model food powders was investigated. Heterogeneity in composition and solubility are important characteristics of food powders. This study was performed regarding the better understanding of wettability in food powder rehydration and the enhanced prediction of this process. Therefore, experimental penetration rates were measured of different mixtures of hydrophobic glass beads and hydrophilic sucrose crystals which are soluble. Additionally, we developed a model dealing with the influence of hydrophobic surface and solubility during the capillary wetting process. Penetration behavior obtained by the model is in fair agreement with the experimental penetration rates. With a decreasing amount of soluble sucrose in the system the increase of sucrose load and viscosity over time of the liquid are reduced. The addition of 5% and 20% of hydrophobic surface to the soluble sucrose powder results in a stabilizing effect and enhances the penetration speed.

References

[1] H. Schubert, Instantisieren pulverförmiger Lebensmittel, Chemie Ingenieur Technik 62 (1990) 892–906.

[2] E.H.-J. Kim, X.D. Chen, D. Pearce, Surface characterization of four industrial spray-dried dairy powders in relation to chemical composition, structure and wetting property, Colloids and Surfaces B: Biointerfaces 26 (2002) 197–212.

[3] B. Cuq, E. Rondet, J. Abecassis, Food powders engineering, between knowhow and science: Constraints, stakes and opportunities, Powder Technology 208 (2011) 244–251.

[4] J. Kowalska, A. Lenart, The influence of ingredients distribution on properties of agglomerated cocoa products, Journal of Food Engineering 68 (2005) 155–161.

[5] C. Gaiani, J.J. Ehrhardt, J. Scher, J. Hardy, S. Desobry, S. Banon, Surface composition of dairy powders observed by X-ray photoelectron spectroscopy and effects on their rehydration properties, Colloids and Surfaces B: Biointerfaces 49 (2006) 71–78.

[6] S.V. Crowley, B. Desautel, I. Gazi, A.L. Kelly, T. Huppertz, J.A. O'Mahony, Rehydration characteristics of milk protein concentrate powders, Journal of Food Engineering 149 (2015) 105–113.

[7] J.J. Fitzpatrick, A. van Lauwe, M. Coursol, A. O'Brien, K.L. Fitzpatrick, J. Ji, S. Miao, Investigation of the rehydration behaviour of food powders by comparing the behaviour of twelve powders with different properties, Powder Technology 297 (2016) 340–348.

[8] A.B. Cassie, S. Baxter, Wettability of porous surfaces, Transactions of the Faraday society 40 (1944) 546–551.

[9] J. Dupas, E. Verneuil, M. Ramaioli, L. Forny, L. Talini, F. Lequeux, Dynamic wetting on a thin film of soluble polymer: effects of nonlinearities in the sorption isotherm, Langmuir : the ACS journal of surfaces and colloids 29 (2013) 12572–12578.

[10] L. Forny, A. Marabi, S. Palzer, Wetting, disintegration and dissolution of agglomerated water soluble powders, Powder Technology 206 (2011) 72–78.

[11] S. Hogekamp, H. Schubert, Rehydration of Food Powders, Food Science and Technology International 9 (2003) 223–235.

[12] E.W. Washburn, The Dynamics of Capillary Flow (1921) 273–283.

[13] I.S. Saguy, A. Marabi, R. Wallach, New approach to model rehydration of dry food particulates utilizing principles of liquid transport in porous media, Trends in Food Science & Technology 16 (2005) 495–506.

[14] D. Benavente, P. Lock, M. Ángeles García Del Cura, S. Ordóñez, Predicting the Capillary Imbibition of Porous Rocks from Microstructure, Transport in Porous Media 49 (2002) 59–76.

[15] I.S. Saguy, A. Marabi, R. Wallach, Liquid imbibition during rehydration of dry porous foods, Innovative Food Science & Emerging Technologies 6 (2005) 37–43.

[16] D.R. Lide, H.V. Kehiaian, CRC handbook of thermophysical and thermochemical data, CRC Press, Boca Raton, 1994.

[17] N.B. Vargaftik, B.N. Volkov, L.D. Voljak, International Tables of the Surface Tension of Water, Journal of Physical and Chemical Reference Data 12 (1983) 817–820.

[18] J. Kammerhofer, L. Fries, J. Dupas, L. Forny, S. Heinrich, S. Palzer, Penetration rates into heterogeneous food systems, Powder Technology (2018) submitted.

[19] F.Á. Mohos, Data on Engineering Properties of Materials Used and Made by the Confectionery Industry, in: Confectionery and Chocolate Engineering, John Wiley & Sons, Ltd, 2017, pp. 617–642.

[20] M. Mathlouthi, P. Reiser, Sucrose: Properties and applications, Blackie Academic & Professional, London, 1995.

[21] VDI Heat Atlas, Springer Berlin Heidelberg, Berlin, Heidelberg, 2010.

[22] M. Asadi (Ed.), Beet-sugar handbook, Wiley-Interscience, Hoboken, N.J, 2007.

[23] J. Kammerhofer, L. Fries, J. Dupas, L. Forny, S. Heinrich, S. Palzer, Impact of hydrophobic surfaces on capillary wetting, Powder Technology 328 (2018) 367–374.

[24] P. Honig, Principles of Sugar Technology, Elsevier Science, Burlington, 1953.

509532 Aspects of Wettability Heterogeneities in Flotation - Investigations with Inverse Gas Chromatography and Colloidal Probe Atomic Force Microscopy

Abstract:

In flotation for the separation of fine particles (approx. 5 µm < x < 200 µm), the main particle property of separation is a differing wettability of the constituents. This most crucial property in the economically important process of flotation is often only and incompletely, quantitatively described in terms of static contact angles of water. A water contact angle larger than 90° is referred to as hydrophobic and hydrophilic vice versa. Looking at Young's equation shows that it is actually the interacting phases' surface energies split in disperse and polar components which describes whether there will be hydrophobic/hydrophilic hydration interactions leading to attraction or repulsion. This, we have previously published for various more or less hydrophobic minerals in correlation to their flotability (c.f. Rudolph & Hartmann, Coll.Surf.A 513, 380-388). In the same paper we also discussed the heterogeneity of wettabilities as assessed with the novel method of inverse gas chromatography. Furthermore we have also reported on hydrophobic force distributions as assessed with atomic force microscopy which are as well due to wettability heterogeneities (c.f. Fritzsche et al., Coll.Surf.A 459, 166-171).

In this paper we present wettability distributions of different substrates and particles with various surface roughnesses/shapes and degrees of hydrophobization as assessed with inverse Gas Chromatography iGC, colloidal probe atomic force microscopy CP-AFM and dynamic sessile drop analyses. We are putting the results in context in unifying models and present methodologies helpful to make use of the fundamental information and their impact on the actual industrial separation process in fundamental process models. Besides froth flotation we also emphasize the impact on selective hydrophobic flocculation process.

508688 Surface Engineering for Designing Superhydrophobic and Superhydrophilic Particulate Solids

Abstract:

Micro-fabrication of flat surfaces for superhydrophobic (contact angle â‰¥ 150^0) and superhydrophilic surfaces (contact angle â‰¤ 5^0) is very common in many advanced applications. However, the technique cannot be directly employed to particulate solids. In this work, superhydrophobic and superhydrophilic surfaces are designed on particulate solids using chemical etching of 120 Âµm glass beads and Ar plasma treatment of 17 Âµm corn starch particles. While the SEM and AFM analysis show that both the surface fabrication techniques created randomly distributed nano-scale surface roughness, the XPS and surface energy (SE) data show that these fabrication techniques are also changed the surface chemistry of the particle. The contact angle (CA) data using Sessile drop method shows that the hydrophilic glass surface (CA ~ 80^0) is transformed into a superhydrophobic surface (CA~150^0) and the hydrophilic corn starch surface (CA ~ 48^0) is converted into more hydrophilic surface with CA~ 20^0 (towards the superhydrophilic surface). The change in wetting property of the glass surface and corn starch surface is implicated to change in surface groups, surface roughness including its distribution and aspect ratio (width to height ratio) of asperities on the particle surface. A comparison between the experimentally determined contact angle and that of theoretically calculated data shows that the adhesion between the water droplet and the glass surface is in Cassie impregnation regime. In contrast, the plasma treated corn starch (in the pellet form) is in the Wenzel wetting regime even after the increase in the surface roughness. The analysis of individual contribution of the surface chemistry and surface roughness shows that wetting behaviour of glass beads CA > 120^0 is mostly controlled by nano-scale surface roughness. However, surface chemistry plays greater role for controlling the wetting behaviour of plasma treated corn starch particles. The present work demonstrated the extent and nature of surface engineering required to design a particle surface which is superhydrophobic or superhydrophilic. The work has immense implications in many advanced applications where wetting of particle surface is critical.

507973 Second Dose during the Breakage Occurred in Initial Flocculation Phase: Does It Work Positively?

Abstract:

Coagulation is a critical unit commonly functions as the first barrier for the removal of particulate and/or natural organic matter during the treatment of drinking water. However, flocs breakage has been found to exert negative effects on the effluent water quality. Although breakage and regrowth of flocs has been studied for several years, little attention was paid to the regrowth ability of broken micro-flocs which mainly formed during the initial flocculation stage. In this paper, we employed a continuous optical monitoring method to investigate the kinetics of flocs formation, and especially more evident flocs properties (strength and recovery factor) were also explored. The results showed that breakage at the beginning of flocculation stage can improve flocs formation to greater sizes when the coagulation dominated by charge neutralization at a predetermined dosage. Furthermore, a second dose of different coagulants during this breakage process improved kaolin particles removal efficiency to varying degrees. Separately, the second dose of $Al_2(SO_4)_3$ enlarged the flocs sizes with increasing dosage, contrarily that of PACl reduced the flocs sizes. Based on the flocs properties, this observation elicits the different coagulation mechanisms between $Al_2(SO_4)_3$ and PACl: hydroxides of $Al_2(SO_4)_3$ formed in larger sizes and entrapped more particles (lower residual turbidity) by sweep flocculation; while PACl formed compact flocs due to stronger charge of single Al. It is worth noting that turbidity removal performance was even better when the overall dosage by second dose only counted for ~ 70% of one-time dosage. Importantly, this study provides a feasible economical dosing strategy for coagulation treatment.

EFFECT OF PARTICLE WETTABILITY ON MINERAL OIL-DISTILLED WATER EMULSION STABILITY

C. Nunez, R. Dabirian, I. Gavrielatos, R.S. Mohan, and O. Shoham, The University of Tulsa, Tulsa, OK, USA

Abstract

Petroleum industry has been encountering oil-water emulsions during production and transportation of crude oil. The presence of natural solid nanoparticles and other surface-active agents together with shear provided by chokes, valves and other devices are the main reasons for the generation of emulsions with a certain degree of stability. The aim of this study is to experimentally investigate the effect of nanoparticles with different wettabilities, namely, hydrophilic and hydrophobic, on the stability of mineral oil-distilled water emulsions.

To study the effects of nanoparticles on separation kinetics of the oil-water dispersion, a Portable Characterization Dispersion Characterization Rig (P-DCR) is used, which enables creation of the emulsions under different conditions. The experiments are conducted with mineral oil and distilled water as fluid phases, and a spherical silica nanoparticle of 20 nm mean diameter at concentrations of 0.01% and 1% by weight. The emulsions were created at a rotational speed of 600 RPM and at three water-cuts of 25%, 50% and 75%.

First, the emulsions are prepared with hydrophobic nanoparticles, which are initially dispersed in the oil phase. The results confirm that highest emulsion stability is obtained with 25% water-cut, and only oil creaming occurs, while the emulsion is very resistant to water coalescence and sedimentation. The data also show that the separation rate improves with increasing water-cut to higher values. The second phase of experiment includes the effects of hydrophilic nanoparticles, where the particles are initially dispersed in the water phase. The data demonstrate that in contrast to the hydrophobic particles, fast separation occurs at 25% water-cut, and increasing water-cut resulted to slower oil creaming process. Emulsions with combination of hydrophilic and hydrophobic particles is the last phase of this study. The data demonstrate that similar behavior as hydrophobic particles are observed, which emphasizes the dominant role of the hydrophobic particles as compared to hydrophilic ones. The experimental data for all cases show that increasing particle concentrations results in more stable emulsions.

Introduction

During oil and gas production process, the separation of the fluids before transportation is a vital issue that needs to be handled. Each separated fluid undergoes different treatment methods for transportation, and inefficient separation can lead to problems such as damaged equipment, inefficient processes, corrosion problems and so on. The presence of certain paraffins, natural solid nanoparticles in the oil, together with produced water and the agitation caused by several factors, generate emulsions with a certain degree of stability (Rodionova et al., 2014). An emulsion can be formed when two immiscible liquids are present in a medium with enough initial agitation to disperse one fluid into small droplets, and the existence of an emulsifier to stabilize the dispersed drops. Water-in-oil emulsions refer to tiny drops of water dispersed in a continuous body of oil. This is usually found in crude oil handling. On the other hand, oil-in-water emulsions are generally found in produced water. It is important to note that the stabilization of the emulsion retards the droplet coalescence and consequently the separation process.

Surface active agents such as asphaltenes, resins and added chemicals are a very common type of emulsifiers found in oil industry. However solid nanoparticles (finely divided sand, clay, etc.), commonly encountered when producing crude from unconsolidated reservoirs, can also act as emulsifiers forming a viscous barrier that inhibits drop coalescence. The agitation needed to form the emulsion can be provided by many sources including pumps and flow restrictions.

The objective of this study is to analyze the kinetics of separation of oil and water, when using nanoparticles to stabilize the emulsion, and observe the effects of changing three major parameters, namely, water-cut (%), wettability, and concentration (wt.%) of the nanoparticles. Having a better understanding of the stabilization effects of nanoparticles on emulsions will lead to the development of more efficient strategies for emulsion separation.

Experimental Program

Instrumentation: Portable Dispersion – Characterization Rig (P-DCR)

Figure 1 shows a picture of the P-DCR. This device was developed at The University of Tulsa in collaboration with Canty Technology. The facility allows the user to mix fluids inside a vessel in the desired proportions, controlling pressure, temperature, stirring time and speed (rpm). The system is also equipped with a backlight system and a high definition camera that allow visualization of the separation process inside the vessel. A fully computerized system is used to monitor the separation and track the fluid interfaces. The volume of fluids introduced in the batch separator was 200 ml for all experiments, which were conducted at ambient pressure and temperature conditions. More details pertaining to the experimental facility can be found in Gavrielatos (2016), Angardi (2016), and Gavrielatos et al. (2017).

Figure 1. *Portable Dispersion Characterization Rig.*

Test fluids and Nanoparticles

For the oil phase, Exxsol D-100 Mineral Oil was selected for these experiments with density 0.805 grs/ml, and a viscosity of 3.1 cP (at 65 °F). Commercial distilled water was used for the water phase. Two types of nanoparticles were employed. A hydrophobic wettability nanoparticle (AEROSIL

R-974), used to be dissolved in the oil phase, and a hydrophilic wettability nanoparticle (AEROSIL 200), to be dissolved in the water phase. Both nanoparticles are silica particles (99.8% silica), with a specific surface of 200 m^2/g, spherical shape and an average diameter of 20 nm.

Experimental Procedure and Test Matrix

The experiments were performed using two nanoparticle concentrations (0.01% and 1%) for both wettabilities (hydrophilic and hydrophobic). For each set of combinations, 3 different water-cuts were used (25%, 50% and 75%). A sonicator was used to dissolve the nanoparticles and a standard procedure was defined. A total time of five minutes ultrasonic agitation was performed for each experiment, but because the dissolution also changes with temperature, this procedure was performed in lapses of 1 minute, with 30 seconds rest time to avoid excessive heating. Consequently, the fluids are added to the P-DCR and stirred for exactly 10 minutes. The separation process is monitored for up to four hours. More information regarding the experimental procedure as well as the determination of the emulsion type is available in Gavrielatos (2016) and Gavrielatos et al. (2017).

The experimental matrix for this study is detailed in table 1:

***Table 1**. Test matrix.*

Experiment #	WC (%)	Hydrophobic NP concentration (%)	Hydrophilic NP concentration (%)
1	25	0.01	-
2	50	0.01	-
3	75	0.01	-
4	25	-	0.01
5	50	-	0.01
6	75	-	0.01
7	25	0.01	0.01
8	50	0.01	0.01
9	75	0.01	0.01
10	25	1.00	-
11	50	1.00	-
12	75	1.00	-
13	25	-	1.00
14	50	-	1.00
15	75	-	1.00
16	25	1.00	1.00
17	50	1.00	1.00
18	75	1.00	1.00

Results and Discussion

The P-DCR allows the user to observe the different interfaces formed. The mineral oil is colored by a lipophilic red dye to distinguish the different fluids and emulsion formed.

A typical separation observation presented by the P-DCR user interface can be seen in Figure 2.

Figure 2. *Typical batch separation image.*

The results obtained show that emulsion stability depends on all three variables namely, water-cut, nanoparticle concentration, and particle wettability. The effect of each variable is investigated separately in different plots.

The effect of different water-cuts, plotted in Figure 3, presents the behavior of the emulsions for the case of hydrophobic nanoparticles, and the lowest concentration (0.01%).

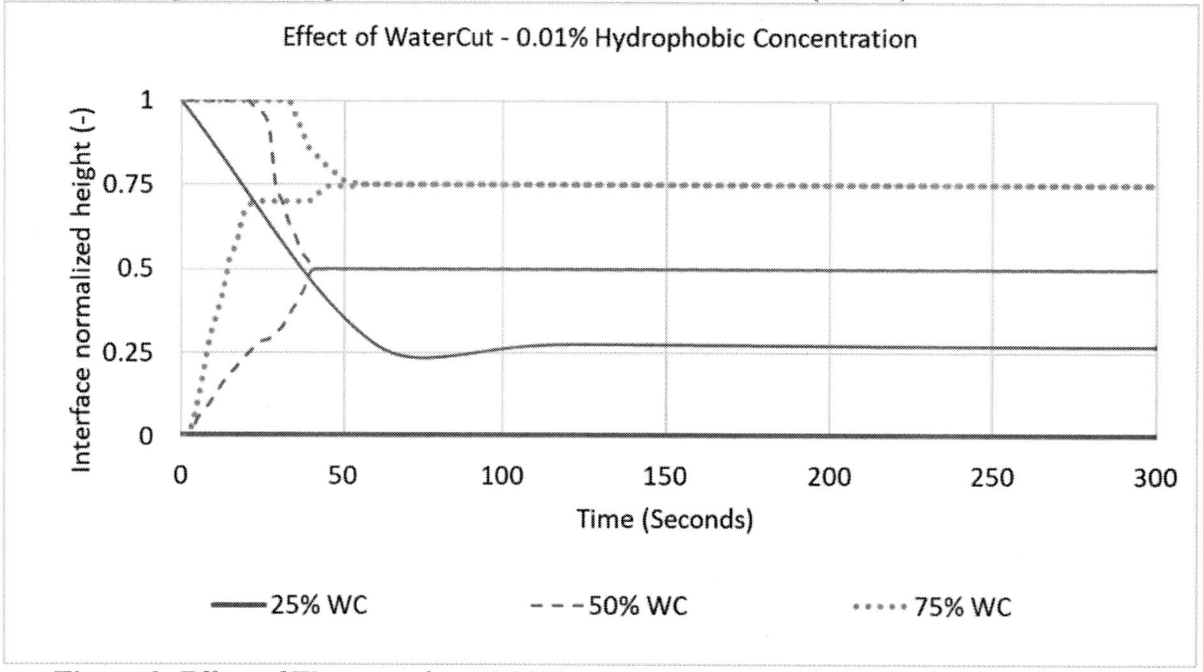

Figure 3. *Effect of Water-cut for a 0.01% concentration and hydrophobic nanoparticles.*

On the Y axis, interface normalized height can be seen and the time is observed in X axis. Continuous line shows the oil-emulsion interface, while dashed line shows the water-emulsion interface. This convention will be held for all next plots as well. In some figures, there is also a black dashed line that shows where the two interfaces are expected to meet in the case of complete separation.

In reference to Fig. 3, the green line shows the behavior for the case of 75% water-cut. The hydrophobic nanoparticles were dispersed in the oil phase, and so, there is a low weight of nanoparticles in the system (0.01% weight/weight of the 25% total volume, for a total of 0.00403 grs). This produces a poor emulsifying effect, and therefore quick oil creaming. Water also comes out of emulsion in a fast manner, obtaining full separation in about 45 seconds.

For the case of 50% water-cut, shown in red line, same trend as previously can be seen. For this case, the weight of nanoparticles in the system is higher than the previous case (0.01% weight /weight of 50% total volume, for a total of 0.00805 grs), but still not enough to stabilize the emulsion over a significant period of time. Fast separation does not allow determination of the kind of emulsion formed. When using the lowest water-cut (25%), the total weight of nanoparticles in the system is high enough to create a stable water in oil emulsion (0.01208 grs in the system). Water drops cannot coalesce because of the action of the emulsifier, although almost all the oil has creamed, as can be seen in figure 10. In spite of very small quantity of oil present in the emulsion (7% oil by volume), it is observed to be oil continuous emulsion as reported by Gavrielatos et al. (2017) due to the stabilizing effect of the hydrophobic nanoparticles.

When higher concentrations are used, of course higher weight of nanoparticles is present in the system, and for all water-cuts, a more stable emulsion can be observed. Figure 4 shows the same conditions as figure 3, but with a nanoparticle concentration of 1%. The hydrophobic nanoparticles have a tendency to retain a significant portion of the oil in the emulsion and the water is present in the emulsion in the form of droplets (dispersed phase).

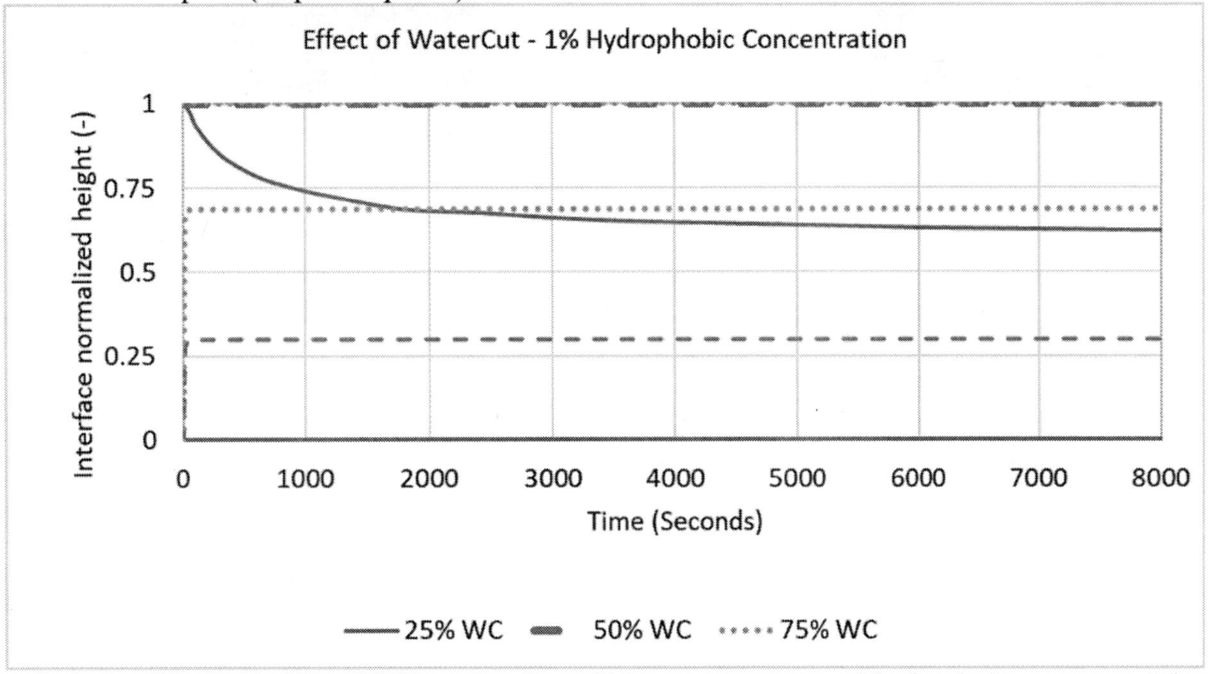

Figure 4. *Effect of Water-cut for a 1% concentration and hydrophobic nanoparticles.*

The higher concentration of emulsifier produces different separation dynamics. For the 75% and 50% water-cut, shown in green and red line, respectively, the emulsions are very stable, showing a different behavior from the ones observed for the low concentration. For both, 50% and 75% water-cut cases, the water sedimenting time does not differ significantly, separating partially in about 40 seconds for both concentrations. For 50% and 75% case, a significant portion of water is drained out of the emulsion resulting in water/oil emulsion. About 8% of the water volume stays inside emulsion for 75% water-cut, and about 40% of water stays in emulsion for 50% water-cut scenario. Regarding the oil phase in both water-cuts, figure 4 shows that no oil creaming occurs even after several hours of observation. The higher nanoparticle concentration is a key point in the stability of the emulsions. In general, the emulsion becomes more stable at higher concentrations. The nanoparticles now are more efficient surrounding the dispersed phase droplets, and so, preventing the flocculation and consequent creaming.

For 25% water-cut case more than 50% of the oil gets separated from the emulsion through the creaming process and all water is retained in the emulsion. The resulting emulsion is still oil-continuous. For this case the emulsion seems to be "tighter", presenting smaller droplets. Taking into account that all the water present in the vessel is part of the emulsion for both nanoparticle concentrations, and that smaller water drops (and so, more droplets) can be seen for the higher concentration, more oil is needed to surround the water droplets due to increased total surface area. This explains the lower volume of oil creaming out of emulsion for the higher concentration case, as shown in figures 5 and 6.

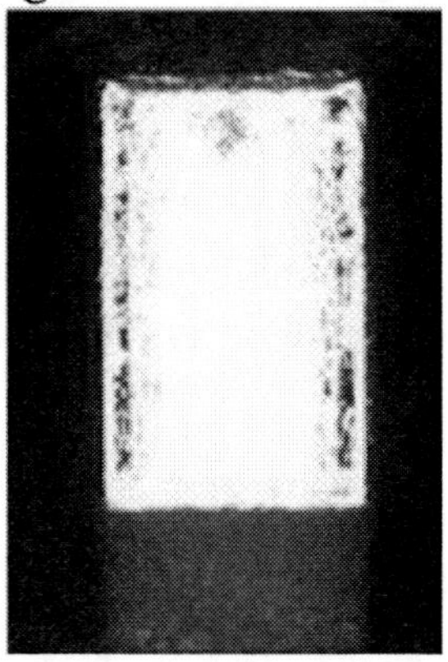

***Figure 5.** 25% WC/hydrophobic / 0.01% conc.*

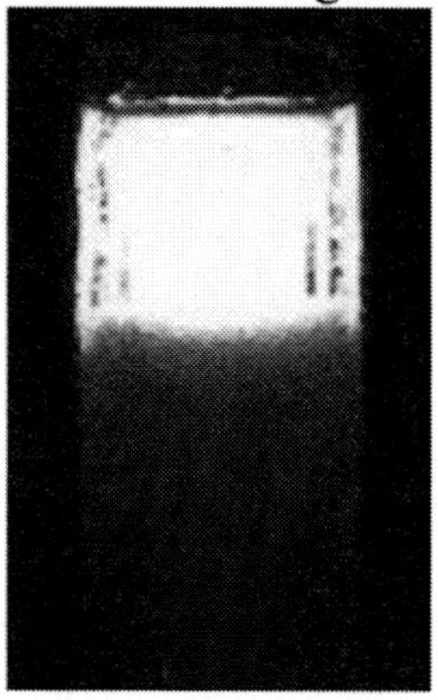

***Figure 6.** 25% WC/hydrophobic / 1% conc.*

The same analysis is performed when the nanoparticle wettability is changed to hydrophilic. Figure 7 presents the effect of different water-cuts when using the low concentration (0.01%).

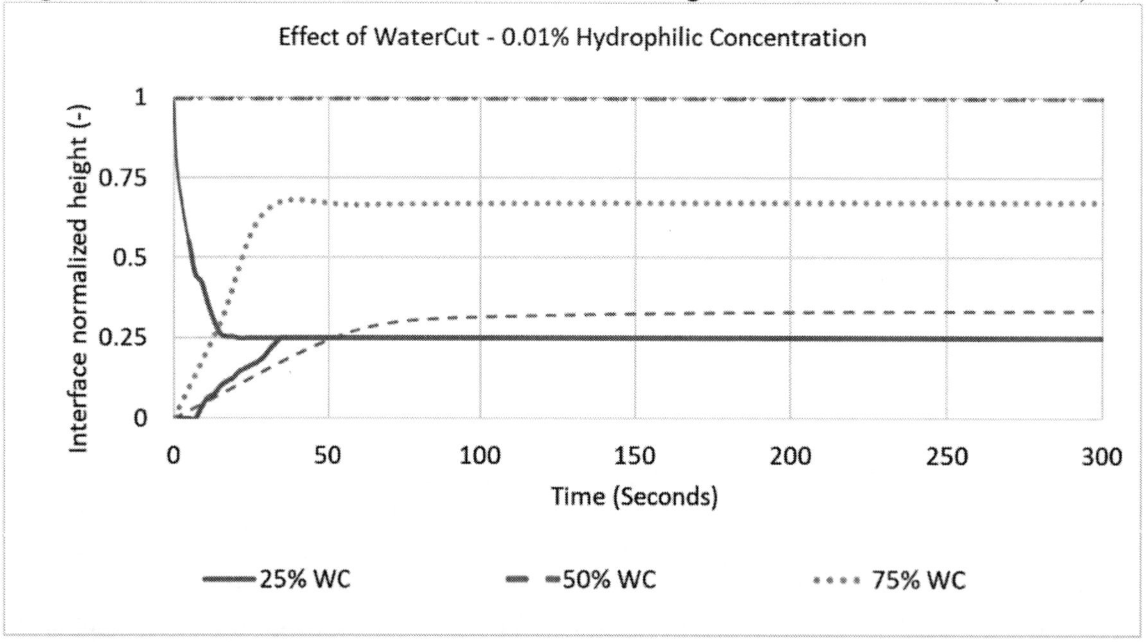

***Figure 7.** Effect of Water-cut for a 0.01% concentration and Hydrophilic nanoparticles.*

In this case, the nanoparticles are dissolved in water. A fair comparison of this figure is against figure 3 (Effect of water-cut for a 0.01% concentration and hydrophobic nanoparticles). The highest weight of nanoparticles in the system is in the highest water-cut (75%) and the lowest weight is in the lowest water-cut (25%). This is exactly the opposite scenario of the hydrophilic wettability type. This fact explains in certain manner the opposite behavior of the fluids. It is interesting to note that a significant portion of water is sedimented out of the emulsion for 75% water-cut case compared to 50% water-cut.

It can be observed that for 25% water-cut complete separation occurs after about 40 seconds. However, for both cases 50% and 75% water-cut, no oil creaming can be observed, and only some percentage of water sediments. The explanation again is found in the weight of particles present in the system. The lowest water-cut contains very little amount of nanoparticles in the system (0.005 grams). Clearly this little amount of emulsifier is not enough to prevent the water drops to flocculate and sediment. However, for the cases of 50% and 75%, the weight of emulsifier present in the system is higher (0.01 and 0.015 grams, respectively). Even with these low concentrations, sufficiently stable emulsions are formed.

Finally, the concentration of hydrophilic nanoparticles was raised to 1% weight/weight, and the results are shown in figure 8.

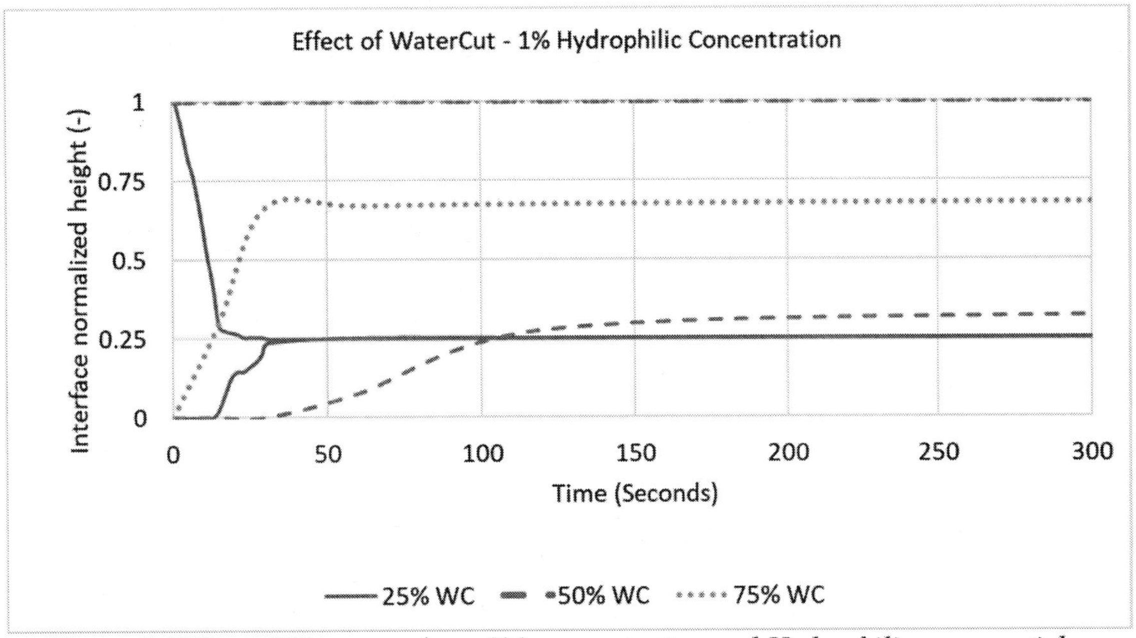

Figure 8. *Effect of Water-cut for a 1% concentration and Hydrophilic nanoparticles.*

It can be seen that the change of concentration for the hydrophilic wettability presents a very similar behavior as seen in the lower concentration case (Fig. 7). This demarks a notable difference in the efficiency of both wettabilities. Even though the weight of emulsifier was raised from 0.005 grams to 0.5 grams (100 times more) for high concentrations, no stable emulsion can be observed for 25% water-cut. For 50% and 75%, this significant increase of nanoparticles amount does not seem to improve the stability of the emulsion, as can be seen comparing figures 9 and 10. In figure 10, the blurry color in the water phase is due to the high nanoparticle concentration, which is expected to facilitate the water sedimentation process.

Figure 9. 75% WC/Hydrophilic/0.01% conc.

Figure 10. 75% WC/Hydrophilic/1% conc.

Using the same results, it is possible to plot in different ways the information obtained, in order to observe the differences when changing the three main variables. Figure 11 shows the effect of using different nanoparticle wettabilities for a 25% water-cut and 0.01% concentration. As stated above, for 25% water-cut, no stable emulsion was observed for both concentrations for the case of hydrophilic wettability (figures 11 and 12), while when using hydrophobic wettability, stable emulsions where observed for both concentrations. It is also notable how the stability of the emulsion increases when higher concentration of nanoparticles is used, as can be observed in the separation interface displayed by the blue lines of figure 11 and figure 12.

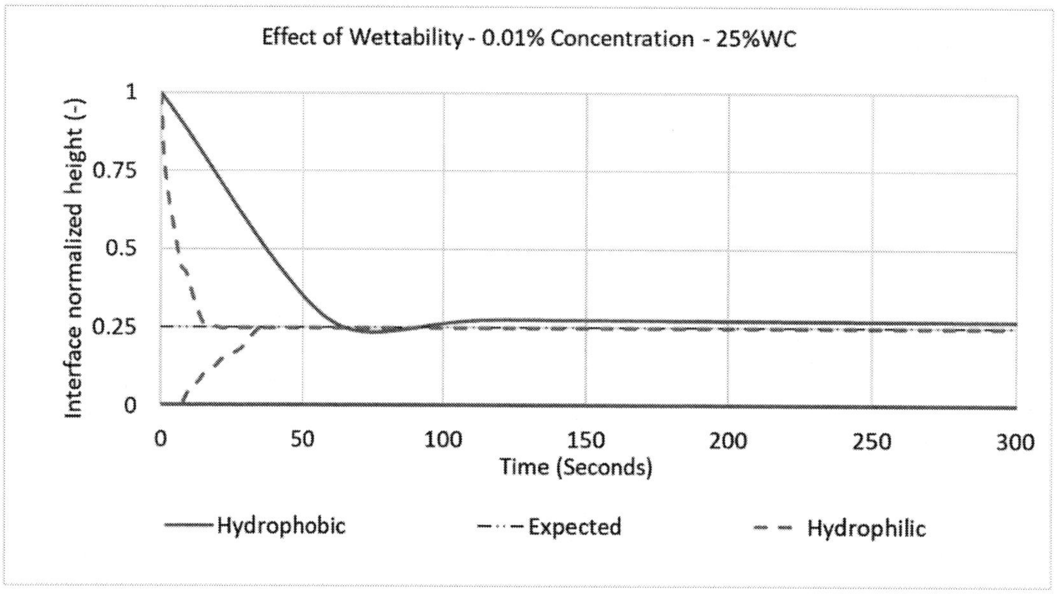

***Figure 11.** Effect of Wettability for a 0.01% concentration and 25% water-cut.*

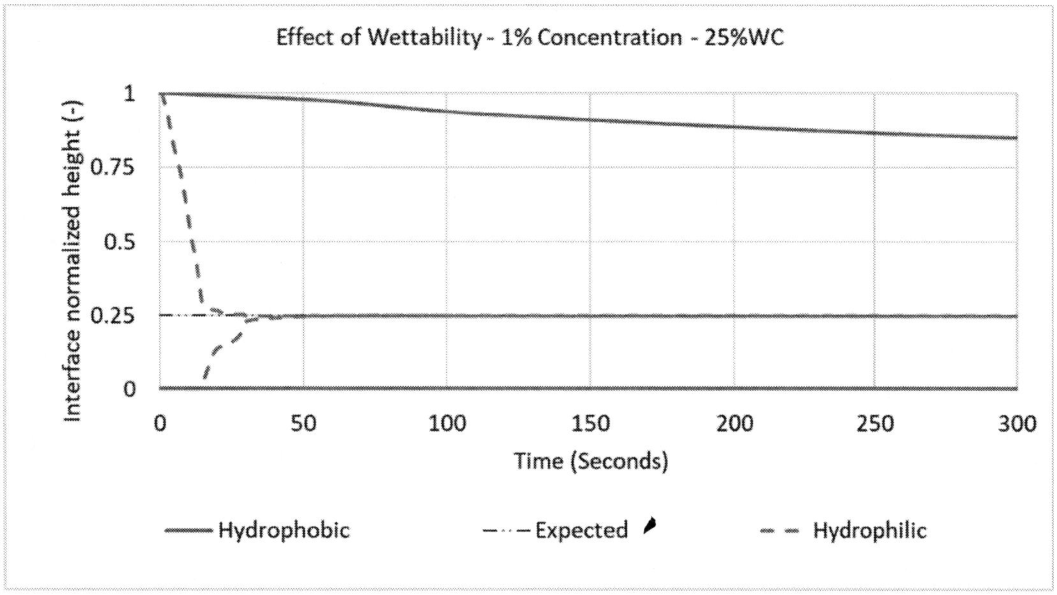

Figure 12. *Effect of Wettability for a 1% concentration and 25% water-cut.*

Next, figures 13 and 14 display the behavior for 50% WC, using lowest and highest concentrations, respectively. It is evident how the hydrophobic nanoparticles (blue lines) are sensitive to the increase of nanoparticle concentration in the system, presenting a notable higher stability in the emulsion formed. For the case of hydrophilic nanoparticles, no significant effect can be observed when increasing the nanoparticle concentration. Both concentrations show a partial separation (about 70%) in about 100 seconds, maintaining the emulsion stable for the rest of the 4 hours of observation (to capture the initial dynamics, data shown only for the first 10 and 5 minutes, respectively).

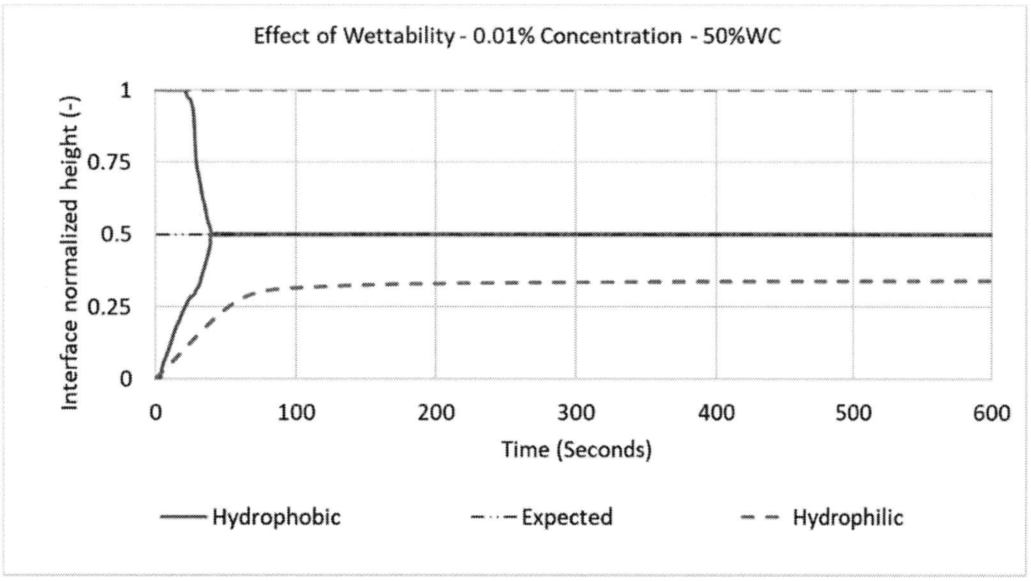

Figure 13. *Effect of Wettability for a 0.01% concentration and 50% water-cut.*

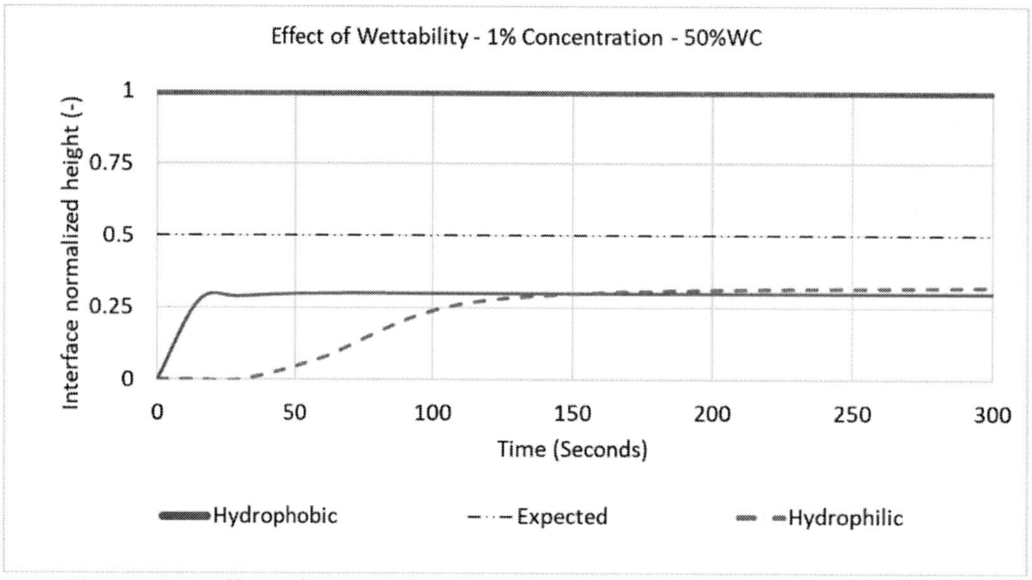

Figure 14. *Effect of Wettability for a 1% concentration and 50% water-cut.*

The exact same trends are observed for the case of 75% water-cut.

Last comparison to be analyzed is the effect of concentration when fixing water-cut and wettability parameters. Figures 15 and 16 show the difference in the behavior when dissolving hydrophobic nanoparticles in oil, and dissolving hydrophilic nanoparticles in water, for different concentrations, respectively. As stated before, while increased emulsion stability can be seen for higher concentrations in the case of hydrophobic nanoparticles (figure 15), no major differences are observed for hydrophilic nanoparticles (figure 16).

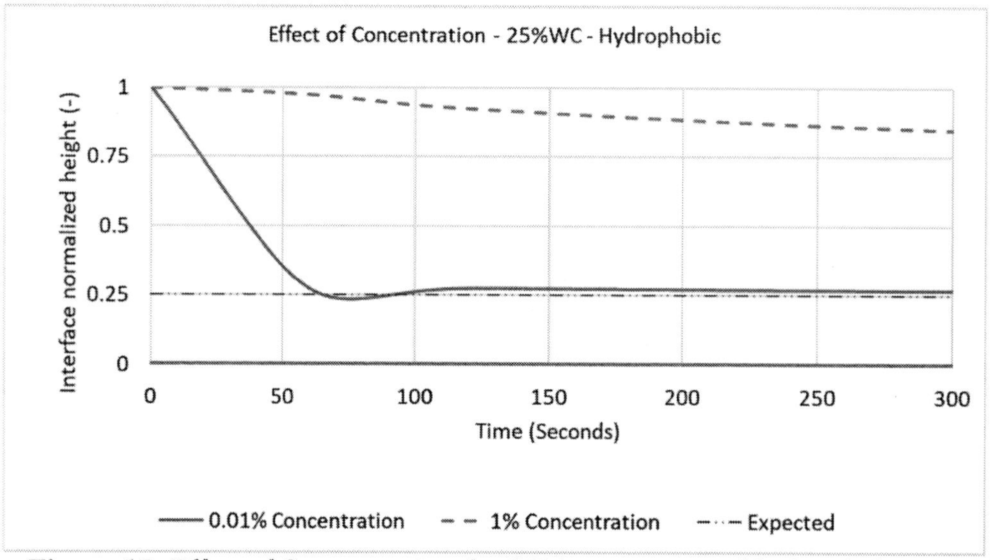

Figure 15. *Effect of Concentration for 25% water-cut and hydrophobic wettability.*

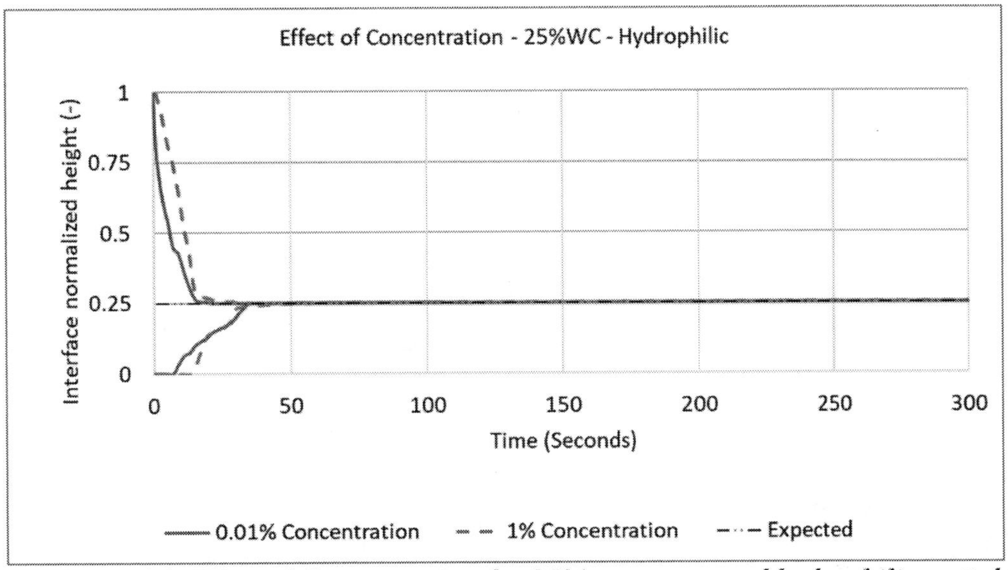

Figure 16. *Effect of Concentration for 25% water-cut and hydrophilic wettability.*

Doing the same comparison, but this time for a 50% water-cut, it can be noted that for hydrophobic nanoparticles, sensitivity to concentration is remarkable, separating in a rapid manner for low concentrations, and stabilizing the emulsion for a higher concentration (figure 17). On the other hand, when comparing concentrations for 50% water-cut, and using hydrophilic nanoparticles, slight differences can be seen when moving from low to high concentrations at the beginning of the experiment, but quickly aligning to a similar asymptotic separation line (figure 18). The emulsion types were estimated using the conductivity test, drop test, and microscopic observation. It is interesting to note that for lower and higher concentrations and 50% water-cut, oil continuous phase was observed (water in oil) for both figures 17 and 18, which needs further investigation.

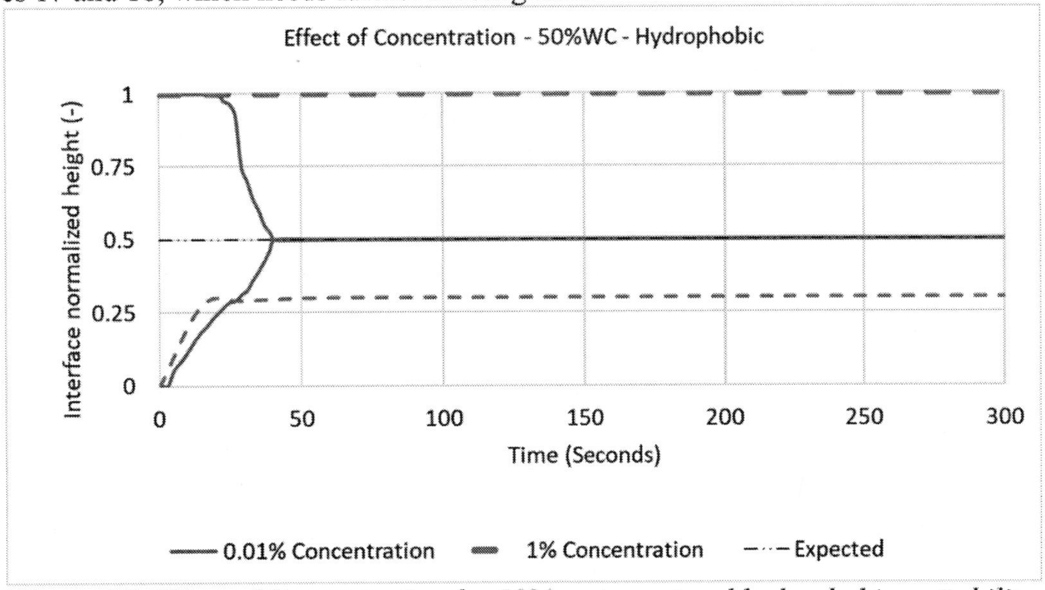

Figure 17. *Effect of Concentration for 50% water-cut and hydrophobic wettability.*

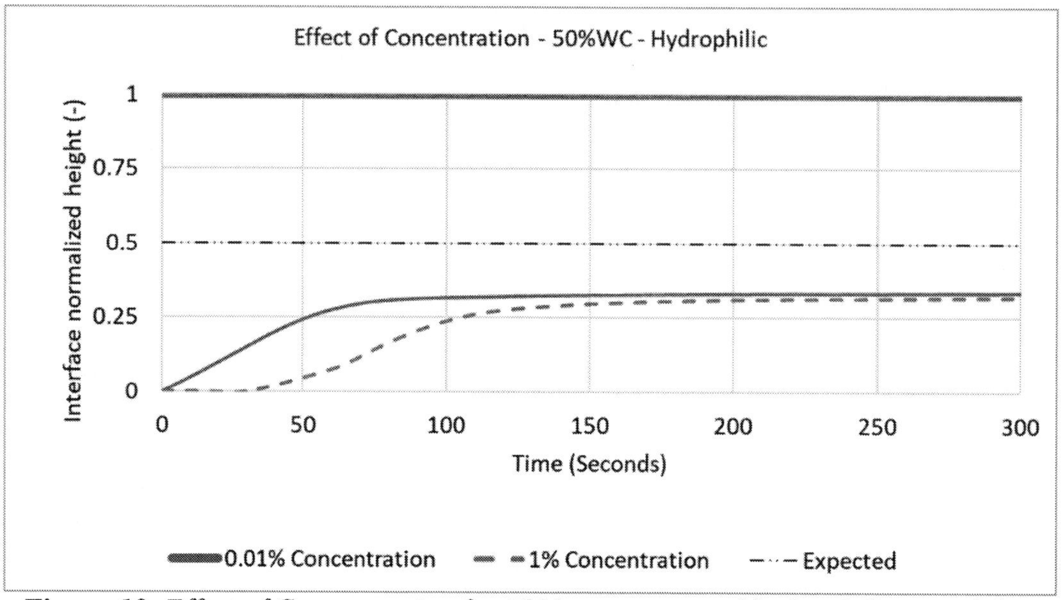

Figure 18. *Effect of Concentration for 50% water-cut and hydrophilic wettability.*

Finally, the separation profiles for 75% water-cut case show similar trends as those discussed for the other water-cuts. For hydrophobic wettability, there is an evident improvement in stability when increasing the concentration from low to high, whereas for the case of hydrophilic nanoparticles, nanoparticle concentration seems to have no effect at all.

When using hydrophobic nanoparticles dissolved in oil, and hydrophilic nanoparticles dissolved in water at the same time, for the same three water-cuts and two concentrations, the results show a big resemblance with the results obtained when using hydrophobic nanoparticles only. This suggests that the effect of hydrophobic nanoparticles is much more important than the hydrophilic nanoparticle effect, as can be observed in figures 19 and 20.

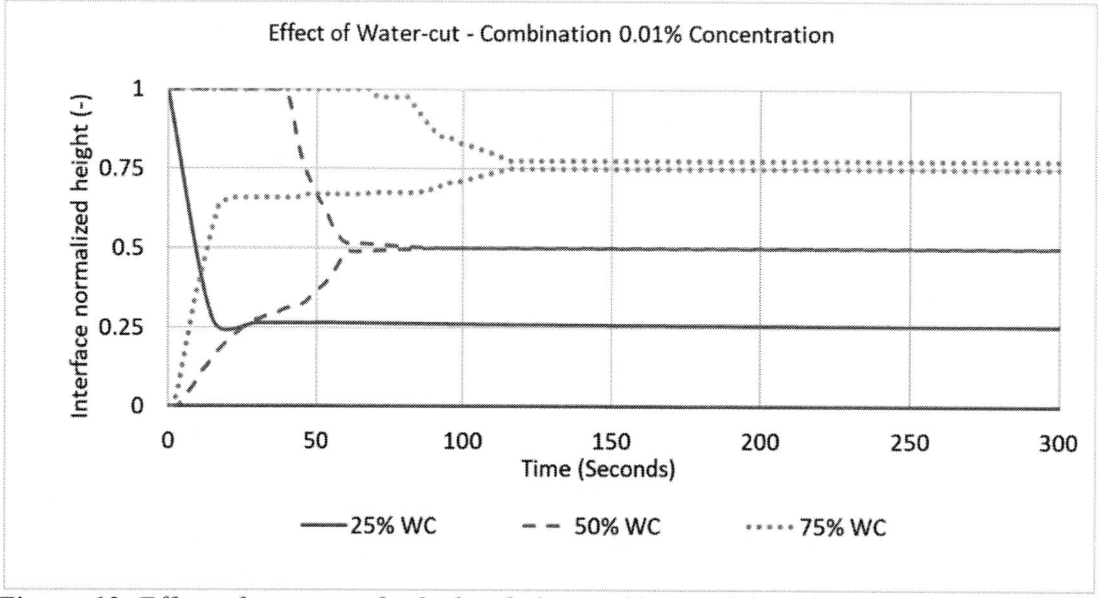

Figure 19. *Effect of water-cut for hydrophobic and hydrophilic nanoparticles, and 0.01% concentration.*

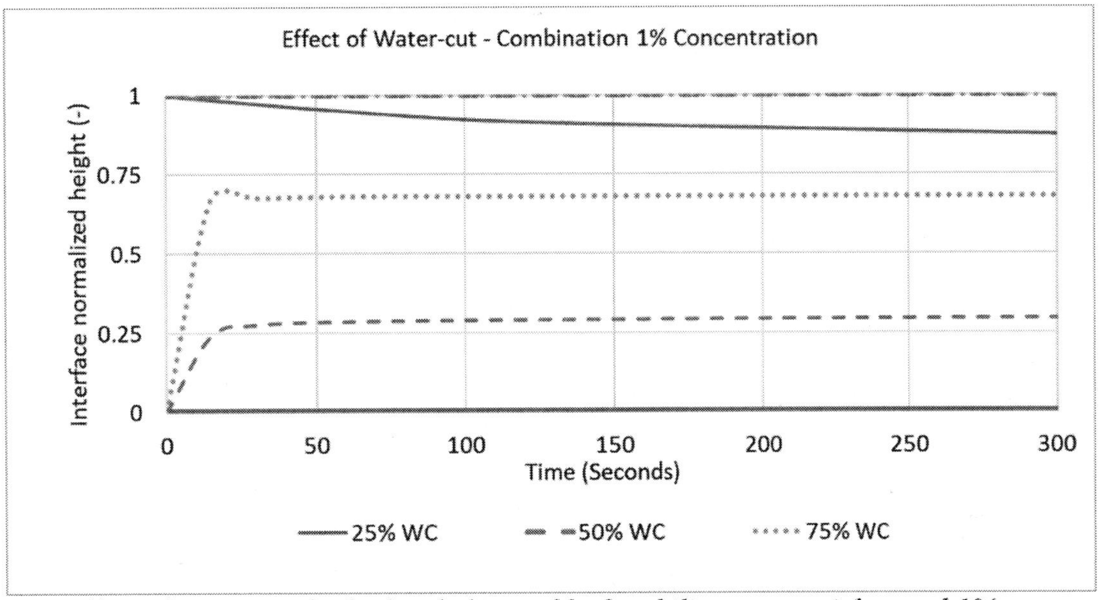

Figure 20. *Effect of water-cut for hydrophobic and hydrophilic nanoparticles, and 1% concentration.*

Conclusions

A state-of-the-art Portable Dispersion Characterization Rig (P-DCR) is used to create emulsions with Exxsol mineral oil, commercial distilled water and hydrophilic and hydrophobic silica nanoparticles as emulsifiers. Stability of these emulsions is evaluated under different conditions such as particle wettability, concentrations and water-cut. The P-DCR is equipped with a camera to track the interfaces of the fluids to investigate the separation kinetics.

The primary conclusions of this study are:

i. The stability of an emulsion depends strongly on wettability of the nanoparticles used as emulsifiers, nanoparticle concentration and water-cut.

ii. For the same concentration, weight of the hydrophilic nanoparticles in the system increases with water-cut, whereas the weight of hydrophobic nanoparticles decreases with water-cut. In general, increased weight of nanoparticles in the system for any wettability produces more stable emulsions.

iii. Compared to hydrophilic nanoparticles, stability of hydrophobic nanoparticle emulsions was shown to be more susceptible to concentration changes for every water-cut used.

iv. When combining nanoparticles of both wettabilities, hydrophilic nanoparticles seem to have no effect on the separation kinetics for all water-cuts investigated. The behavior of this emulsion is very similar to the case with hydrophobic nanoparticles alone.

References

1. Rodionova G., Pettersen B., Kelesoglu S., and Sjoblom J. (2014), "Preparation and characterization of reference fluid mimicking behavior of North Sea heavy crude oil". Fuel 135, 308-314.
2. Gavrielatos I. (2016) "Effect of nanoparticles on oil-water emulsion stability". MS Thesis, The University of Tulsa, OK, USA.
3. Gavrielatos I., Mohan R., and Shoham O. (2017), "Effect of intermediate wettability nanoparticles on oil-water emulsion stability". Journal of Petroleum Science and Engineering 152, 664-674
4. Gavrielatos I., Dabirian R., Mohan R.S., and Shoham O. (2018), "Oil/water emulsions stabilized by nanoparticles of different wettabilities", Journal of Fluids Engineering (under review).
5. Gavrielatos I, Dabirian R., Mohan R.S., and Shoham O. (2017), "Separation kinetics of oil/water emulsions stabilized by nanoparticles" FEDSM 2017 – 69112.
6. Angardi V. (2016), "Effect of aqueous phase composition and surfactant on oil-water dispersion stability". MS Thesis, The University of Tulsa, OK, USA.
7. Allouche J., Tyrode E., Sadtler V., Choplin L., and Salager J.L. (2004) "Simultaneous conductivity and viscosity measurements as a technique to track emulsion inversion by the phase-inversion-temperature method". Langmuir 20, 2134-2140.
8. Tyrode E., Allouche J., Choplin L., and Salager J.L. (2005), "Emulsion catastrophic inversion from abnormal to normal morphology. 4. Following the emulsion viscosity during three inversion protocols and extending the critical dispersed-phase concept". Industrial & Engineering Chemistry Research 44, 67-74.
9. McClements D.J. (2007), "Critical review of techniques and methodologies for characterization of emulsion stability". Critical Reviews in Food Science and Nutrition 47, 611-649, 2007. DOI: 10.1080/10408390701289292.
10. Binks B.P. and Lumsdon S.O. (2000), "Effect of oil type and aqueous phase composition on oil-water mixtures containing particles of intermediate hydrophobicity". Phys. Chem. Chem. Phys. 2, 2959-2967.

508669 Effect of Oil Filtration on Oil-Water Dispersion Characterization

Abstract:

Oil-water emulsions make up a significant percentage of the worldâ€™s crude oil production. Emulsions generate operational problems in the Petroleum Industry, such as incremental pressure drop, decline in oil production rate, and equipment corrosion, among others. Crude oil naturally includes solid nano-particles and other surface-active agents, such as asphaltenes, resins, inorganic salts, bases, and acid end groups. The agents can act as stabilizers for oil-water emulsions when present in the interface and promote interfacial viscosity, which hinders the droplet-droplet coalescence. The objective of this paper is to investigate the effect of the removal of surface active agents on the stability of the oil-water emulsion under different water cuts and crude oil filtration media pore size.

A state-of-the art Portable Dispersion Characterization Rig (P-DCR) is utilized to study the effect of crude oil filtration on oil-water dispersion characterization. The P-DCR is a batch separator that includes a separation cell where the oil and water are introduced, sheared, and allowed to separate over time. The device is equipped with a high-resolution camera, and data acquisition software that tracks and analyzes the separation profiles. Oklahoma crude oil and distilled water are used as fluid phases, which are sheared with rotational speed of 600 rpm for a total time of 10 min. The experiments are collected with four water cuts, namely, 15%, 25%, 50%, and 75%. A filtration flask attached to a vacuum pump was used for the filtration process. Filter papers with pore sizes of 8 and 20-25 μm were used as filtration media for the removal of the surface active particles from the crude oil.

The experimental data show that the fastest separation occurs with increasing water-cuts for both filtered and unfiltered crude oil. The filtration of crude oil results in a faster separation as compared to unfiltered cases for low water cuts, namely, 15% and 25%. However, the effect of crude oil filtration on the separation rate is negligible for higher water cuts due to the lower concentration of the surface-active agents in the emulsion. The experiments confirm that the smaller pore size filtration media results in an increased removal of surface active particles, resulting in ahigher separation rate for lower water cuts.

„THE INFLUENCE OF NANOBUBBLES ON THE INTERACTION FORCES BETWEEN ALUMINA PARTICLES AND CERAMIC FOAM FILTERS“

Ditscherlein, L.; Peuker, U.A.; Institute of Mechanical Engineering and Mineral Processing, Agricolastrasse 1, 09599 Freiberg, Germany*

**corresponding author: lisa.ditscherlein@mvtat.tu-freiberg.de*

The knowledge of interaction forces and thus influencing parameters for various industrial processes like flotation, filtration or agglomeration is of great interest. As it is seen for many water-based systems, ordinary DLVO-forces do not explain the interactions between hydrophobic surfaces [1, 2]. Besides VAN DER WAALS interactions (vdW) there is another attractive force, up to tenfold stronger than vdW, which influences the process significantly. Since the inventions of the surface force apparature and the atomic force microscope [3, 4], hydrophobic interactions are the scope of many investigations [5-11] and many theories about their origin were proclaimed [12-14]. A promising approach is a division of hydrophobic interactions in two parts: a short-range true hydrophobic interaction (1-2 nm) caused by a structuring of water molecules and a long-range interaction (up to several hundreds of nm) caused by capillary bridging due to small cap-shaped bubbles, so-called nanobubbles [15-17]. A structuring of water molecules near the hydrophobic surfaces occurs because of a reorientation of the dipoles parallel to the surface. Consequently, a water depletion layer is generated. In turn, nanobubbles are formed during immersion, because of a gas solubility gradient (temperature or solvent-exchange induced) or by perturbation. As a result of the depletion layer, dissolved gas molecules are enriched at this layer and act as a gas reservoir for the nanobubbles. Their long-term stability, for years subject of discussions, can be explained by a dynamic equilibrium between in- and outflux of gas if the liquid is supersaturated as well as a pinned contact line.

Investigations on nanobubbles are usually done on atomic rough surfaces, that is to say samples with root mean square (rms) values below 5 nm. Surfaces in industrial process are much rougher, e.g. rms > 200 nm, so adhesion models and/or phenomena seen on smooth surfaces might be irrelevant for technical processes like flotation, filtration or agglomeration. VdW interaction models for smooth and slightly rough (rms < 20 nm) surfaces are numerous [18-22], but only You et al. [23] and Fritzsche et al. [24] introduce models for technical rough surfaces yet. On the other hand, there are hardly any models that describe short range hydrophobic interactions on smooth [25] and technical rough surfaces [24] and until now no models for long-range hydrophobic interactions are available. Hydrophobic forces are not fully understood, and in case of rough surfaces the impact of both short- and long range interaction has to be estimated.

In this study we focus on technical rough ceramic foam filter samples that are typically used for metal melt filtration. Since the process cannot be investigated under real conditions, we developed a water-based model system that mimics the same wetting conditions as in the real system [26] to examine adhesion mechanisms between particulate inclusions and filter surface. Filter materials and alumina particle (model inclusion) are silanized so that the contact angle between a water droplet and the surface is about 104°. Force spectroscopy is done using the atomic force microscope XE-100 from Park Systems. The particle is glued on a tipless cantilever AIO-Al-TL from budget sensors with a water-insoluble epoxy resin; the spring constant of the cantilever was determined to 1.55 N/m. For each sample 64 force distance curves on a 10 x 10 µm² area on at least 6 different parts of the substrate surface are measured to get statistical valid results. A MatLAB® routine recalculates the raw data (e.g. baseline correction etc.) and determines the adhesion force that is the minimum of the retrace for each force distance curve. Adhesion forces are then normalized on the particle diameter because of the linear relationship between diameter and adhesion force. Topographical scans on the samples are done using a ContAl-G cantilever from Budget Sensors with a spring constant of 0.21 N/m for nanobubble stability tests and the determination of roughness values.

As it can be seen in Figure 1, the overall force distribution for a rough alumina sample ranges from 0.06-14.62 mN/m. This width of distribution is explainable by different mechanisms: small forces are explainable by VAN DER WAALS interactions in liquid media on rough surfaces (0-20 % of the overall distribution). A structuring of water molecules might be disturbed due to the inhomogeneous morphology and should influence the adhesion between the surfaces on smoother parts of the substrate (20-60 % of the overall distribution). Large adhesive forces are explainable by capillary bridging due to nanobubbles. The overall distribution can be split in force curves with and without snap-in (threshold snap-in: 10 nm), whereby a snap-in indicates capillary bridging. Here, it can be seen that adhesion without snap-in results in smaller adhesion forces. But there is also a transition zone visible that can be explained nanobubble generation after colloidal probe perturbation (in case of no snap-in) and not fully accessible nanobubbles (in case of snap-in).

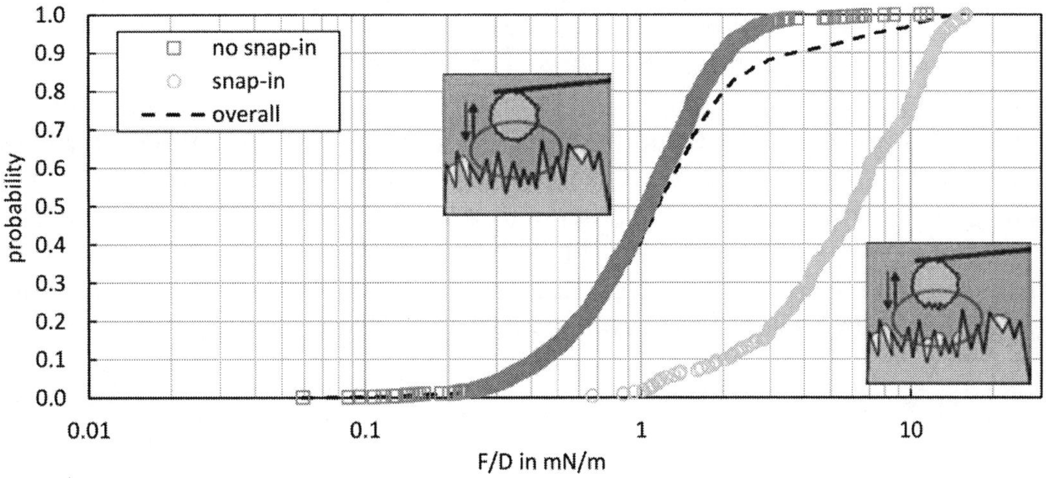

Figure 1: force distribution of normalized adhesion forces measured between a hydrophobic alumina particle and a hydrophobic alumina surface

The adhesion force depends on the snap-in force as it can be seen in Figure 2a: The higher the snap-in force the larger is the adhesion.

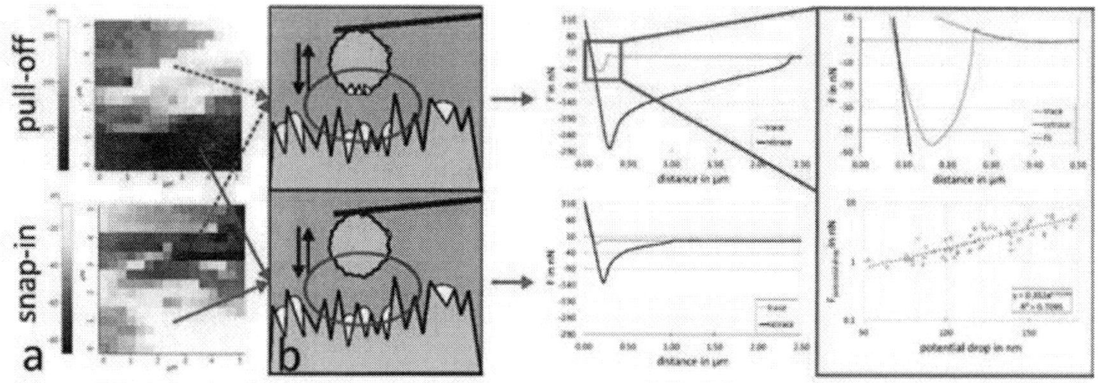

Figure 2: 2D map of adhesion forces and corresponding snap-in forces (a) proposed mechanisms, exemplary force distance curves and relation between force and decay length of the potential drop (b)

Interestingly, the force distance curves with snap-in differ regarding their behaviour just before the particle jumps into contact (Figure 2b). Curves with larger snap-in/adhesion force always show a slight repulsion which has an exponential nature. An explanation might be a trapped bubble on the interacting particle area. During approach, the water film between the bubbles has to get disrupted before a capillary bridge is formed. Film rupture is impaired if both soft surfaces (nanobubbles on top of the rigid surfaces) are deformed. In case of melt filtration, this leads to the necessity of enough provided energy, kinetic energy respectively, to overcome this repulsive force.

Another influencing property of hydrophobic surfaces is their roughness. As it is seen in air or for hydrophilic surfaces, higher roughness results in a decrease of adhesion force because of a smaller contact area (Figure 3 left). Hydrophilic TiO_2 (rms of a 3x3 µm² area: 0.3 µm) shows the largest adhesion forces followed by spinel (rms = 0.6 µm) and alumina (rms = 0.8 µm). Although AC1_0, a sample made of alumina and additives, is the smoothest sample (rms = 0.2 µm) there are only small VAN DER WAALS forces detectable. This is because of the slightly different chemical composition of the additives. For hydrophobized samples, the different chemism can be neglected due to the silane layer onto the surface. Hydrophobic systems are more complex [27]. The hydrophobized ceramics that are used as filter materials with varying roughness show an increase of adhesion with increased roughness.

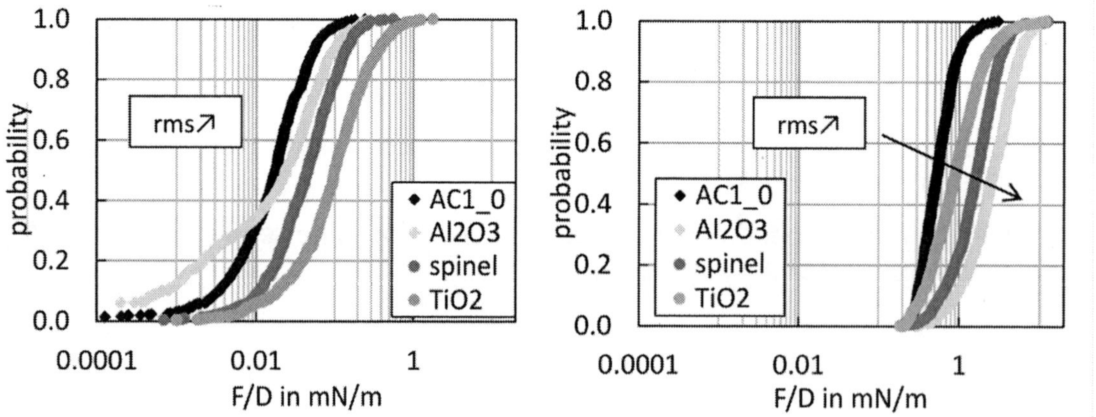

Figure 3: normalized force distributions of the four investigated oxide surfaces for the hydrophilic case (a) and the hydrophobic case (b)

This change compared with good wetted surfaces is caused by small bubbles inside of asperities and pores: The process of immersion with water (or in the real case of melt filtration) is too fast to wet pores, and so there are nano- and also mesobubbles distributed all over the surfaces, depending on the pore size and distribution. They sit stable inside of the asperities and cannot be easily depinned up to scanning forces of 40-50 nN. In a previous study, we have shown that bubble depinning forces needed on smooth hydrophobic surfaces are decreased (below 5 nN) due to the less pinned three phase contact line.

To optimize melt filtration or other deep bed filtration processes, we propose rough filters with a rms of 0.8 µm because of the increase of adhesion due to small bubbles and the higher stability of them inside of pores/asperities. In case of pre-existing bubbles on inclusions or agglomerates, sufficient kinetic energy must be provided to overcome the repulsion just before a snap-in.

The authors would like to thank the German Research Foundation (DFG) for supporting the studies in subproject B01 which is part of the Collaborative Research Center CRC 920.

[1] J. Israelachvili and R. Pashley, "The hydrophobic interaction is long range, decaying exponentially with distance," *Nature,* vol. 300, pp. 341-342, 1982.

[2] H. K. Christenson and P. M. Claesson, "Direct measurements of the force between hydrophobic surfaces in water," *Advances in Colloid and Interface Science,* vol. 91, pp. 391-436, 2001.

[3] D. Tabor and R. H. S. Winterton, "DIRECT MEASUREMENT OF NORMAL AND RETARDED VAN DER WAALS FORCES," vol. 312, pp. 435-50 plate, 1969.

[4] G. Binnig, C. F. Quate, and C. Gerber, "Atomic force microscope," *Physical Review Letters,* vol. 56, pp. 930-933, 1986.

[5] P. Attard, "Nanobubbles and the hydrophobic attraction," *Advances in Colloid and Interface Science,* vol. 104, pp. 75-91, 2003.

[6] T. D. Blake and J. A. Kitchener, "Stability of aqueous films on hydrophobic methylated silica," *Journal of the Chemical Society, Faraday Transactions 1: Physical Chemistry in Condensed Phases,* vol. 68, pp. 1435-1442, 1972.

[7] A. Carambassis, L. C. Jonker, P. Attard, and M. W. Rutland, "Forces measured between hydrophobic surfaces due to a submicroscopic bridging bubble," *Physical Review Letters,* vol. 80, pp. 5357-5360, 1998.

[8] Y. Wang and B. Bhushan, "Boundary slip and nanobubble study in micro/nanofluidics using atomic force microscopy," *Soft Matter,* vol. 6, pp. 29-66, 2009.

[9] X. Cui, C. Shi, L. Xie, J. Liu, and H. Zeng, "Probing Interactions between Air Bubble and Hydrophobic Polymer Surface: Impact of Solution Salinity and Interfacial Nanobubbles," *Langmuir,* 2016/06/13 2016.

[10] W. A. Ducker, Z. Xu, and J. N. Isrealachvili, "Measurements of hydrophobic and DLVO forces in bubble-surface interactions in aqueous solutions," *Langmuir,* vol. 10, pp. 3279-3289, 1994.

[11] M. L. Fielden, R. A. Hayes, and J. Ralston, "Surface and capillary forces affecting air bubble-particle interactions in aqueous electrolyte," *Langmuir,* vol. 12, pp. 3721-3727, 1996.

[12] E. Ruckenstein and N. Churaev, "A possible hydrodynamic origin of the forces of hydrophobic attraction," *Journal of Colloid And Interface Science,* vol. 147, pp. 535-538, 1991.

[13] Y. I. Rabinovich and B. V. Deryagin, "DIRECT MEASUREMENT OF FORCES OF ATTRACTION OF HYDROPHOBIZED QUARTZ FIBERS IN AQUEOUS KCl SOLUTIONS," *Colloid journal of the USSR,* vol. 49, pp. 598-602, 1987.
[14] J. L. Parker, P. M. Claesson, and P. Attard, "Bubbles, cavities, and the long-ranged attraction between hydrophobic surfaces," *Journal of Physical Chemistry®,* vol. 98, pp. 8468-8480, 1994.
[15] M. A. Hampton and A. V. Nguyen, "Nanobubbles and the nanobubble bridging capillary force," *Advances in Colloid and Interface Science,* vol. 154, pp. 30-55, 2010.
[16] D. Lohse and X. Zhang, "Surface nanobubbles and nanodroplets," *Reviews of Modern Physics,* vol. 87, 2015.
[17] J. R. T. Seddon and D. Lohse, "Nanobubbles and micropancakes: Gaseous domains on immersed substrates," *Journal of Physics Condensed Matter,* vol. 23, 2011.
[18] B. Derjaguin, "Untersuchungen über die Reibung und Adhäsion, IV - Theorie des Anhaftens kleiner Teilchen," *Kolloid-Zeitschrift,* vol. 69, pp. 155-164, 1934.
[19] H. C. Hamaker, "The London-van der Waals attraction between spherical particles," *Physica,* vol. 4, pp. 1058-1072, 1937.
[20] E. M. Lifshitz, "The Theory of Molecular Attractive Forces between Solids," *Journal of Experimental and Theoretical Physics USSR,* vol. 29, pp. 94-110, 1956.
[21] R. P. Jaiswal, G. Kumar, C. M. Kilroy, and S. P. Beaudoin, "Modeling and validation of the van der Waals force during the adhesion of nanoscale objects to rough surfaces: A detailed description," *Langmuir,* vol. 25, pp. 10612-10623, 2009.
[22] S. You and M. P. Wan, "Mathematical models for the van der waals force and capillary force between a rough particle and surface," *Langmuir,* vol. 29, pp. 9104-9117, 2013.
[23] S. You and M. P. Wan, "Modeling and experiments of the adhesion force distribution between particles and a surface," *Langmuir,* vol. 30, pp. 6808-6818, 2014.
[24] J. Fritzsche and U. A. Peuker, "Modeling adhesive force distributions on highly rough surfaces," *Powder Technology,* vol. 289, pp. 88-94, 2016.
[25] C. J. Van Oss, M. K. Chaudhury, and R. J. Good, "Interfacial Lifshitz-van der Waals and polar interactions in macroscopic systems," *Chemical Reviews,* vol. 88, pp. 927-941, 1988.
[26] J. Fritzsche, J. Teichmann, F. Heuzeroth, K. G. Van Den Boogaart, and U. A. Peuker, "Impact of wetting to the agglomeration of dispersed particles in an aqueous medium," *Advanced Engineering Materials,* vol. 15, pp. 1299-1306, 2013.
[27] L. Ditscherlein, A. Schmidt, E. Storti, C. G. Aneziris, and U. A. Peuker, "Impact of the Roughness of Alumina and Al2O3-C Substrates on the Adhesion Mechanisms in a Model System," *Advanced Engineering Materials,* 2017.

509311 Effect of Physical Properties of Coal Gasification Fine Ash on Its Wettability

Abstract:

Optical contact angle measurement was used to measure the contact angle of different size of fine ash particles. The fractal characteristics of fine ash was studied from size and surface porous structure based on fractal theory and effect of fractal dimension on wetting ability of fine ash was also investigated. Scanning Electron Microscope (SEM), Fourier Transform Infrared Spectroscopy (FTIR), X-Ray Diffraction (XRD) and humidity measuring instrument and other experimental measurements have been applied to study the effect of physical characteristic on fine ash wettability. The research shows that wettability increases with the fractal dimension of particle size and particle surface. With the particle size decreasing, surface roughness and structure irregularity decrease and the amount of spheroidal particle increases. The content of unsaturated functional group increases and the hydrophilic mineral substance decreases with particle size decreasing, resulting in the decease of its wettability. For particles larger than 40Î¼m, the wettability increases with the increase of moisture content. On the contrary, for particles smaller than 40Î¼m, the wettability increases at first and then decreases with the increase of moisture content.

508097 Analysis and Control of Adhesion Behavior of Sewage Sludge Combustion Ashes at High Temperature

Abstract:

In recent energy recycle system of sewage sludge combustion, fine ash particle deposition and growth on the surface of heat exchanger, heater and wall decrease the energy recycle efficiency and hindered stable operation of combustion plants. These phenomena were generated by the adhesiveness and stickiness increase of ash particles at high temperature. Since the sewage sludge combustion ash included higher contents of alkali metals, phosphate and chloride, the stickiness of ashes promoted at high temperature compared with coal combustion ashes. It seems that eutectics phase with relatively low melting points are contained in ash, and generated liquid phase at high temperature. Since phosphate concentration in sewage sludge combustion ash is generally higher than that in other biomass and coal combustion ash. However, the action mechanism of phosphate on adhesion phenomenon has not been discussed.

In this work, 4 kinds of sewage sludge combustion ash from commercial plant, of which two samples were collected during stable operation in 2 different plants, and other 2 samples were collected after ash adhesion and growth trouble in different plants, were studied. By using 4 samples, after particles characterization, such as particle size distribution and chemical component analysis, tensile strengths of each ash powder bed at high temperature condition ranging from 673 to 1073 K were measured by split type tensile strength tester. In order to analyze the increasing mechanism of adhesion behavior of ash particles, thermodynamic equilibrium calculations applied for liquid phase formation behavior. The results suggest that the concentration of phosphate in ash would influence the liquid phase formation at high temperature, the adhesion phenomenon would be influenced consequently. Ashes with higher contents of phosphate tend to generate K_2O-SiO_2 series eutectic compounds, of which the melting points are relative low. On the contrary, ashes with lower contents of phosphate, the formation of low melting point compound was restrained, thus the adhesiveness of ashes decrease. Based on the calculation results, we estimated the useful element, alumina, to hinder liquid phase formation, and the effect of alumina addition on tensile strength of ash layer at high temperature were examined.

508729 Mixed Layer Formation in a Blast Furnace and Its Effect on the Performance

Abstract:

Mixed layer formation in a blast furnace and its effect on the performance

Dianyu E,[1][*] Qinfu Hou,[1] and Aibing Yu[1,2]

[1]ARC Research Hub for Computational Particle Technology, Department of Chemical Engineering, Monash University, Clayton, VIC 3800, Australia

[2]Centre for Simulation and Modelling of Particulate Systems, Southeast University - Monash University Joint Research Institute, Suzhou 215123, PR China

Abstract

Gas flow distribution in a blast furnace (BF) plays a significant role in BF smooth operation, productivity and thermal efficiency. It is affected by the distribution of burden materials composed of alternating coke and ferrous ore layers. While moving downward coke and ferrous layers of different sizes can mix and form the so-called mixed layers. Generally, the porosity is lower and hence the pressure drop is higher in the mixed layers. These variations can change the gas flow distribution and BF performance. Previous work tried to quantify the effect of material properties and process conditions on the formation of mixed layers and the resulting local porosity variation. However, these studies were often conducted under simplified conditions. Few was dedicated to the formation of mixed layers in a BF and its effect on BF performance. This work studies the formation of mixed layers in an experimental BF by using a combined computational fluid dynamics and discrete element method approach. The effect on BF performance is evaluated under different operational conditions including different size ratios of coke and iron ore particles, burden distribution and batch weight, and discharge rates. The results are helpful to optimize burden charge for improving BF performance.

Keywords: Blast furnace, Mixed layer, Discrete element method, Computational fluid dynamics

---

[*]Corresponding author. Email: dianyu.e@monash.edu.

508066 Anti-Agglomeration Strategy Study of High Sodium Coal Gasification in a Commercial Circulating Fluidized Bed Gasifier

Abstract:

High-sodium coal is mainly located in Australia, Germany, the United States, China, Indonesia and other places. The utilization of high-sodium coal has attracted worldwide interest in recent years. Gasification seems to be an ideal route for high-sodium coal utilization because of high reaction activity, high efficiency and low pollution emissions. Among various gasification technologies, circulating fluidized bed (CFB) gasification is one of the most popular gasification technologies. Some advantages of CFB gasification are that it can use low-cost pulverized coal, has satisfied adaptability of a wide range of solid fuels, can use poor quality coals and emits lower amounts of pollutants. However, the high-sodium content in coal ash may cause serious ash-related problems, such as agglomeration, fouling and contamination. Therefore, it is important to understand the agglomeration mechanism and provide anti-agglomeration strategy for the successful utilization of high-sodium coal.

Institute of Engineering Thermophysics (IET) has built up 28 sets of CFB gasifiers. During the commercial operation process, high-sodium Indonesia coal was used as feedstock in Guangxi Province. In order to reduce the operating risk, SiO_2-Al_2O_3-Na_2O ternary phase diagram calculation and lab scale blended coal gasification experiments were carried out. The calculation result of SiO_2-Al_2O_3-Na_2O ternary phase diagram indicates that Indonesia coal ash is located in the liquidus region and has serious agglomeration tendency. Via phase diagram optimization analysis, proper blending coal was selected for utilization of high-sodium Indonesia coal, and the thermodynamic prediction was in accordance with the fundamental experimental results. 75% of Indonesia blended coal gasification was conducted in one set of commercial CFB gasifier. The system temperature and pressure are stable, and no slagging phenomena occur. The cold gas efficiency is 71.96%, the carbon conversion is 86.90% and the gas heat value is 5.59 MJ/Nm^3. In addition, more than 320 days of long-term and stable operation was achieved. The results indicate that blended coal gasification is an efficient method to alleviate the agglomeration tendency of high sodium coal. The essence of the strategy is to change the ash properties via chemical reactions, and phase diagram can provide some clue for the identification of blending coal and suitable conditions.

EXPERIMENTAL INVESTIGATION AND FORCE BALANCE MODELLING OF WET PARTICLE COLLISIONS

Britta Buck, Hamburg University of Technology, Hamburg, Germany
Yali Tang, Eindhoven University of Technology, Eindhoven, The Netherlands
Niels G. Deen, Eindhoven University of Technology, Eindhoven, The Netherlands
Hans Kuipers, Eindhoven University of Technology, Eindhoven, The Netherlands
Stefan Heinrich, Hamburg University of Technology, Hamburg, Germany

Abstract

Knowledge about collision dynamics of particles is fundamental for describing and modelling particulate processes. If liquids are additionally involved in the collisions, the rebound dynamics strongly depend on the liquid properties. However until now, wet particle collisions are still not fully understood. Hence, influence of liquid properties, particle attributes and impact conditions on rebound behavior of dry particles from wet walls is investigated experimentally and a force balance model is described, which can predict collision dynamics on a physical basis. The results of this proceeding were already published in [1].

Introduction

Particulate processes are characterized by intense contacts between particles and between particles and the apparatus walls. In addition to the particles often liquids are involved in the process as droplets or layers on the surfaces, e.g. as injected liquids in agglomeration and granulation or wetness of spray dried particles. These liquids have a fundamental influence on the collision dynamics as well as on agglomeration mechanisms of the particles. Although these micro-mechanics are crucial for understanding and modelling of wet particulate systems, until now the collision dynamics of wet particles and walls are still not fully understood. Especially, a physics-based model is missing, which is validated by extensive single particle collisions. Hence, the authors proposed a force balance model considering contact forces F_c, gravitational forces F_g, viscous forces F_{vis} and capillary forces F_{cap} in normal and tangential direction [1]. Therefore, Newton's laws of motion are solved numerically:

$$m_P \frac{dv}{dt} = \sum F_i \qquad (1)$$

$$I_P \frac{d\omega}{dt} = F_{c,t} R_P + M_{vis} \qquad (2)$$

With particle mass m_P, velocity v, moment of mass inertia I_P, angular velocity ω, particle radius R_P and momentum due to viscous dissipation M_{vis}.

To compare experimental and numerical results collision dynamics are analyzed using the so called coefficient of restitution (CoR) e, which is often used for this purpose (e.g. [2,3,4]). The coefficient of restitution of a particle – wall collision is defined as ratio of rebound velocity v_R to impact velocity v_0:

$$e = \left|\frac{v_R}{v_0}\right| = \sqrt{1 - \frac{E_{diss}}{E_{kin,0}}} \qquad (3)$$

If the collision is oblique, thus featuring a collision angle, the coefficient of restitution can be described for normal, tangential and rotational direction:

$$e_n = \left|\frac{v_{R,n}}{v_{0,n}}\right|; \qquad e_t = \frac{v_{R,t}}{v_{0,t}}; \qquad e_\omega = \frac{\omega_R}{\omega_0} \qquad (4); (5); (6)$$

Materials and Methods

For experimental investigation of collision dynamics a dry particle is made to collide with a wall, covered by a liquid layer of defined thickness. The collision is captured by two synchronized high-speed cameras, while the layer thickness is controlled by a confocal thickness sensor (compare figure 1). For normal collisions, the particle is dropped onto the wall, for oblique collisions without initial rotation the particle is shot obliquely onto a horizontal wall by pressurized air, while for oblique collisions with initial rotation of the particle it was rolled down an angled ramp. A more detailed description of each collision setup is found in our respective publications [5,6,7]. During oblique collisions the normal impact velocity was kept constant at 1 m/s.

As particles glass spheres of 0.91 mm and 1.74 mm diameter and γ-Al_2O_3 particles of 1.74 mm diameter were used. The wall was made of glass for all experiments. As liquid layers water were used as well as 35 wt% and 50 wt% Glycerol-water solutions to investigate different viscosities and a Tween 20-water solution (60 mg/l) with about half the surface tension of water were applied to the wall.

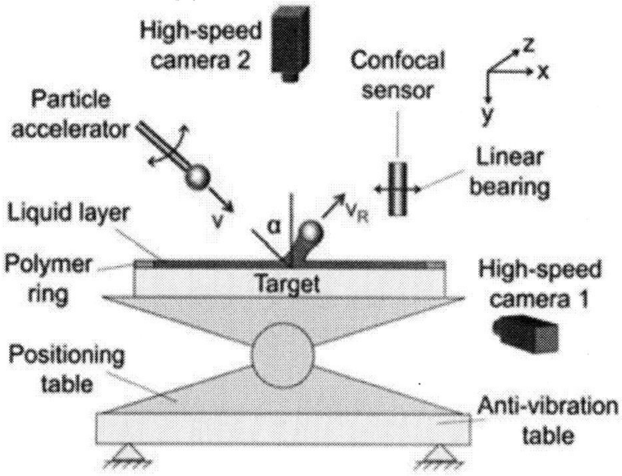

Figure 1: Schematic of experimental setup

The force balance model was calculated for the same materials. Refer to the original paper [1] for a detailed list of input parameters. The contact forces are modeled using the work of Tsuji et al. [8], while the viscous forces of White [9] and Lian et al. [10] are used for the normal direction and the model of Goldman et al. [11] is applied in tangential direction. The capillary force is considered only in normal direction using the model of Kralchevsky and Nagayama [12].

Results and Discussion

Normal collisions

Figure 1 shows results for the normal coefficient of restitution during purely normal collisions of a dry glass particle with a wet glass wall. Experimental findings are compared to the results of the force balance model. For dry collisions the normal coefficient of restitution is constant over the complete range of collision velocity investigated (black filled data in figure 2a). This trend can also be represented by the force balance model. If a liquid layer is applied to the wall, the normal coefficient of restitution features a characteristic trend with collision velocity: If the collision velocity is below a critical "sticking velocity" the particle sticks to the liquid layer, because all initial kinetic energy is dissipated during the collision. Hence, the normal coefficient of restitution is zero. With increasing collision velocity above the sticking velocity the normal coefficient of restitution first increases steeply and then levels out on a constant value at high (infinite) collisions velocities. This value however, is always smaller than the dry coefficient of restitution. An increase in layer thickness leads to a strong decrease of the normal coefficient of restitution. Both the trend with layer thickness and impact velocity is nicely represented by the force balance, qualitatively and quantitatively.

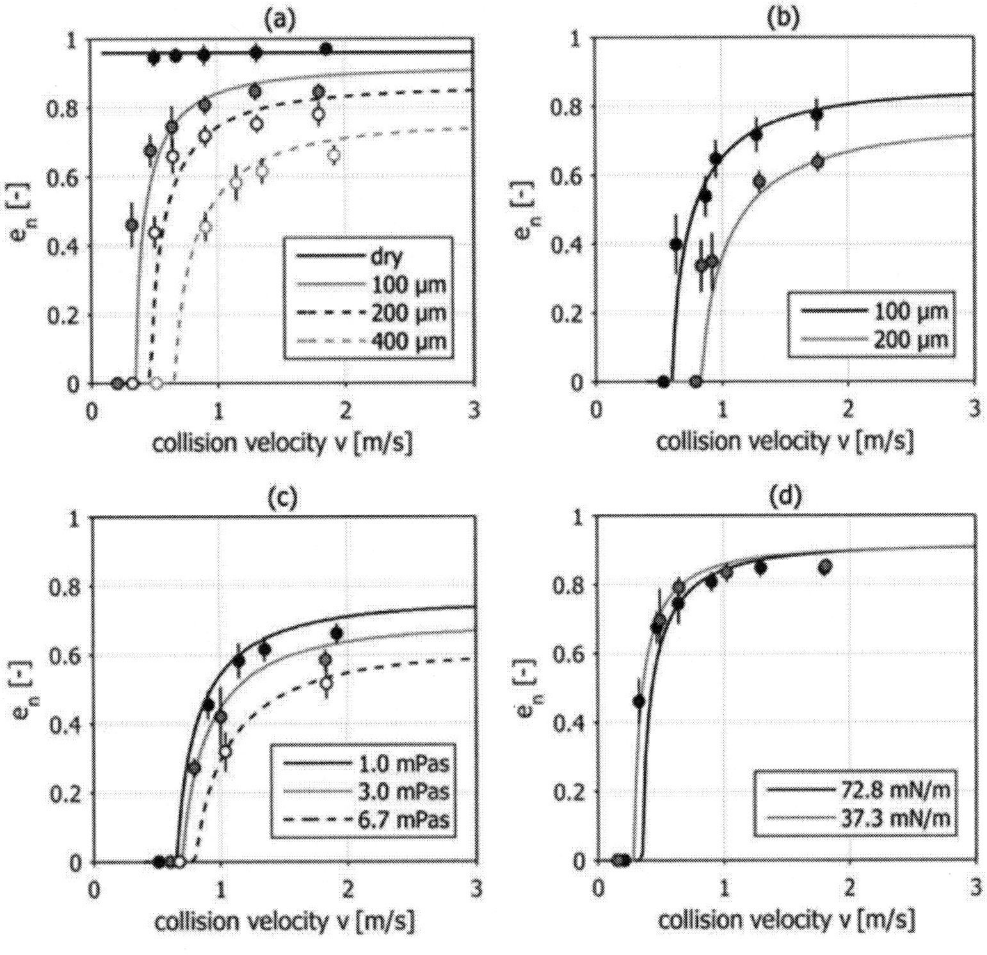

Figure 2: Comparison of experimental results (circles) and force balance model (lines) regarding normal coefficient of restitution for a 1.8 mm particle and varying layer thickness of water (a), a smaller glass particle (b), varying viscosity at 400 µm layer thickness (c) and different surface tension at 100 µm thickness (d). Adapted from [1].

Figure 2 (b) shows the results for similar collisions, but of smaller particles. Comparing the results in (a) and (b) at the same layer thickness, a smaller particle features smaller coefficients of restitution than a larger particle, since it is farther immersed in the liquid.

However, the trends are the same and the influence of different particle size can also be represented by the force balance model. Influence of viscosity is shown in figure 2 (c), where a decrease of the normal coefficient of restitution with increasing viscosity can be seen, which is also predicted by the force balance model. Surface tension showed no influence on the normal coefficient of restitution experimentally (compare (d)), although the force balance model predicts a small increase of the normal coefficient of restitution with decreasing surface tension. This difference is however small and lies inside standard deviations of the experimental results.

Oblique collisions without initial rotation

Results for oblique collisions without initial rotation of the particle are discussed in figure 3. The collisions were performed for γ-Al_2O_3 particle impacting a glass wall covered by a 100 µm layer of the Tween 20 solution. The normal coefficient of restitution is found independent of collision angle experimentally as well as predicted by the force balance.

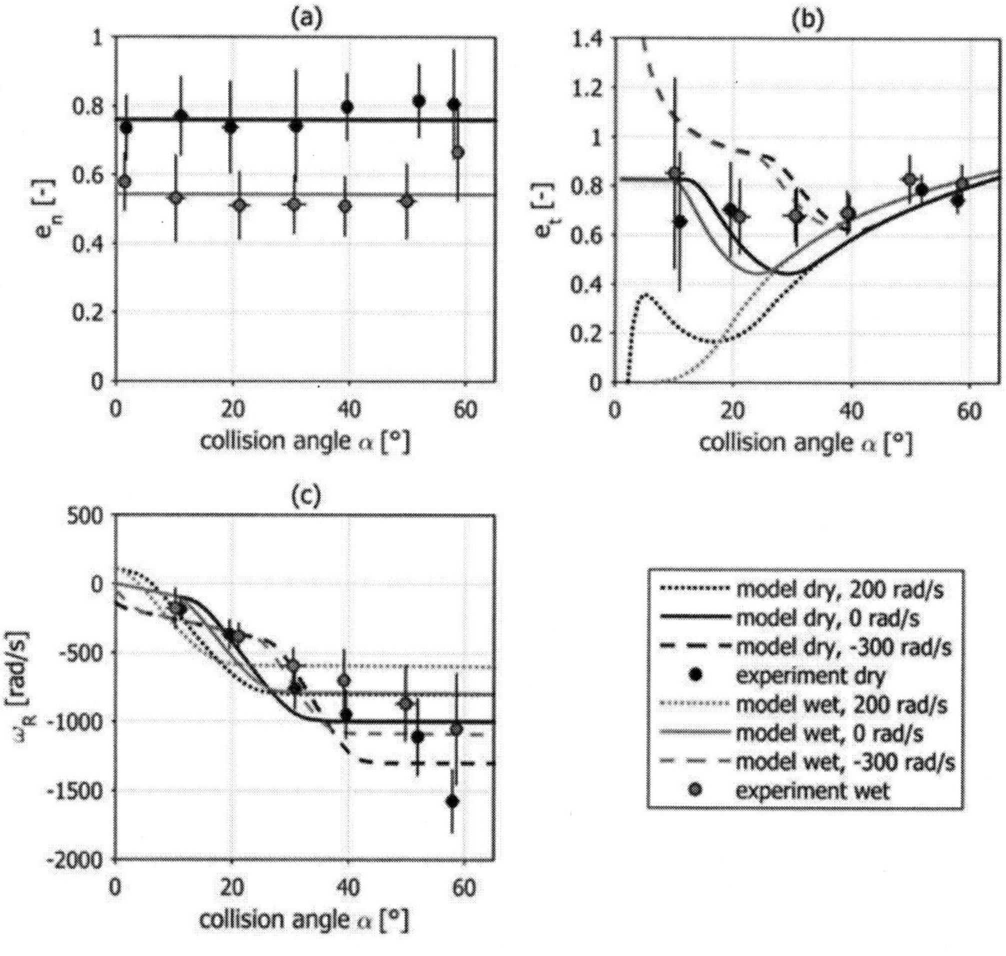

Figure 3: Comparison of experimental results (circles) and force balance model (lines) for a γ-Al_2O_3 particle impacting obliquely with no initial rotation onto a dry wall or a wet wall covered by a 100 µm thick liquid layer. (a) normal CoR, (b) tangential CoR, (c) particle rotation after rebound. Adapted from [1].

In contrast, the tangential coefficient of restitution is found strongly dependent on the collision angle. Since during experiments initial rotation of the particle varied between approximately -300 rad/s and 200 rad/s, the simulations were also conducted using initial rotation velocities of -300 rad/s, 0 rad/s and 200 rad/s. At small collision angles the experimental results of the tangential coefficient of restitution features high standard deviations, which results from the varying initial rotation, as the force balance model proves. At higher collision angles (30°-60°), the tangential coefficient of restitution slightly increases and initial rotation of the particles has only negligible influence. This different behavior at small and large collisions angles results from a change of rolling/sticking contact at small angles to a sliding contact between particle and wall at high collision angles.
Rotation velocity is also dependent on the impact angle. With increasing collision angle the rotation after rebound increases in negative direction. This trend is a result of tangential movement being converted to rotation due to the tangential contact forces between particle and wall [7]. The tangential impact velocity increases with growing collision angle ($v_n = \mathrm{const.}$), thus the tangential forces increase and more energy is converted to rotation. The force balance model also predicts an increase of rotation in negative direction, but with a much more complex trend, than found experimentally so far. This complex trend is a result of the transition of rolling/sticking contact to sliding contact [1], which was mentioned before.

Oblique collisions with initial rotation

Figure 4 compares force balance model and experiments for particles impacting with initial rotation on a wet wall, covered by a 100 µm thick layer of the Tween 20 solution. Due to the setup of the particle rolling down a ramp, initial rotation increased with impact angle. Hence, the force balance was calculated for each collision with the specific initial conditions of the experiment. Additionally, the trend with constant initial rotation but varying collision angle is investigated using the force balance model.
The normal coefficient of restitution is constant for all collision angles and initial rotation of the particles both in experiment and force balance model. Consequently, in addition to the independence of the normal coefficient of restitution regarding collision angle, found in the previous section, it is also independent of the initial rotation of the particle.
The tangential coefficient of restitution again shows different results. For particles impacting with mathematically negative initial rotation, as in these experiments, the tangential coefficient of restitution is larger than 1. Hence, rotational energy must have been converted to translational movement during the collision. The force balance model fits the experiments comparably well, but additionally shows a complex trend with various turning points of the tangential coefficient of restitution with initial rotation and collision angle.
Rotation of the particle after rebound again shows an increasing trend in negative direction with increasing collision angle. However, the exact values are a combined effect of changing collision angle and initial rotation of the particle. Nevertheless, the force balance model is able to predict the rotation velocities after rebound. Furthermore, a complex trend is found by the model for rotation in dependence of collision angle, with turning points at the same angles as for the tangential coefficient of restitution.
The rotational coefficient of restitution features values below 1 in the experiments. Hence, rotation is reduced during the collision. This was to be expected, since rotation had to be converted to tangential movement for the tangential coefficient of restitution to be larger

than 1. Comparing force balance model and experimental results, again good agreement is found, while the exact trend of the rotational coefficient of restitution, predicted by the force balance, is more complex than expected from the experimental results. Since rotation after rebound and rotational coefficient of restitution are linked, the same turning points in the trend with collision angle are found via force balance model.

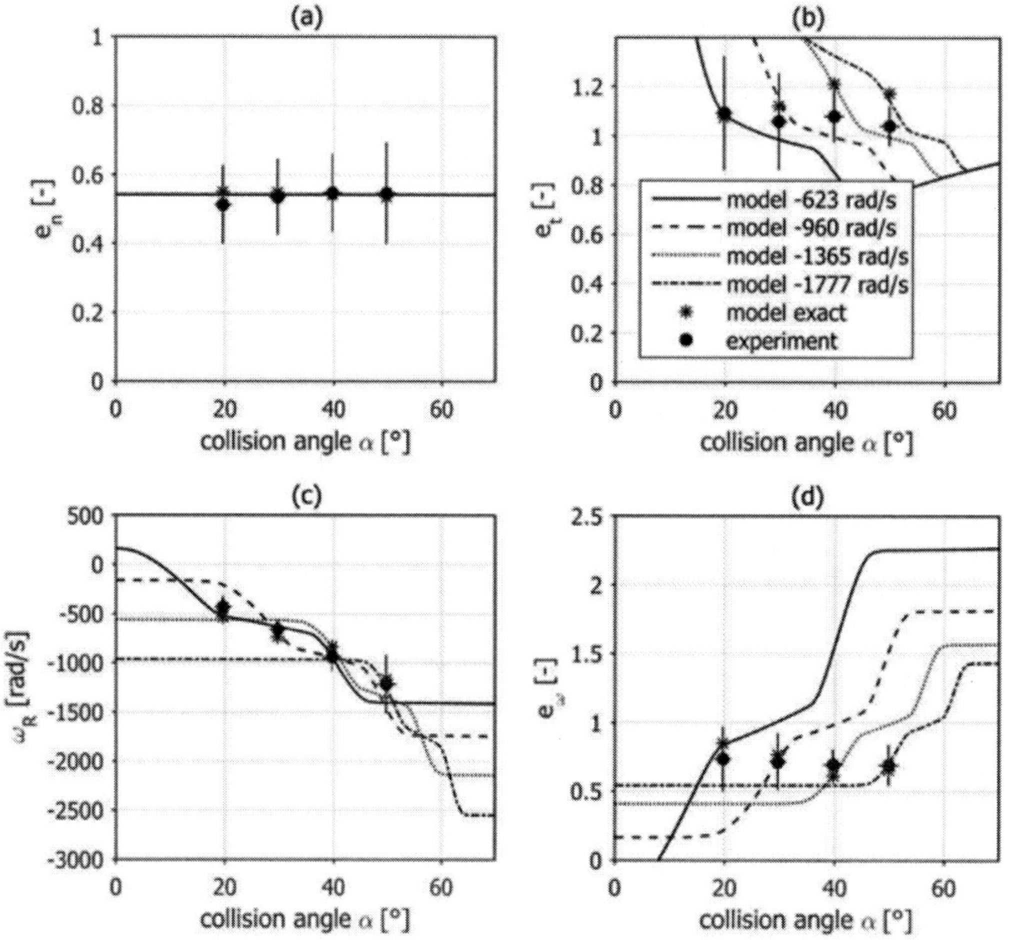

Figure 4: Comparison of experimental results (circles) and force balance model (lines) for a γ-Al_2O_3 particle impacting obliquely with initial rotation onto a wall covered by a 100 µm thick liquid layer. (a) normal CoR, (b) tangential CoR, (c) particle rotation after rebound, (d) rotational CoR. Adapted from [1].

Conclusion

A force balance model was proposed, which predicts the coefficient of restitution for dry particles impacting a wet wall [1]. It was validated using normal and oblique collision experiments. Good agreement was found. Additionally the force balance model showed some complex dependencies of tangential coefficient of restitution and rotation of the particle after rebound regarding collision angle and initial rotation.

Acknowledgement

We gratefully acknowledge for the financial support: German Research Foundation (DFG), Germany. Project number HE 4526/9-2.

References

1. Buck, B., Y. Tang, N.G. Deen, J.A.M. Kuipers and S. Heinrich (2018), „Numerical investigation of collision dynamics of wet particles via force balance," Chemical Engineering Research and Design, 132, pp. 1143-1159.
2. Davis, R.H., J.-M. Serayssol and E.J. Hinch (1986), "The elastohydrodynamic collision of two spheres," Journal of Fluid Mechanics, 163, pp. 479-497.
3. Antonyuk, S., S. Heinrich, N. Deen and H. Kuipers (2009), "Influence of liquid layers on energy absorption during particle impact," Particuology, 7 (4), pp. 245-259.
4. Gollwitzer, F., I. Rehberg, C.A. Kruelle and K. Huang (2012), „Coefficient of restitution for wet particles," Physical Review E, 86, 011303.
5. Crüger, B., V. Salikov, S. Heinrich, S. Antonyuk, V.S. Sutkar, N.G. Deen, and J.A.M. Kuipers (2016), "Coefficient of restitution for particles impacting on wet surfaces: An improved experimental approach;" Particuology, 25, pp. 1-9.
6. Crüger, B., S. Heinrich, S. Antonyuk, N.G. Deen and J.A.M. Kuipers (2016), "Experimental study of oblique impact of particles on wet surfaces," Chemical Engineering Research and Design, 110, pp. 209-219.
7. Buck, B., Y. Tang, S. Heinrich, N.G. Deen, and J.A.M. Kuipers (2017), "Collision dynamics of wet solids: Rebound and rotation," Powder Technology, 316, pp. 218-224.
8. Tsuji, Y., T. Tanaka and T. Ishida (1992), "Lagrangian numerical simulation of plug flow of cohesionless particles in a horizontal pipe," Powder Technology, 71 (3), pp. 239-250.
9. White, F. M. (2006), "Viscous Fluid Flow," 3rd Edition, McGraw-Hill.
10. Lian, G., Y. Xu, W. Huang and M. J. Adams (2001), "On the squeeze flow of a power-law fluid between rigid spheres," Journal of Non-Newtonian Fluid Mechanics, 100 (13), pp. 151-164.
11. Goldman, A., R. Coxa and H. Brenner (1967), "Slow viscous motion of a sphere parallel to a plane wall - I motion through a quiescent fluid," Chemical Engineering Science, 22 (4), pp. 637-651.
12. Kralchevsky, P. A. and K. Nagayama (2001), "Chapter 11 - capillary bridges and capillary-bridge forces", in: P. A. Kralchevsky, K. Nagayama (Eds.), "Particles at Fluids Interfaces and Membranes - Attachment of Colloid Particles and Proteins to Interfaces and Formation of Two-Dimensional Arrays", Vol. 10 of "Studies in Interface Science", Elsevier, pp. 469 – 502.

Interaction forces in relation to agglomerate size of hydrophobic particles

Knüpfer P., Peuker U.A; TU Bergakademie Freiberg,

Institute of Mechanical Process Engineering and Mineral Processing,

Freiberg

In order to improve the quality of metal casting products it is very necessary to remove non-metallic inclusion. An efficient way is the application of a ceramic foam filter which separates the inclusions from the liquid metal during the casting procedure. Such a filter for aluminum melts is shown in fig 1.

Figure 1: ceramic foam filter made of alumina, size 50x50x20 mm

The Collaborative research center 920 deals with the development of multifunctional filters for metal melt filtration. One root of this assignment was an accident of a German high speed train called "Inter City Express" (ICE) back in 2008. After analysis of the damage by Klinger et al. [1], it was clear that a fatigue fracture due to non-metallic inclusions was responsible for the damage of an axle of the train. Furthermore, they found that the inclusions consisted of agglomerates of primary alumina particles and grew up to a critical size of more than 70 µm.

Liquid metals are extremely hot (steel 1600 °C, aluminum 700 °C) and opaque, which make investigations of the filtration and agglomeration extremely costly and complex. In addition, metal melts have a very high surface tension, which leads to a non-wetting behavior on both the filter surfaces and the solid inclusion particles. Based on these factors, the agglomeration of non-wetting particles is investigated in a room temperature model system. Water is used as model melt because of its unusual high surface tension of 72 mN/m. Hydrophobic particles are used for the agglomeration experiments, which are dispersed in a stirred tank filled with aqueous solutions. With the aid of a dynamic image analysis apparatus, the change in particle size is measured over time. The main target is to find the mechanisms present during agglomeration of particles in a poor wetting liquid and the limiting parameters of the process. Additionally, various measurements via atomic force microscopy are possible and useful to analyze and describe the occurring sub-processes of non-wetting agglomeration.

In choosing appropriate hydrophobic model particles, a large number of materials is possible. Most synthetic materials and polymers like polystyrene (PS) or polytetrafluorethylene (PTFE) show a hydrophobic character (planar contact angle of water over 90°). But the behavior of such inert surfaces in water, especially the charge distribution at hydrophobic interfaces are highly discussed, yet not completely understood in literature [2]. In the experiments, alumina particles are used, which are coated with a hydrophobic silane coating (Dynasilane F8261 from Evonik, Germany) in the same way as described in a previous study [3]. Apart from its hydrophobicity, this particle system shows further advantages. The particles are opaque in contrast to the polymer particles at small particle sizes, which makes them easy to detect via dynamic image analysis. Furthermore, the influence of wetting will be directly visible, if the uncoated material is also part of investigation, having the same particle size distribution as the hydrophobic material. The alumina particles were classified in a fraction of about 2-20 µm.

The particles are added to a stirred vessel having a volume of 4 l. A scheme of the experimental setup is shown in fig. 2. The vessel is closed on the top and contains baffles to develop a turbulent stream while stirring with a 6-blade-propeller.

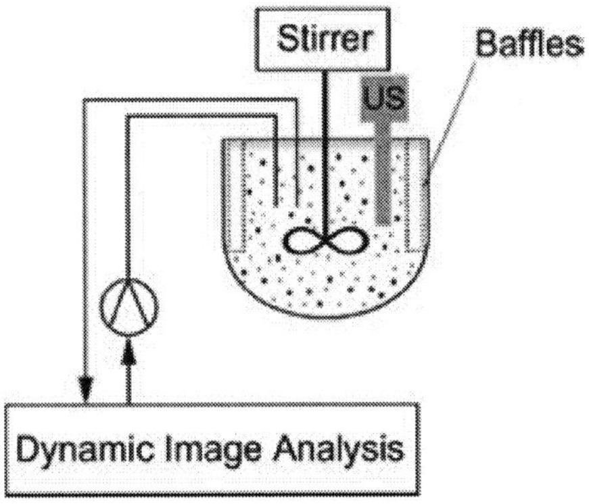

Figure 2: experimental setup to measure the kinetics of agglomeration

After adding the particles to the water-filled vessel, a sampling system directs the agglomerates to the dynamic image analysis. As a function of time, the agglomerate size distribution is recorded and leads to a representation of agglomeration kinetics. After a certain time, agglomeration and breakup will form an equilibrium and a constant agglomerate size evolves. A focus on the wetting properties of the particles show a significant difference in agglomerate size. The hydrophilic agglomerates are much smaller than the hydrophobic ones. The hydrophilic particles were agglomerated in a saline solution of 0.5 M NaCl. This leads to a total compression of the diffuse double layer around the particles and electrostatic repulsion can be neglected. Only van der Waals forces dominate the attraction and constitute the interaction forces.

Hydrophobic particles are hard to disperse in water, so different methods were used for dispersion. At first, the particles were forced into the water due to treatment via an ultrasonic sonotrode. Here the particles grew up to larger agglomerates, which is only explainable due to hydrophobic forces. These forces are heavily discussed in literature and different kinds (long and short range) of interaction forces can be distinguished. The most common explanation of the origin of these forces is due to capillary bridges. In this scenario it is assumed that gas remains trapped inside the cavities of the rough particles.

Another method consists of initial dispersion of the particles in ethanol and subsequent addition of this suspension to the water phase. This solvent exchange leads to a nucleation of surface nanobubbles at the particle surface [4], which induce capillary forces between the particles [5].

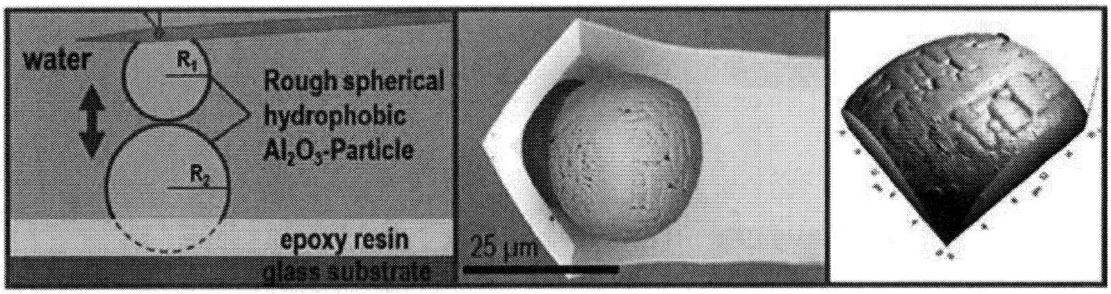

Figure 3: left: scheme of the experimental setup for measuring the interaction forces between two rough colloidal particles via CP-AFM, middle: image of an colloidal probe cantilever preparation with a 25 µm sized alumina particle, right: topography of the embedded alumina particle – measured via contact mode AFM

The interaction forces were measured via colloidal probe atomic force microscopy (CP-AFM). Thereby a spherical colloidal particle with a size of 15-50 µm is glued to a tipless cantilever. This preparation allows the direct force measurement between the CP-particle and a discretionary interface. In this case, interaction forces were measured between a Al_2O_3 CP-particle (20-25 µm) and a fixed grounded Al_2O_3-particle (20-35 µm) with similar roughness. The experimental setup is shown in figure 3. Force-distance curves were recorded while bringing the particles in contact over a thousand times. The maximum detected attractive forces can be evaluated via a MATLAB ® routine. This method is necessary because of the heterogeneity of the particles and allows an assemblage of all interaction forces, resulting in a force distribution. For hydrophobic particles, the attractive forces are higher, which can be explained by the additional hydrophobic interaction. After applying a solvent exchange, the particle interactions further increase due to the formation of surface nanobubbles. In consequence, the agglomerate size distribution is directly connected to the results of the AFM measurements, which can be clearly seen in figure 4.

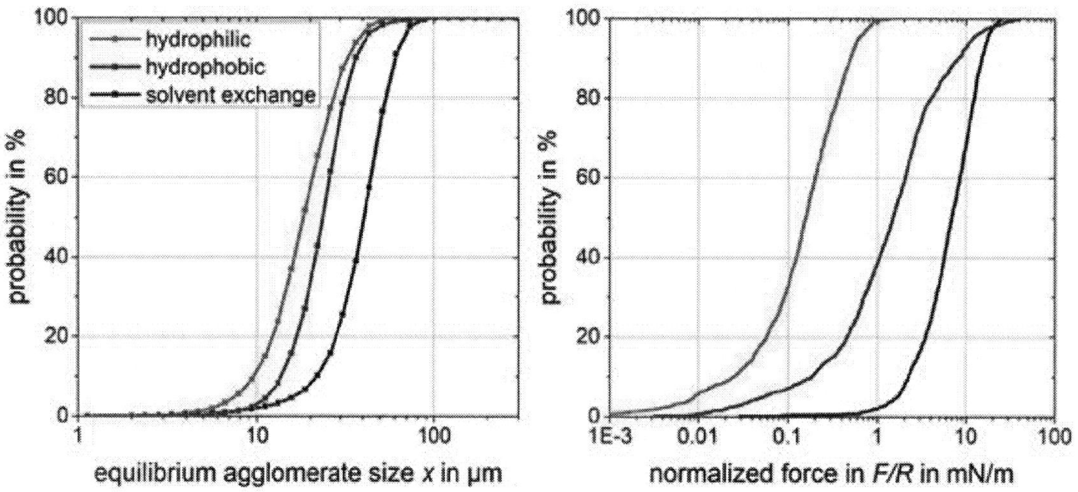

Figure 4: left: agglomerate size distribution after 20 min agglomeration of 2-20 µm primary alumina particles with hydrophobic and hydrophilic surface properties, additionally the equilibrium size distribution after applying the solvent exchange, right: interaction force distribution between alumina particles with the same physico-chemical properties

The investigations with the room temperature model system gives insights into the mechanisms of poor wetting agglomeration behavior. It is therefore believed that metal melts behave in a similar fashion and capillary forces are present either between the inclusion particles themselves or the inclusions and filter surfaces.

Acknowledgement

The authors gratefully thank the German Research Foundation (DFG) for supporting the Collaborative Research Center CRC 920, subprojects B04.

References

[1] C. Klinger, D. Bettge, Axle fracture of an ICE3 high speed train, Engineering Failure Analysis 35 (2013) 66–81.

[2] R. Zimmermann, U. Freudenberg, R. Schweiß, D. Küttner, C. Werner, Hydroxide and hydronium ion adsorption — A survey, Current Opinion in Colloid & Interface Science 15 (2010) 196–202.

[3] P. Knüpfer, J. Fritzsche, T. Leistner, M. Rudolph, U.A. Peuker, Investigating the removal of particles from the air/water-interface – Modelling detachment forces using an energetic approach, Colloids and Surfaces A: Physicochemical and Engineering Aspects 513 (2017) 215–222.

[4] M.A. Hampton, B.C. Donose, A.V. Nguyen, Effect of alcohol–water exchange and surface scanning on nanobubbles and the attraction between hydrophobic surfaces, J Colloid Interface Sci 325 (2008) 267–274.

[5] M.A. Hampton, A.V. Nguyen, Nanobubbles and the nanobubble bridging capillary force, Adv Colloid Interface Sci 154 (2010) 30–55.

508201 Characterizing Nanoparticle Agglomerates in a Nanopowder Fluidized Bed

Abstract:

Nanoparticles in a gaseous environment agglomerate in a hierarchical fashion; this leads to peculiar hydrodynamics when fluidizing nanopowders. From production until powder processing in the gas phase, nanoparticles typically go from individual particles to aggregates, simple agglomerates, and complex agglomerates. Even though the structures at each level have unique properties, they are commonly assessed as a whole. Additionally, the effect of external factors on the morphology of these structures during gas processing is not well understood and challenging to study due to the limited techniques for in situ analysis of the dynamic phenomena.

Here, we study different ceramic oxide nanoparticles. In our lab, we use such particles to functionalize them via atomic layer deposition in a fluidized bed, e.g. for catalysis. In order to optimize this process, it is important to know the structure of the agglomerates. We describe the structural characteristics of each hierarchical level of complex agglomerate formation obtained from two in situ techniques: small-angle neutron scattering for the smaller scales, and a settling tube coupled with a boroscope for the larger scales. The first scale, namely aggregates, are open structures with a fractal dimension of about 1.5, which then form simple agglomerates with a fractal dimension close to 3, that later cluster into complex agglomerates that present a fractal dimension of about 2. In addition, we characterize the contact mechanics of the nanopowders agglomerates using atomic force microscopy. Finally, we will discuss the entrainment of nanosized clusters from the fluidized bed.

508691 Granule Structure and Formation Mechanisms of Single Drop Formed Granules from Binary Mixtures

Abstract:

In wet granulation, single drop granulation (where one drop forms one granule) can simplify the complexity and reveal the fundamentals of the interaction between the liquid and the powder bed. Combining drop penetration time, granule formation mechanisms and the characterization of granule properties is a novel study to achieve a comprehensive map of the single drop granulation process, which is relevant in particulate applications like pharmaceuticals, detergents, agricultural chemicals and catalysts.

In this work, single drop impact of liquid on a static powder bed was studied to investigate the granule formation mechanism, droplet penetration time, and the characterization of granule properties (morphology, internal structure and content uniformity). Water was used as the liquid binder and two pharmaceutical powders, microcrystalline cellulose (MCC) (mean particle size ~54.8 Î¼m, contact angle with water ~36Â°) and acetaminophen (APAP) (mean particle size ~9.74 Î¼m, contact angle with water ~89°), were mixed to make heterogeneous powder beds with varying compositions. The complete drop impact and penetration was recorded with a high-speed camera. The granule morphology, its internal structure and content uniformity were characterized by the prism method with image analysis, micro-CT and UV-vis spectrometry, respectively. Besides the current MCC and APAP system, MCC and APAP with other particle sizes and relative contact angles will also be studied. It is believed that the mean particle size of the powder bed is predominant in influencing the granule formation mechanism, drop penetration time, and granule properties.

509632 Forming Lumps and Jets when Pouring Particles through a Liquid Interface

Abstract:

Abstract.

The dispersion of dehydrated powders in water is a key step in many process industries. During the rehydration process, the powder grains can form heterogeneous powder lumps that are wet outside and dry inside [1, 2]. It is often difficult to predict and control the conditions leading to lump formation since dissolution is hindered subsequently by entrained air. Phenomena associated with the balance of forces responsible for controlling whether small objects float at an interface or sink have been the subject of recent studies [3, 4]. A theoretical condition for the formation of granular jets was also proposed [5], but a comprehensive experimental validation is still lacking.

The focus of this contribution is therefore on studying experimentally the behaviour of a powder poured on a static liquid interface and the formation of a granular jet, understanding the effect of grain flow rate, contact angle, density and size. The wetting properties of the system were varied by using ethanol-water mixtures. Two types of grains, of different density, were used and the size of the grains varied between 50 microns and 1 mm.

The conditions under which powder grains form a â€˜jetâ€™ through the interface, or alternatively float to form a powder island, were quantitatively investiaged. The relationship between the kinetic energy of the grains generated as they flow through a circular orifice under gravity and the surface energy created by the grains at the interface per unit time was also quantitatively studied. The experimental results indicate that the powder island size grows monotonically with decreasing grain density and increasing contact angle. These finding can be used to promote powder dispersion and avoid lumps.

References

[1] J. Dupas, L. Forny, M. Ramaioli, Powder wettability at a static air-water interface, Volume 448, p. 51â€“56, *JCIS*, 2015.

[2] J. Dupas, E. Verneuil, M. Van Landeghem, B. Besson, L. Forny, M. Ramaioli, F. Lequeux and L. Talini, Glass transition accelerates the spreading of polar solvents on a soluble polymer, *Phys. Rev. Lett.* 112, 188302, 2014.

[3] M. Abkarian, S. Protière, J. M. Aristoff, and H. A. Stone, Gravity-induced encapsulation of liquids by destabilization of granular rafts., *Nature Commun.*, 4, 2013.

[4] D. Vella, P. Metcalfe, and R. Whittaker, Equilibrium Conditions for the Floating of Multiple Interfacial Objects, *J. Fluid Mech.*, 549, 215-224, 2005.

[5] G. Lagubeau, Interfaces à grains, et autres situations de mouillage nul, *PhD Thesis*, 2010 (https://www.pmmh.espci.fr/fr/gouttes/Equipe_files/TheseGuillaume.pdf

Particle Measurement in High-Temperature Gas based on Light Scattering

Li-feng LU[1,3], Xiao-lin WU [1,3], Zhong-li JI [1,2], Ming-xing WANG[1,3]

[1] Department of Chemical Engineering, China University of Petroleum, Beijing, P. R. China

[2] Department of Mechanical and Transportation Engineering, China University of Petroleum, Beijing, P. R. China

[3] Beijing Key Laboratory of Process Fluid Filtration and Separation, China University of Petroleum, Beijing, P. R. China

Abstract: In order to achieve the particle measurement at 650 °C, a single particle measurement method and device suitable for high-temperature gas were developed. The influence of gas temperature on optical sensor of optical particle counter (OPC) was investigated in detail based on geometric optics theory and heat transfer theory. The thermal expansion of aerosol tube along the airflow direction was caused by the heat conduction between the tube and the gas, resulting in the position of the light apertures to shift to obscure the partial light beam. The invar material can be adopted to eliminate the above mentioned disadvantages. According to the principle of gas state equation, the optical parameters (refractive index) of the gas at different temperatures change with the change of its state parameters. Based on the geometrical optics theory and the geometric mathematical relationships, the theoretical and experimental dynamic models of the optical measuring volume with the change of the gas temperature were established. The established models were then used to analyze the influence of gas temperature on the optical measuring volume. The results showed that the influence of the temperature on the optical system is negligible. Meanwhile, particle detection experiments were conducted at different temperature to achieve a good performance of OPC.

Keywords: High-temperature gas; Optical measuring volume; Dynamic model

1 Introduction

With the continuous improvement of environmental protection and the safety requirements for industrial production, the technology of high-temperature purification is drawing more and more attention, especially in the fields of clean coal power generation, coal gasification, petrochemical industry and metallurgy [1]. For example, nuclear power plants often produce a large number of low radioactive solid wastes, and the combustible wastes usually account for a considerable proportion. The incineration and pyrolysis are effective treatment techniques for the combustible waste, during which the temperature of flue gas is 600~800 °C. With the aim to purify the flue gas, the German Karlsruhe Nuclear Research Institute, the French nuclear power station and the Japanese nuclear power station have installed the silicon carbide ceramic filter tube, aluminum silicate fiber and alumina ceramic filter tube at gas temperatures above 650 °C, respectively, and investigated their properties properties [2-4]. These filters have a direct influence on the purification quality of gas and the particle concentration in the downstream. In an industrial cracker system, the particle concentration is required to reach 0.014 mg/m^3 after the filters; the Integrated Gasification Combined Cycle (IGCC) requires the particle size to be smaller than 1 um with a particle concentration of less than 2.2 mg/m^3 and in the range of 1~10 um with less than 5.8 mg/m^3, and to remove the dust size of larger than 10 um [5]. In a Shell Coal Gasification Process (SCGP), the particle concentration is required to be less than 20 mg/m^3 [6]. In order to detect the concentration and size of the particles in pipes in real time for calculating the separation efficiency of evaluating the purification performance of the filtration system, the particle detection technology is needed to be improved [7].

The widely-used particle detection devices in industry are mainly based on the principle of light scattering [8-10], and the particle size and concentration are obtained through an inversion of the scattering light intensity and particle size, which are suitable for the conditions of below 200 °C. The method of first cooling and then detecting can be adopted for the particle detection under higher-temperature conditions, but droplet condensation and sublimation will inevitably occur and result in an incorrect particle samples.

With the aim to achieve the particle detection at 650 °C, an optimization was conducted on the existing particle detector (OPC) in this paper. And based on geometric optics theory [11] and heat transfer theory, a detailed research was performed on the influence of gas temperature on OPC optical sensor, including the dynamic change of the aerosol tube and the optical measuring volume (OMV). Finally, the experiments were conducted on the on-line particles measurement of different temperatures.

2 Experimental setup and method

Figure 1 gives the experimental setup for detecting particles in high-temperature gas, mainly including the heating unit, optical detection unit and gas purification and processing unit.

The heating unit consists of the vibration particle feeder and the air heater. The gas brings the particle out of the feeder, and is heated up to 1000 °C by the heater, which is monitored quickly by a PID temperature control instrument with easy operation.

The optical detection unit is used to achieve the experimental measurement of the characteristics of the OPC optical system and the on-line detection of particles, including the optical sensor and the OPC control host. The optical sensor is shown in Figure 2, which is composed of an aerosol tube and optical system. Aiming to avoid the influence of high-temperature gas on optical system, the aerosol tube (shown in Figure 3) has been developed, and is in a distance of 15 mm to the optical lens. In addition, the insulation coating with lower thermal conductivity (-80-1700 °C) is adopted to treat the outer wall of the aerosol tube. A K-type micro thermocouple was bent and installed into the light aperture to monitor the gas temperature at the center of the aerosol tube in real time without affecting the airflow and optical paths. Figure 4 shows the optical system of the sensor, which is composed of an incident light path and an observed scattering light path. Both the paths are arranged in 90° with the same optical lens and elements [12-14], and the intersected area is the center of the aerosol tube to form an OMV [15]. The intersected area is enough small and allows particles to pass through it one by one [16, 17]. The OPC host consists of the light source system (wavelength: 300-700 nm) [18], the photoelectric detection system and the signal analysis and processing system. In the particle detection, particle is illuminated by the incident light when passing through the OMV, and the scattering light is collected and transmitted to the photoelectric detector to achieve the conversion of a pulse voltage signal from an optical signal. Finally, the tested particle size is calculated based on the calibration curve of particle size and voltage amplitude [14, 19].

The gas purification and processing unit is composed of a vacuum pump, a flow controller, a dust collector and a heat exchanger. The high-temperature gas form the aerosol tube is completely cooled during the heat exchanger, the flow rate of which is adjusted by the flow controller. In order to avoid the damage of particles on the pump, the dust is captured and collected by a filter in the dust collector.

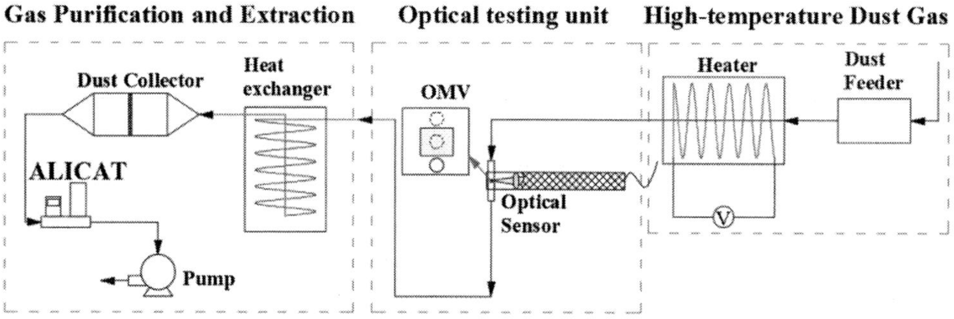

Figure 1. Experimental setup for detecting particles in high-temperature gas

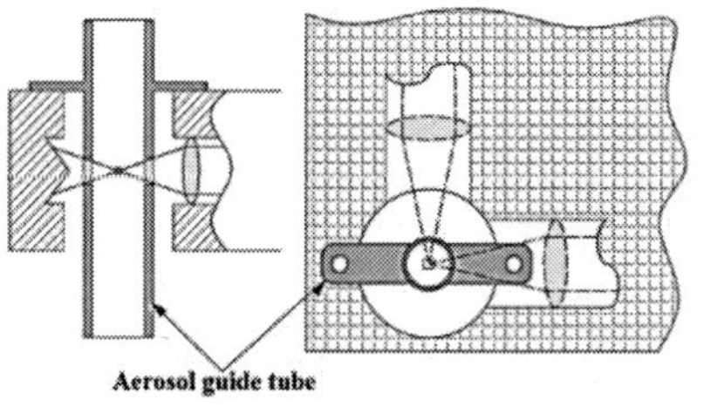

Figure 2. Optical sensor

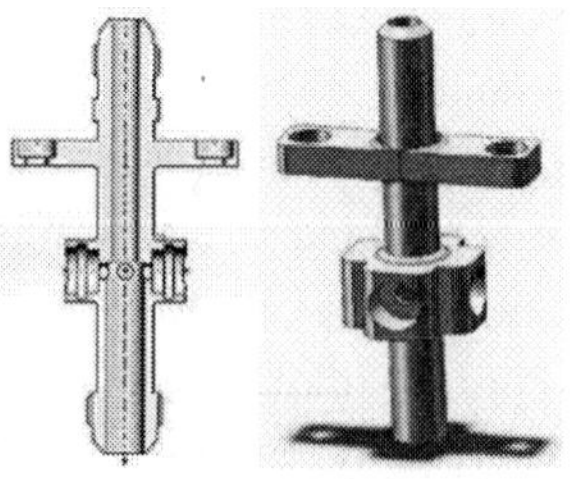

Figure 3. Aerosol tube for high-temperature gas

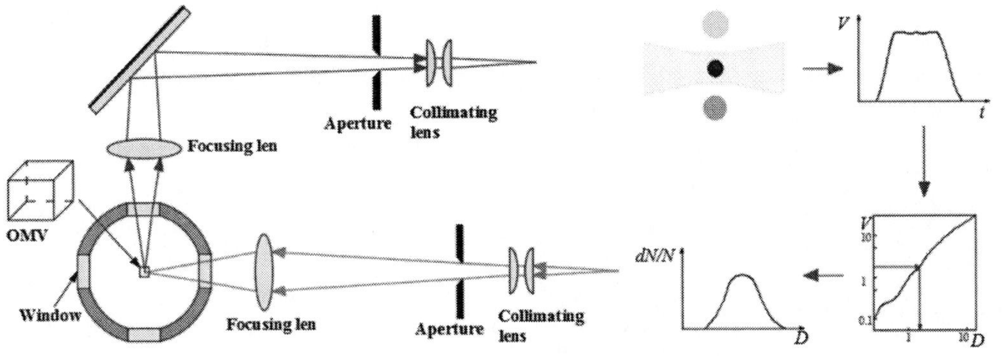

Figure 4. Optical system and principle of OPC

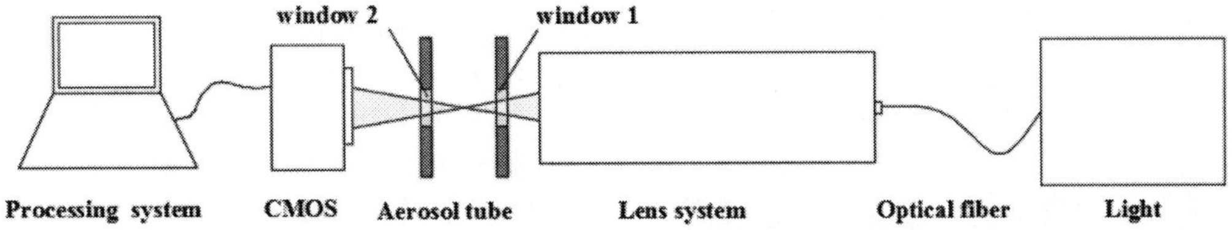

Figure 5. Evaluation and analysis device for light beam

3 Analysis and discussion

3.1 Influence of temperature on aerosol tube

The aerosol tube shown in Figure 3 has four apertures along the center axis, in which the glasses with a thickness of 2 mm were installed, as shown in Figure 4. When the gas flows in the tube, it can reduce the damage of high-temperature gas to the optical system elements. But the heat resistance of the tube has a direct influence on the imaging of the optical system at the center of the aerosol tube.

Figure 5 gives an evaluation and analysis device for light beam, including the light source system, the light path system, the aerosol tube and the light beam analyzer (CINOGY CMOS-1201, Germany), which can be equipped at the optical testing unit as shown Figure 1. For the gases at different temperatures, the light passes through the glasses in the aerosol tube and images on the CCD of CMOS-1201 to detect the spot of the light beam in real time.

After eliminating the background noise, the spot image at different temperatures can be acquired. When the aerosol tube is made of 304 stainless steel, the measurement results are shown in Figure 6. It can be found that the shape of the light spots is distorted and shows the imperfection of the imaging with an increase in gas temperature. That's due to thermal expansion along the airflow direction caused by the change of gas temperature, which results in the shift of light apertures position to obscure the partial light beam. A bad result must be acquired for smaller particle size and lower particle count efficiency because of the decreased incident light. Meanwhile, the shape of the OMV was changed to greatly reduce the performance of OPC detection by a failure of eliminating the error of edge detection in optical system.

With the aim to eliminate the above mentioned disadvantage, the invar material with a linear expansion coefficient of 1.6×10^{-6} m/°C was adopted for high-temperature experiment. The results are shown in Figure 7. At different temperatures, no deformation was found in the shape of light spots, indicating that the material can meet the testing requirement for high temperature up to 650 °C with a good heat resistance.

Table 1. Comparison of the application conditions of the two materials

Material	Light divergence angle /°	Gas temperature /°C
304 (Stainless steel)	<10	<200
4J36 (Invar)	<30	<700

The application conditions of the two materials are

summarized in Table 1. Compared with the stainless steel, the invar material is of great significance for the development of high-temperature detection device.

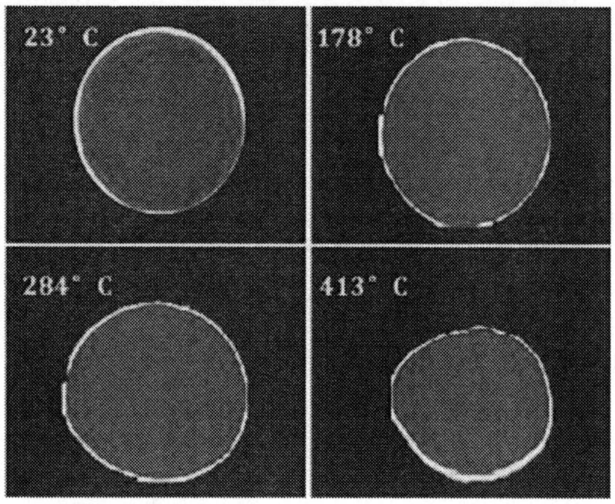

Figure 6. The spot images measured at four different temperatures of gas for 304 stainless steel.

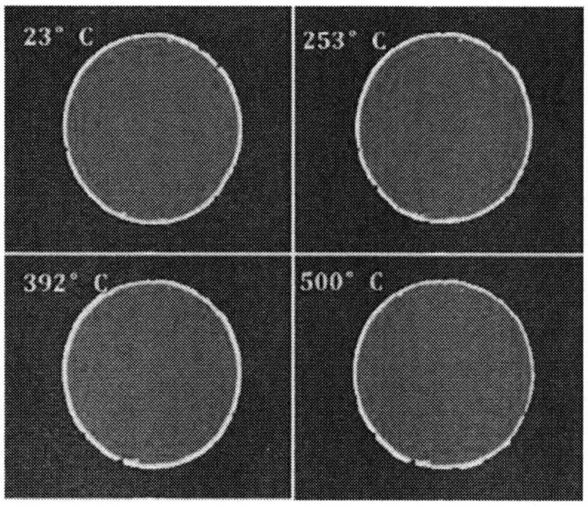

Figure 7. The spot images measured at four different temperature of gas for 4J36 (Invar).

3.2 Influence of temperature on optical system

In general the optical sensing system in the particle detection device is only applied at atmospheric temperature, but can also satisfy the temperature up to 150 °C. Under high-temperature condition, the gas volume is going to expand with the reduction in the gas density. The transmission characteristic of light beam could be changed, resulting from the decrease in the optical performance of the sensor. In order to investigate the influence of temperature on optical system, the specific causes are analyzed by theoretical method, and the extent of its influence is determined through experimental verification.

3.2.1 Theoretical method

By calculating the refractive index of the gas at different temperatures, the change of the OMV with the temperature can be obtained.

a) Calculation of gas parameters

The gas medium is air, so its refractive index can be calculated by the formula proposed by Rueger at the working temperature, given by:

$$n_{ph} = \frac{273.15}{1013.25} \cdot P \cdot N_{sph} \cdot \left(a_0 + \frac{a_1}{T} + \frac{a_2}{T^2} + \frac{a_3}{T^3} + \frac{a_4}{T^4}\right) \times 10^{-6} + 1 \quad (1)$$

Where P is the gas pressure; a_0, a_1, a_2, a_3 and a_4 are the correlation coefficient and constant; N_{sph} is the gas refractive index at atmospheric temperature, given by:

$$N_{sph} = (n_{sph} - 1) \times 10^6 = 287.6155 + \frac{1.62887}{\lambda^2} + \frac{0.01360}{\lambda^4} \quad (2)$$

The wavelength range is 400-700 nm. The gas refractive indexes at different temperatures in the range of 25-650 °C are calculated by Eq. 1 and 2. The results are shown in Figure 8. It can be found that the refractive index of air decreases with the increase in the gas temperature. And it can be inferred that the light divergence angle at high temperature is going to increase compared with that at atmospheric temperature. Consequently, the position of the beam focus is offset, leading to larger optical measuring volume, as shown in Figure 9.

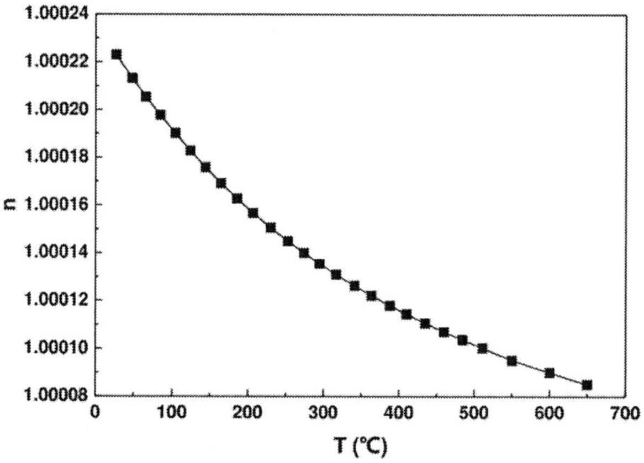

Figure 8. Gas refractive index at different temperature

In order to analyze the above unfavorable influence, a model of the OMV and gas temperature is needed to be built, mainly including the functions of the determination of the optical parameters and the calculation of the dynamic variable.

b) Calculation of optical parameters

On the basis of the Lorentz-Lorenz formula and actual gas state equation, the relationship between gas pressure and refractive index is given by:

$$\frac{n^2-1}{n_0^2-1}=\frac{\rho}{\rho_0}=\frac{PT_0Z_0}{P_0TZ} \tag{3}$$

Where T_0, P_0, ρ_0, Z_0 and n_0 are the gas temperature, pressure, density, compression factor and refractive index under atmospheric condition, respectively; T, P, ρ, Z and n are the gas temperature, pressure, density, compression factor and refractive index under high-temperature condition, respectively.

The light divergence angle is calculated by the refraction law shown below:

$$n_0' \sin\alpha = n \sin\beta \tag{4}$$

Where n_0' and α are the air refractive index and light divergence angle under atmospheric condition; β is the light divergence angle under high-temperature condition.

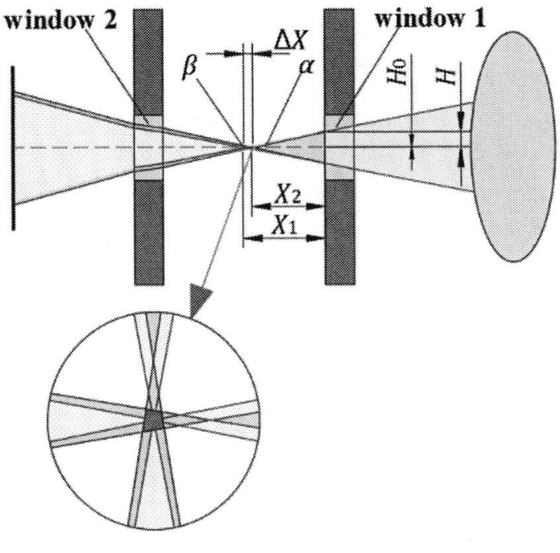

Figure 9. Schematic of the enlargement of the OMV

c) Calculation of dynamic variable

Based on the geometric optical theory and the mathematic relationship, the distance between the light focus position and the glass window 1 is triangular with the equivalent spot radius of the inner side of the glass window 1. According to X_1 and X_2 (the distance between the light focus position and the inner side of the glass window 1 under atmospheric condition and high-temperature condition), the theoretical model of the OMV with gas temperature was built. X_1 and X_2 are calculated by:

$$X1=\frac{H-H_0}{\tan\alpha} \tag{5}$$

$$X2=\frac{H-H_0}{\tan\beta} \tag{6}$$

The change of the light focus position (ΔX=$X1$-$X2$) can be calculated by Eq. 5 and Eq. 6.

The change of the equivalent spot diameter of the OMV can be calculated based on the triangle relation, as shown below:

$$\Delta D = 2\Delta X \tan\beta \tag{7}$$

Eq. 6 and Eq. 7 are the formulas of the theoretical dynamic model of the OMV with the change of the gas pressure.

Using the model, the optical parameters at different gas temperatures in the range of 25 ~ 700°C were calculated with three light divergence angles (4.02°, 9° and 13°, tested under atmospheric condition, respectively). The results are presented in Figure 10. It can be found in Figure 10 that ΔX increases with the gas temperature with approximate two patterns. In addition, ΔD changes by a smaller margin (less than 1 um). However, ΔD and ΔX satisfy a trigonometric relation. It can be also noted that the enlargement rate of the OMV volume (δ) has a small increase with the increase in gas temperature. Under the constant temperature, δ is going to increase for the larger light divergence angle, but it is still relatively smaller (<1%) compared with the OMV at atmospheric condition. Consequently, the influence of the gas temperature on the OMV can be ignored based on the above theoretical analysis.

3.2.2 Experimental method

In order to verify the reliability of theoretical analysis, an experimental dynamic model of the OMV with gas temperature variation was established based on the image analysis and the mathematical geometry relations, including the measurement of the change of light spot, the calculation of the geometric relation and the calculation of the relative variable.

By the refraction law, the changes of light spot on the inner and outer side of the glass window 2 under high-temperature condition (ΔD and $\Delta D'$) meet ΔD=$\Delta D'$. Taking the light focus position as the central axis, a triangle relation was formed by the distances between the light -

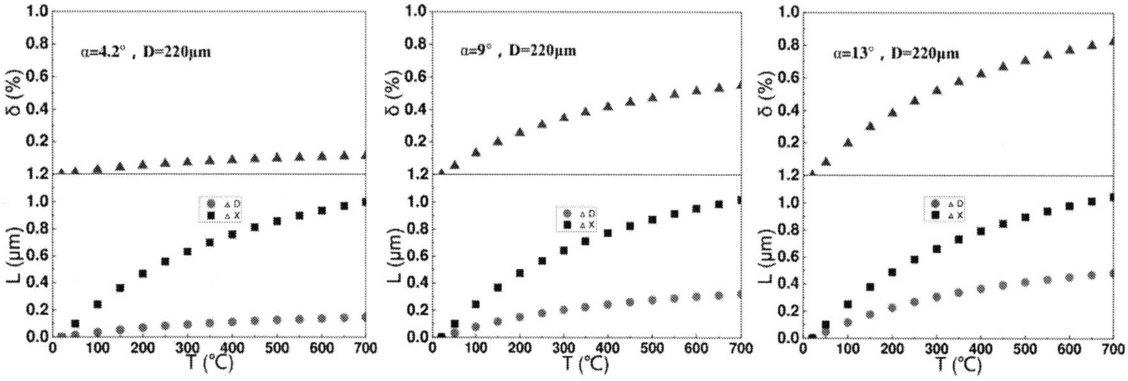

Figure 10. Results calculated based on theoretical dynamic mode with three light divergence angles (4.02°, 9° and 13°)

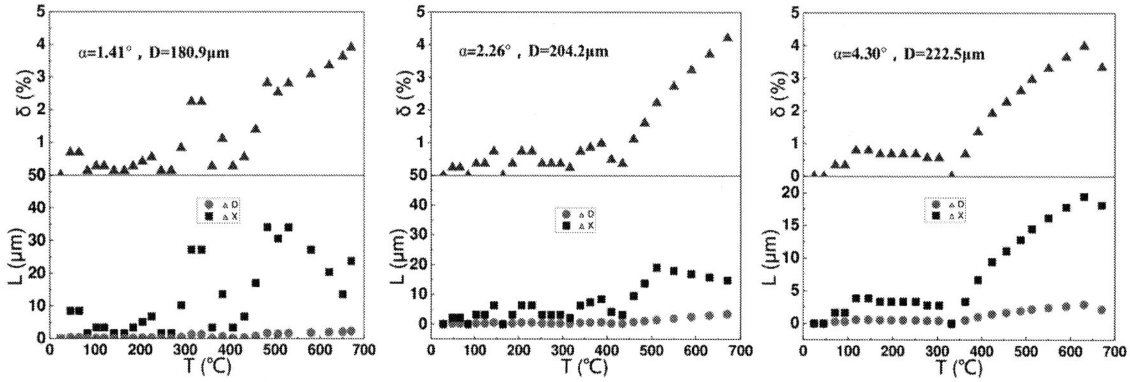

Figure 11. Results calculated based on experimental dynamic model

focus position and the inner side of the glass windows 1 and 2 (x and y), the difference of the light spot equivalent radius on the inner side of the glass windows 1 and 2 (H and h, $h = H - \frac{\Delta D}{2}$) and the equivalent radius of the OMV (H_0). Therefore, we have the following equations.

$$\frac{H - H_0}{h - H_0} = \frac{y}{x} \quad (8)$$

$$x + y = k \quad (9)$$

$$\beta = \arctan\left(\frac{h - H_0}{x}\right) = \arctan\left(\frac{H - H_0}{y}\right) \quad (10)$$

where β is the light divergence angle under high-temperature condition; k is the distance of the inner sides of the glass windows 1 and 2.

The changes of ΔX and ΔD under high-temperature condition to those under atmospheric condition are shown as follows:

$$\Delta X = \frac{y - x}{2}, \text{ or } \Delta X = y - \frac{k}{2}, \text{ or } \Delta X = \frac{k}{2} - x \quad (11)$$

$$\Delta D = 2\Delta X \tan\beta = \Delta D' \quad (12)$$

Eq. 11 to Eq. 12 are the formulas of the experimental dynamic model.

Using the above established experimental dynamic model, the ΔX, ΔD and δ of four cases in Table 2 were calculated, and the results are shown in Figure 11. It can be found that these three parameters increase obviously with the gas temperature increasing, but cannot meet the two function growth mode as the results of theoretical model. That may be caused by the reasons listed below: 1) the fluctuation of gas flow in the pipeline causes the unstable gas temperature field; 2) the beam quality analyzer is slightly influenced by other vibration sources; 3) the aerosol tube is affected by the gas flow fluctuation. But according to the experimental results, the maximum values of ΔX and ΔD are very small (less than 50 μm and 5 μm at the temperature below 700 °C, respectively) compared with the equivalent diameter of the OMV. Consequently, the ratio of the volume of the OMV is less than 5%. To sum up, the gas temperature has greatly weakened influence on the OMV, which can be ignored in actual technical development.

Table 2. Optical parameters of experimental model

number	light divergence angle /°	spot equivalent diameter /μm
1	1.41	180.9
3	1.94	285.1
4	2.26	204.2
5	4.30	222.5

3.2.3 Experiments

The performance of the optical sensor is investigated at different temperatures up to 650°C. The experimental results showed that the temperature of the optical sensor is around 50°C with the gas temperature in the range of 200-300 °C, and is going to increase about 10 °C at the gas temperature of 400-500 °C. The temperature is in the controllable range with a maximum value of 70 °C under the gas temperature of 650 °C. Therefore, the developed optical sensor can be used for the detection of particles in high-temperature conditions for a long time.

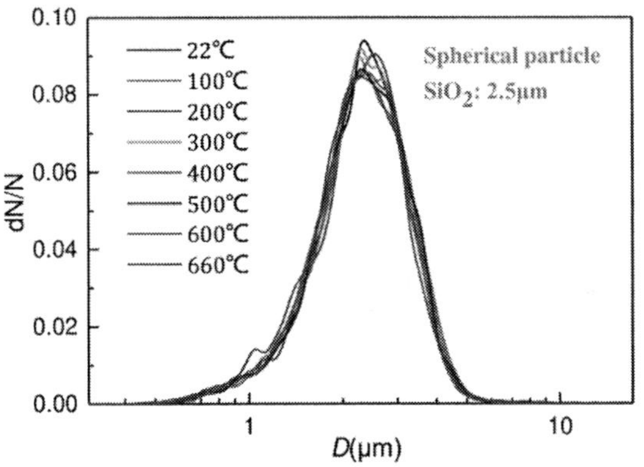

Figure 12. The results obtained under high-temperature condition

In the experiment of particle detection, the standard silica spherical particles with a strong temperature resistance ($D_{particle}$=2.5 μm) were tested at different gas temperatures. The testing time is 60 s. Figure 12 shows the test results. It can be found that all of the particle size distributions are in good agreement and have a good normal distribution, which meets the test results of single particle size, and the peak particle size is about 2.5 μm. Consequently, the optical sensor worked well for the particle detection under higher-temperature condition.

4 Conclusion

The emission of high-temperature flue gas causes serious air pollution and partial waste of available heat. In order to meet the relevant emission standards, additional purification units (high-temperature ceramic filters, etc.) are installed in the industrial process to collect the dusts in the flue gas. As the purification units are important for controlling the dust concentration, it is necessary to calculate the separation efficiency through a real-time measurement of the dust concentration in the flue gas in the upstream and downstream of the purification equipment.

In present work, a single particle measurement method and device suitable for high-temperature condition were developed. In order to complete the particles measurement at 650 °C, the beam images in optical sensing system under different temperature conditions were measured in real time by CMOS sensor. It was found that the beam images were deformed when the light divergence angle increased to more than 10°, and the deformation became more serious as the temperature increased. Depending on the principle of heat transfer, thermal expansion along the airflow direction caused by heat conduction between the duct and the gas is the main cause of this phenomenon. By changing the aerosol tube material t o one with an ultra-small linear expansion coefficient (Invar: 4J36), the optical system can be applied to the condition with temperature up to 700 °C, and in particular the light divergence angle extends to 30°.

According to the principle of gas state equation, the optical parameters (refractive index) of the gas at different temperatures change with the change in its state parameters. According to the results calculated by the formula of Ruegur's atmospheric refractive index, it was found that the refractive index of the gas decreases with the increase in temperature, which leads to the increase in the divergence angle of the beam in the aerosol tube, and shift of the position of the optical beam. Based on the geometrical optics theory and the geometric mathematical relationships, the theoretical dynamic model of the optical measuring volume with the change of the gas temperature was established. By calculating the amount of the beam focus offset and the volume of the optical measuring volume at different temperatures, it was found that the optical measuring volume increases with the temperature increasing. However, when the gas temperature was 650 °C, the increase rate of optical measuring volume was extremely small, so the influence of the temperature on the optical system can be negligible.

The performance of the whole detection system was tested by using 2.5 μm standard spherical silica particles. The satisfying measuring results were obtained and they could be applied to the actual on-site measurement at the temperature of 650 °C.

Acknowledgement

The authors thank the supports of "The National Key Research and Development Program of China (NO.2016YFB0601100)"; and would also like to thank the laboratory and staff of the Beijing Key Laboratory of Process Fluid Filtration and Separation, China University of Petroleum.

Reference

[1] J. Meyer, H. Umhauer, K. Reuter, A. Schiel, G. Kasper, M. Förster, Concentration and size measurements of fly ash particles from the clean gas side of a pressurised pulverised coal combustion test facility, Powder Technology, 180 (2008) 57-63.

[2] H. Leibold, F. Dirks, V. Rüdinger, Particulate emissions from a LLW incinerator and off-gas cleaning with a new type of ceramic candle filter, Waste Management, 9 (1989) 87-94.

[3] B.L. Perkins, Incineration facilities for treatment of radioactive wastes: a review, Incinerators, (1976).

[4] D. Fino, N. Russo, A. G. Saracco, V. Specchia, Multifunctional Filter for Treatment of the Flue Gases from Municipal Waste Incinerators, Industrial & Engineering Chemistry Research, 44 (2005) págs. 9542-9548.

[5] Scheibner, Wolters, Schumacher hot gas filter long-term operating experience in the nuon power buggenum IGCC power plant, (2002).

[6] J.M. Wang, Shell coal gasification technology and its application in China, Coal Quality Technology, (2006).

[7] Q. Xu, Z. Ji, L. Yang, Performance assessment of high temperature flue gas on-line particle size analyzer, Ciesc Journal, 63 (2012) 3506-3512.

[8] B.Y.H. LIU, V.A. MARPLE, K.T. WHITBY, N.J. BARSIC, Size distribution measurement of airborne coal dust by optical particle counters, Aihaj, 35 (1974) 443-451.

[9] R. Maus, F. Löffler, Size analysis of airborne bacteria by an optical particle counter and an impactor, Journal of Aerosol Science, 25 (1994) 141-142.

[10] B. Sachweh, H. Büttner, F. Ebert, Improvement of the resolution and the counting accuracy of an optical particle counter by fast digital signal recording, Journal of Aerosol Science, 20 (1989) 1541-1544.

[11] E.W. Max Born, Principles of Optics, Cambridge University Press, Cambridge, 1999.

[12] S. Hochrainer, Further developments of the particle counter sizer PCS-2000, Journal of Aerosol Science, 31 (2000) 771-772.

[13] B.S. Li, Z.L. Ji, L. Mei, An approach for correcting particle size distributions measured by the optical particle counter welas 2000, Aip Conference, 2010, pp. 686-689.

[14] J. Stier, M. Quinten, Simple refractive index correction for the optical particle counter PCS 2000 by Palas, Journal of Aerosol Science, 29 (1998) 223-225.

[15] H. Umhauer, Particle Size Distribution Analysis by Scattered Light Measurement Using an Optically Defined Measuring Volume, Journal of Aerosol Science, 14 (1983) 765-770.

[16] B. Sachweh, H. Umhauer, F. Ebert, H. Buttner, R. Friehmelt, In situ optical particle counter with improved coincidence error correction for number concentrations up to 107 particles cm-3 quantification and elimination of the effect of particle shape and structure, Journal of Aerosol Science, (1998) 1075-1086.

[17] D.I.J. Raasch, D.I.H. Umhauer, Errors in the Determination of Particle Size Distributions Caused by coincidences in optical particle counters, Particle & Particle Systems Characterization, 1 (1984) 53-58.

[18] H. Buettner, Measurement of the Size of Fine Nonspherical Particles with a Light-Scattering Particle Counter, Aerosol Science & Technology, 12 (1990) 413-421.

[19] D. Holve, S.A. Self, Optical particle sizing for in situ measurements Part 2, Applied Optics, 18 (1979) 1632-1645.

508245 An Experimental Investigation of Single Droplet Drying Above Boiling Point

Abstract:

An experimental method has been developed to enable dried particle structure, properties and drying rate to be predicted based on droplet drying history, particularly focusing on effects driven by boiling including bubble nucleation, growth and collapse. The method involves a single droplet drying rig which will be used to investigate drying kinetics above boiling point and across a broad range of drying conditions, drop size and composition space. The drying rig was improved as compared to previous apparatuses reported in literature, in that it is designed to measure droplet drying rates and morphological changes at high temperatures and under conditions that more closely resemble those found in industrial applications. The aim is to conduct drying experiment for droplets as small as 300μm diameter, in an air stream which temperature may be as high as 350°C.

Drying kinetics and morphological evolution will be determined, particularly focusing on the second drying stage as well as inflation/deflation periods. Initially, the constructed drying rig is checked and evaluated by conducting experiments on droplets of pure water. Afterwards, experiments will be conducted on solutions of different materials such as Hydroxypropyl Methylcellulose (HPMC), sodium sulphate and sucrose. An assessment of morphologies, transformations and drying behaviour under isothermal drying would be completed at a wide space of process condition, initial water content, initial drop size and composition space. The investigation will look to link material properties, particle size, initial moisture content and drying history to morphology. Results will be compared with single droplet drying models reported in literature.

An initial investigation has focused on mapping the morphological drying behaviour of HPMC, sodium sulphate and sucrose. It has shown that the three materials show different behaviour; sucrose showed deflation and inflation cycles, sodium sulphate droplets supersaturated and crystallized while HPMC showed inflation and rupture. Differences in drying behaviour is due to difference in material and transport properties. Therefore, the next stage of this work is focused on conducting some physical properties measurement at different water concentrations and temperature. In addition, different materials will be dried on the rig and the drying behaviour will be determined. The experimental data from this work will be used to develop, a universal morphology regime map which will link material properties, particle size, initial moisture content and drying history to morphology.

508482 Characterising Powder Flow in Dynamic Processes

Abstract:

Flow inconsistency of cohesive powders is a major issue for process industries. It is often difficult to predict and control flow of cohesive powders to enable efficient and reliable process operations particularly for dynamic processes such as mixing and filling. Many quantitative and qualitative flow measurement techniques exist that are used for measuring powder flowability under quasi-static conditions. However, there are few available techniques capable of assessing powder flowability under dynamic conditions. The FT4 Powder Rheometer (Freeman Technology), which drives a rotating blade along the vertical axis through a column of powder contained within a cylindrical vessel, is one of the few available devices that is capable of measuring powder flow in a dynamic state [1]. The flow resistance of the rotating blade as it penetrates the powder bed is characterised by flow energy, however, there is lack of understanding in how to relate this measurement to process design.

This work focuses on determining the relationship between the torque and shear stress in the FT4. The effect of particle size on the external flow profile around the blade has been investigated experimentally for glass beads using a high speed camera. The material investigated shows a decrease in downward specific flow energy with increasing particle size; indicating that the increased shear resistance in the bulk for smaller particle sizes exists under dynamic conditions. Particle Image Velocimetry (PIV) analysis indicates an increase in the volume of particles displaced by the blade with increasing particle size for both upward and downward blade movement. An increase in impeller tip speed leads to a reduction in flow energy for this material, and a reduction in the volume displaced by the blade. The prevailing stresses and strains as well as the volume of particles displaced by the blade are estimated by Discrete Element Method (DEM). DEM allows for the determination of a relationship between the flow energy and/or shear stress with particle properties and blade velocity and direction. Furthermore, Positron Emission Particle Tracking (PEPT) analysis of the FT4 has been carried out to determine the radial and axial displacement of particles with respect to the bed height, and to validate the DEM results.

508671 Impact ofÂ Non-Spherical Projectiles on Granular Media

Abstract:

Recently, several studies involving numerical and experimental methods have focused on the study of impact dynamics in both dry and wet granular media. An example of such instances can be an animalâ€™s foot impacting sand during its locomotion. Understanding the physics behind the interaction between granular media and animal feet can be helpful in optimizing locomotion parameters for bio-inspired robots. However, most of the studies involving impact dynamics considered spherical projectiles under different conditions, while characteristic models should involve more complex shapes. Impacting different geometries with conserved density, volume, and velocity on a granular bed may experience contrasting drag forces upon penetration. This is due to the difference in the surface areas coming in contact with the granular media. In this work, non-spherical geometries of following shapes were considered: cylinder, triangular prism, hexagonal prism and cuboid. These geometries were impacted on a loosely packed non-cohesive dry granular bed with the same impact velocities. The effect of surface area in contact with the granular media was analyzed for impact, with the help of penetration depth measurements. We also tested whether the power law for spherical projectiles is applicable for non-spherical projectile impacts.

508664 Computational and Experimental Shear Cell Study with Rigid Cylindrical Particles

Abstract:

Flow properties of particles are commonly assessed through shear cells. Both the Schulze ring shear tester (SRST) and Jenike shear tester (JST) measure the interparticle stresses which are important to the design of many granular flow processes. This work uses precision-cut rigid cylinders of various aspect ratios and sizes to compare the measured stresses between both testers. It is found that the shearing results for both testers are similar, which is important for future work involving flexible biomass materials.

Previous work has sought to use shear cell results to validate various aspects of discrete element method (DEM) simulations. The SRST is recreated in a DEM simulation with rigid glued-sphere cylindrical particles to assess the contact detection method by comparing the experimental and computation results. Also, a parameterization study is performed to examine the effects of coefficient of restitution, coefficient of friction, and other particle properties on the measured steady-state stress. Overall, this serves as an efficient validation of the contact detection method which will be used to simulate complex particle shapes.

508865 Defluidization Behaviour of Industrial Reactive Powders at High Temperature

Abstract:

Fluidized bed reactors are used in a wide range of industrial chemical processes, and they are particularly useful in high temperature systems thanks to their ability to provide high heat transfer rates and rapid solids mixing which lead to isothermal and more controllable conditions. Their applications spread the environmental, chemical, energy and process industries. For example, they are a key technology in the petroleum industry, catalytic chemicals synthesis processes, combustion and gasification of solid fuels (coal, wastes and biomass), metals productions and many more.

Use of fluidized beds in combustion-type operations may lead to particles sintering and bed agglomeration, which again may result in defluidization, unscheduled process downtime and additional costs. However, the details of the physical mechanisms behind such phenomena are still not entirely understood. Despite the countless researches that have been done so far, such processes still disclose unrevealed features and challenge the researchers worldwide.

The objective of this paper is to assess the effect of high temperature on the flow properties and fluidization behaviour of mixtures of carbon coke and two different titanium ores, namely synthetic and natural rutile. To this end, a simple approach for the early detection of agglomeration in the regimes of incipient fluidization -where the carbon combustion causes defluidization- will be followed. Such a monitoring approach is based on the simultaneous measurements of local temperatures and the pressure drop across the particles bed.

507739 Manufacturing and Characterization of Spherical Blend PBT-PC Particles for Additive Manufacturing

Abstract:

Polymer powder-based additive manufacturing technologies often yield excellent mechanical and functional part qualities. One of these technologies is selective laser sintering (SLS), a process employing micron-sized polymer particles, which are selectively fused by a laser [1]. While hardly any boundaries regarding part design exist, there are quite some restrictions concerning the variety of commercially available SLS materials. For the time being, the most widely used polymeric material for SLS is polyamide 12 (PA12) with a market share of roughly 95% [2]. Hardly any blend materials are available; mostly , the processing of multi-material systems via SLS is achieved by simply mixing different polymer powder materials [3].

In this contribution, a new approach to blend particles is introduced. Via co-grinding and subsequent rounding, manufacturing of spherical particles consisting of semi-crystalline polybutylene terephthalate (PBT) and amorphous polycarbonate (PC) thermoplastic is possible. The polymers are jointly comminuted and mixed in a planetary ball mill and the product particles are rounded in a heated downer reactor [4,5]. The thereby achieved intermixed, but not transesterified, PBT-PC composite particles are characterized regarding size and morphology via electron microscopy. Intermixing of the polymers is investigated via staining techniques and Raman spectroscopy. Thermal properties of the blend particles are analyzed via DSC, also in regard to transesterification in SLS process-like conditions. Furthermore, flowability and size distribution of the blend particles are examined. This way, we show the high potential of our process route [4] for the production of novel polymer blend particle systems for SLS.

References:

[1] I. Gibson, D. Rosen, B. Stucker, Additive manufacturing technologies, 2. ed., Springer, New York, NY, 2015.

[2] T.T. Wohlers, Wohlers Report 2016, Wohlers Associates, Fort Collins, Col., 2016.

[3] D. Drummer, K. Wudy, F. Kühnlein, M. Drexler, Polymer Blends for Selective Laser Sintering: Material and Process Requirements, Phys. Procedia. 39 (2012) 509–517. doi:10.1016/j.phpro.2012.10.067.

[4] J. Schmidt, M. Sachs, C. Blümel, B. Winzer, F. Toni, K.-E. Wirth, W. Peukert, A novel process route for the production of spherical LBM polymer powders with small size and good flowability, Powder Technol. 261 (2014) 78–86. doi:10.1016/j.powtec.2014.04.003.

[5] J. Schmidt, M. Sachs, S. Fanselow, M. Zhao, S. Romeis, D. Drummer, K.E. Wirth, W. Peukert, Optimized polybutylene terephthalate powders for selective laser beam melting, Chem. Eng. Sci. 156 (2016) 1–10. doi:10.1016/j.ces.2016.09.009.

506658 Creating Tuneable Agglomerates via 3D Printing

Abstract:

It is essential to offer a universally accepted way to systematically evaluate agglomerate breakage propensity and mechanisms. Discrete Element Method (DEM) simulation is a commonly used tool but is limited by the lack of identical, controlled agglomerates to test and validate simple models, let alone replicate the complex structure of real industrial agglomerates.

Our research presents a systematic approach to produce 3D printed agglomerates with tuneable properties. Agglomerates with different mechanical properties and structures were designed and printed by an Objet500 Connex 3D printer. The tensile and shear strength of inter-particle bonds was characterized using Instron tester. Breakage results of agglomerates with simple cubic structures and the design of agglomerates with irregular structures were presented. This 3D printing method will allow more rigorous testing of agglomerate breakage models. Discrete Element Method (DEM) has been used to simulate the agglomerate crushing using the Timoshenko Beam Bond Model (TBBM) with bond properties matching the 3D printed agglomerates. For both two agglomerate structures used in this research, the simulation and experiment showed similar breakage patterns. However, the simulation results significantly underestimated the compressive loads of spherical random structured agglomerates. In order to improve the simulations of compressive granule breakage, the influences of bond geometry definitions and non-linear behaviour of the bond contacts need to be explored.

508699 Effect of Particle Size Distribution and Wettability on Penetration Behavior, Granule Formation and Granule Properties in Single Drop Granulation

Abstract:

In wet granulation, single drop granulation (where one drop forms one granule) can simplify the complexity and reveal the fundamentals of the interaction between the liquid and the powder bed. Combining drop penetration time, granule formation mechanisms and the characterization of granule properties is a novel study to achieve a comprehensive map of the single drop granulation process, which is relevant in particulate applications like pharmaceuticals, detergents, agricultural chemicals and catalysts.

In this work, single drop impact of liquid on a static powder bed was studied to investigate the granule formation mechanism, droplet penetration time, and the characterization of granule properties (morphology, internal structure and content uniformity). Water was used as the liquid binder and two pharmaceutical powders, microcrystalline cellulose (MCC) (mean particle size ~54.8 Î¼m, contact angle with water ~36Â°) and acetaminophen (APAP) (mean particle size ~9.74 Î¼m, contact angle with water ~89°), were mixed to make heterogeneous powder beds with varying compositions. The complete drop impact and penetration was recorded with a high-speed camera. The granule morphology, its internal structure and content uniformity were characterized by the prism method with image analysis, micro-CT and UV-vis spectrometry, respectively. Besides the current MCC and APAP system, MCC and APAP with other particle sizes and relative contact angles will also be studied. It is believed that the mean particle size of the powder bed is predominant in influencing the granule formation mechanism, drop penetration time, and granule properties.

508017 Atomically Deposited Sintering Aids: Assessing the Effects of Alumina Particle ALD on the Sintering and Performance of SOFC Electrolytes and Dental Ceramics

Abstract:

The crystal structure of zirconium oxide (ZrO_2) can be tailored by the incorporation of aliovalent dopants, most commonly, yttrium oxide (Y_2O_3). The concentration of dopant dictates the thermodynamically stable crystal phase where 3 mol%
Y
conducting cubic phase. To achieve suitable mechanical properties and ionic conductivity, Y-TZP and YSZ ceramics must be near theoretical density, requiring high sintering temperatures. Previous researchers have successfully lowered the required Y-TZP and YSZ sintering temperatures through incorporation of aluminum oxide (Al_2O_3) particles. While this technique has proven to reduce the sintering temperature in both systems, the presence of particulate Al_2O_3 inclusions in densified parts reduces the homogeneity of the material which may have deleterious effects on densification and microstructure. In this work, we report on the use of atomic layer deposition (ALD) to conformally coat individual Y-TZP and YSZ particles with the desired amount of Al_2O_3. Constant-rate-of-heating (CRH) experiments were conducted using a horizontal push-rod dilatometer to extract the activation energy of sintering and also gain mechanistic insights into the active diffusion mechanisms as a function of ALD film thickness. Transmission electron microscopy (TEM) and scanning electron microscopy (SEM) were used to characterize the as-deposited thin Al_2O_3 films and the effects of the Al_2O_3 films on grain growth and microstructure during densification. Ionic conductivity and flexural strength tests were conducted to determine the effect of Al_2O_3 thin films on relevant material properties. Al_2O_3 incorporation by ALD is shown to have unique phenomenological effects on the densification of Y-TZP and YSZ as compared with conventional Al_2O_3 incorporation using particle-based approaches.

508644 Effect of Particle Size Distribution and Operating Parameters on Conduction and Convection Heat Transfer Mechanisms in Rotary Drums

Abstract:

Granular materials undergo process steps that include transportation, drying, heating and chemical or physical conversion. In several instances, these materials need to be heated or cooled during processing, and rotary drums (kilns) are the most commonly used process equipment for this purpose. It is not always possible to conduct experiments using different materials and consider all process conditions to study the thermal mechanism due to the inherent complexity of these processes. Therefore, computational modeling and simulations have been extensively used to understand the heat transfer mechanism in rotary drums. To maintain simplicity and to overcome the challenges of simulating polydispersed particles, a common practice is to assume that the polydispersed particles with a narrow size distribution behave like monodispersed particles thermally. However, no effort has been made to validate this assumption. This work focuses on understating the effect of polydispersed particles on the conduction and convection heat transfer mechanisms in indirectly heated rotary drums. It also aims to investigate the effect of rotation speed, fill level, particle size and their interaction effects on the conduction and convection heat transfer mechanisms in rotary drums. For this, the discrete element method (DEM) (using MFIX-DEM, an open-source computational fluid dynamics solver suite) simulation technique is used to analyze the thermal behavior of spherical silica beads, which are the most common catalyst support material.

507974 Selective Hydrogenation of Citral over Supported Pt Catalysts on Various Substrates

Abstract:

Highly dispersed platinum (Pt) nanoparticles (NPs) were deposited on various substrates by atomic layer deposition (ALD) for the reaction of citral hydrogenation. The substrates included multi-walled carbon nanotubes (MWCNTs), silica gel (SiO_2), commercial Î³-Al_2O_3, and ALD-prepared porous Al_2O_3 particles (ALD-Al_2O_3). Based on TEM and CO-chemisorption analysis, the Pt NPs on all substrates were around 1.3 nm, and Pt/MWCNTs, Pt/ALD-Al_2O_3 and Pt/Î³-Al_2O_3 showed similar Pt dispersion (75%-85%). But the Pt dispersion on SiO_2 was low (63%). In the reaction of citral hydrogenation, Pt/SiO_2 showed the highest activity due to the strong acidity of SiO_2 and the conversion of citral reached 82% after 12 hr with a selectivity of 58% of unsaturated alcohols (UA), including geraniol and nerol. Among all catalysts, Pt/MWCNTs showed the highest selectivity of UA (65%) due to its unique structure and electronic effect. In the cycling test, Pt/MWCNTs and Pt/ALD-Al_2O_3 catalysts were more stable than Pt/SiO_2 which could be attributed to the different interactions between the catalysts and the solvent.

508345 Numerical Simulation of Particle Classification in a Classifier Based on Coanda Effect

Abstract:

Efficient and precise classification of particles has been an important issue in many industrial applications as well as in academia. A relatively new particle classifier based on 'Coanda effect' was developed and reported to have many favorable features (Liang 2010). They include simultaneous multiple classification, simple design with no moving parts, and easy scale up to higher production rates. However, this classifier has attracted little attention so far, and our understanding of underlying mechanism associated with classification performance needs to be improved.

The purpose of this study is to investigate the detailed fluid flow and particle transport characteristics in a Coanda classifier using CFD (Computational Fluid Dynamics). The main focus is placed on the effects of working conditions (inlet flow velocity, particle diameter) and numerical method for turbulence dispersion of particles on the classification performance. As a numerical method, three different approaches are considered: k-ε model, k-ε model combined with DRW (Discrete Random Walk) model, LES (Large Eddy Simulation). In the first approach, standard k-ε model is adopted to obtain the mean flow and then particle trajectory is calculated based on the mean flow without consideration of turbulence dispersion. However, in the second approach using DRW model, the instantaneous flow velocity is constructed under the assumptions of isotropic turbulence in order to include turbulence effect. Finally, in LES, the instantaneous flow field is dynamically and accurately obtained at each time step. Numerical simulations are carried out for two-dimensional mean flow in case of k-ε model, but for three-dimensional unsteady flow in LES. Particles are tracked using a Lagrangian approach based on one-way coupling. We consider talc particles with density of 2,860 kg/m^3 and diameter of 1 to 20 μm.

The Coanda classifier is mainly composed of three inlet channels, three outlet channels, and Coanda block. Inlet channels are divided into one main inlet and two control inlets. The main air jet containing particles is supplied with the velocity of 35 m/s through the main inlet. Two control air jets with velocities of 32 m/s (lower) and 40 m/s (upper) are also supplied through the control inlets for enhanced control.

The trajectories of moving particles obtained from the mean flow (k-ε model) are analyzed for different diameters. The main jet containing particles deflects toward the curved Coanda block due to Coanda effect as expected. Fine particles ($d \leq 2\mu m$) follow the flow streamlines and enter the lower outlet channel. However, coarse particles ($d \geq 8\mu m$) are detached from the streamlines and go into the upper outlet channel. This observation agrees well with the previous result (Liang, 2010).

Even though the mean flow and corresponding particle trajectories provide useful information, realistic particle passage do not take a form of single curve but they change in time due to flow turbulence. Therefore, particle trajectories are simulated again, turbulence dispersion being considered with DRW model. Due to turbulence dispersion, particles of the same diameter are scattered around the particle pathline obtained from the mean flow. As a result, particles of the same diameter can not be collected into a single outlet. For example, some particles (about 5%) of 1 μm enter the middle outlet even though most of them (about 95%) enter the lower outlet designed for collecting fine particles. This means that flow turbulence deteriorates the accuracy of classification.

Finally, LES simulations are performed to understand dynamic behaviors of fluid flow and particle transport. Instantaneous flow fields such as vorticity contours clearly show unsteady fluid motions inside the classifier. It is also shown that the generation of unsteady vortices leads to the scattering of particle trajectory.

A more detailed analysis including the classification efficiency will be given in the final presentation.

Acknowledgements

This research was supported by the National Research Foundation of Korea (NRF) under grant number 2017M2A8A4018482 and 2015R1D1A1A0105 9675.

References

Liang, S. (2010). *Chem. Eng. Comm.*, **197**, 1016-1032.

503491 Numerical Simulation of the Secondary Air Distribution Layer in the Gas Solid Fluidized Bed Based on Flunt Software

Abstract:

Due to the continuity beneficiation in the practical production, the gangue would accumulate on the distributor forming the gangue layer, which was called Secondary Air Distribution Layer and has vital effect on the gas distribution. In this paper, the ventilation characteristics of the secondary air distribution layer was analyzed. In addition, FLUENT software was used to study the effect of the modeled layer for the fluidity characteristics (e.g., pressure drop, bed height, volatility, etc.) in the gas-solid fluidized bed with the complex layer. The results showed that the complex layer could promote the secondary distribution of the airflow in the bed, which effectively inhibited the production of large bubbles and reduced the bed pressure drop fluctuation. Thus, itâ€™s of significance to provide synergy effect on the fluidization stability in the fluidized bed. Based on the above analysis, the introduced complex structure would be convenient to achieve the coal separation effectively.

507949 CFD-DEM Simulation of the Fluidization of Non-Spherical Particles in Fluidized Bed

Abstract:

In chemical and energy industries, fluidization concerning non-spherical particles (such as rod-like particle, disk-like particles, ellipsoidal particles, etc.) is often encountered, but research in this area is minimal. Yet understanding fluidization of non-spherical particles is crucial in many relevant industrial process. To investigate the fluidization of non-spherical particles, a CFD-DEM coupling algorithm is employed in this research to simulate a bubbling fluidized bed. The non-spherical particles are modeled by super-ellipsoids, multi-sphere models, and other particle models. The simulations results show that, the fluidization of non-spherical particles has significant differences from that of the spherical particles. The local void fraction and particle's projected face area perpendicular to fluid flow direction are directly related to the orientations of the non-spherical particles, which has a significant influence on the particle drag force and fluidization. Due to the significance of particle orientation, we further investigate the factors that influence the distribution of particle orientations. The results show that particle orientation depends on the fluidization time, fluidization velocity, and structure of the fluidized bed.

508320 CFD and DEM Simulation of the Cold Spray Process for Surface Coating with Fine Particles

Abstract:

The coating and functionalization of technical surfaces has been become increasingly important in the last two decades for various applications, since of its manifold possibilities for process optimization, as friction reduction, corrosion resistance or improved wetting behavior. Cold spray is a thermal spraying method for surface coating. It was developed in the late 1980s in Novosibirsk [1] and has gone through continuous optimization on certain applications since then. For example showed our previous study a reduction of the friction properties on steel pins, microstructured with fine Titania particles by an adapted spraying method [2]. In the cold spray process particles are dispersed in a gas stream, which is heated up to several hundred degrees and accelerated to supersonic velocities in a Laval nozzle. The particles impact on a substrate with about 800 m/s and form a strong bonding due to the high kinetic energy which is transferred to deformation and heat in the contact zone. In contrast to other thermal spraying methods, the temperatures occurring during the process are far below the melting temperatures of the used materials, which makes the process suitable also for thermal sensitive powders. The quality of the produced coating layers depends highly on the properties of the powder and substrate as well as on the nozzle design and the process parameters as initial temperature and particle velocities. Due to the high gradients, the high occurring velocities and the small size of the particles, typically in the range of 1 to 50 µm, the contact zone of impacting particles can only be measured indirectly and with destructive methods. Therefore, still not all phenomena which lead to a strong bonding in the contact area and a high quality coating are completely understood. For the study of these mechanisms two different simulation approaches have been used:

a) Simulation of particle and gas flow with Computational Fluid Dynamics (CFD)

For a better understanding of the spray process, the nozzle, the adjacent flow areas and the substrate were computed with a CFD method. The trajectories of single particles were calculated by an Euler-Lagrange approach. For the flow the k-Ï‰ SST model was used. Local temperatures in the substrate were calculated by the law of heat conduction. The impact conditions dependent on the nozzle geometry and the particle properties. With the local temperature distribution in the substrate and in the particle, precise impact conditions could be estimated. A study has been conducted to vary the parameters of the spray process, temperature and pressure as well as the particle properties as shape factor, density and size. Moreover the nozzle geometry has been varied in the simulation to optimize the spray process. The simulation results are compared to images of coated surfaces obtained by Scanning Electron Microscopy (SEM) for validation.

b) Simulation of particle-surface impact behavior with Discrete Element Method (DEM)

The DEM is widely used for investigations of particular processes. The advantage of this method is the detailed calculation of particle-particle or particle-wall contacts to obtain the important parameters as forces and deformations. The contact force in the particle wall contact is also dependent on the surface morphology. Therefore the surface topography of two real technical surfaces, a polished and an unpolished Titanium substrate were measured by Scanning Probe Microscopy (SPM) imaging with a nanoindenter (TI Premier, Hysitron). The surface topography was discretized and implemented in a DEM model. A parameter study has been conducted to determine the distribution of forces and contact angles occurring at the particle impact. Different particle sizes and impact velocities were studied.

The cold pray process dependents on various parameters which affect each other. Since of the high gradients and the small scales occurring during the process experimental measurements are often difficult. With this numerical study different parameters could be considered uncoupled of the other parameters. The results can be used for a better process understanding and an optimization of commercial cold spray systems.

[1] Alkhimov, A. P., Papyrin, A. N., Kosarev, V. F., Nesterovich, N. I., Shuspanov, M. M.: Gas dynamic spraying method for applying a coating; US Patent, 5302414; 1994

[2] Buhl, S., Schmidt K., Sappok D., Merz, R., Godard, C., Sauer, B., Kerscher, E., Kopnarski, M., Antonyuk, S., Ripperger, S.: Surface structuring of case hardened chain pins by cold-sprayed micro particles to modify friction and wear properties; Particuology 21, 2015

508568 Two-Fluid Validation of Constituive Models for the Simulation of Cylindrical Particles

Abstract:

Two-fluid simulations are a popular tool in the modeling of large processes. The constitutive models that close the governing equations are based in granular kinetic theory and typically assume that the particles are spherical. In reality, very few processes utilize spherical particles, hence there is a need for constitutive models that describe more complex shapes. Previous work used discrete element method (DEM) simulations to create a solid stress[1,2] and collisional dissipation rate[3,4] model for cylindrical particles that can replace the current spherical models. These new models have been added to MFiX and this work focuses on their validation. Cylindrical particles of various aspect ratios and solid densities were used in cratering and hopper discharge experiments. Simulations are designed to recreate the experiments in order to compare the results for crater depth and hopper discharge rate. The model effectiveness is assessed with regards to assumptions made concerning particle rotation and preferential alignment.

1. Guo Y, Wassgren C, Ketterhagen W, Hancock B, James B, Curtis J. A numerical study of granular shear flows of rod-like particles using the discrete element method. *J Fluid Mech*. 2012:1-26.

2. Berzi D, Thai-Quang N, Guo Y, Curtis J. Stresses and orientational order in shearing flows of granular liquid crystals. *Phys Rev E*. 2016;93(4):40901.

3. Buettner KE, Guo Y, Curtis JS. Using the Discrete Element Method to develop Collisional Dissipation Rate Models that Incorporate Particle Shape. *AIChE J*.

4. Berzi D, Thai-Quang N, Guo Y, Curtis J. Collisional dissipation rate in shearing flows of granular liquid crystals. *Phys Rev E*. 2017;95(5):50901.

509002 The Comparsion of Coarse Grained CFD-DEM for Simulating the Dense Bubbling Fluidized Bed" Requires Immediate Action

Abstract:

CFD-DEM is an effective tool to study the hydrodynamics in dense bubbling fluidized bed. However, the expansive cost of calculating the particles motion is not easy for its application in industrial application or dense particles system. Thus several coarse grained models (CGM) were put forward to simplifying the calculation of particles and decreasing the particles motion, such as PSM, SPA, MSPA,CGEE, EMMS-DPM where a group of particles are set as a parcel or coarse particle to reduce the computation. In this study the experiment and simulation test conducted to test five coarse grained methods. The result indicated that the models increasing the gravity force, pressure gradient force and drag force linearly with the coarse grained ratio (CGR) can be effective to ensure similar macro behavior between original system and coarse graining system. The simulation data of the solid fraction distribution and bed expansion ratio reflected good agreement as experiment. But the bubbles structure became less clear as increasing the CGR. Furthermore, the power spectrum density (PSD) and Wavelet analysis of the pressure signal were used and showed that SPA and EMMS-DPM gave better agreement result in term of dominant frequency and energy ratio at mesoscale feature. Finally the computational efficiency for different models was compared and the models of EMMS-DPM, PSM and CGEE have higher computation efficiency than SPA and MSPA relatively if the collision model of spring-damping were adopted.

509153 Stability Analysis of Uniform Gas Solids Flow

Abstract:

Uniform distribution of gas solids flow should be ensured to gain sufficient performance of fluidized beds. The highly chaotic entropy dissipation across solids bed will trigger non-uniform fluidization, although distributor plays an important role in obtaining homogenous gas stream. Hence, total solids pressure loading (Î¦$_T$), the pressure drop ratio between solids bed and distributor, is introduced here to present their interaction. The boundary between uniform and non-uniform fluidization can be detected by stability analysis based on Prigogineâ€™s minimum of entropy production principle, and this criterion for *N* parallel paths can be simplified into calculating the second derivative of pressure drop through one path with respect of Î¦$_T$. Furthermore, a phase diagram, illustrating the effects from operational parameter (Î¦$_T$), geometrical characteristics of distributor (C_d) and property of solids (*Ga*) on instability of uniformity, is attained to draw some useful conclusions of robust uniform fluidization.

509309 Numerical Simulation of Particle Sedimentation Related to Nuclear Safety By CFD-DEM Algorithm

Abstract:

Numerical Simulation of Particle Sedimentation related to Nuclear Safety by CFD-DEM Algorithm

Byoungcheol Hwang[1], Kiyofumi Moriyama[1], and Hyun Sun Park[1*]

[1]Division of Advanced Nuclear Engineering (DANE), POSTECH, Pohang, Korea

**corresponding author: hejsunny@postech.ac.kr*

Abstract

The Discrete Element Method (DEM) is a technique to treat particle-particle interactions, and is used in many industries such as civil engineering, powder processing, pharmacy, and so on. Coupling of this method with Computational Fluid Dynamics (CFD) is needed in problems where the particles interacts with fluid significantly, such as bubble column reactor, mineral processing, and so on.

The simulation of important phenomena in nuclear power plant accidents that involves significant core damage and melting, i.e. so called severe accidents, is also one of the topic where the CFD-DEM is useful. After the Fukushima accident in 2011, the severe accident analysis and preparation of counter measures to mitigate the accident consequences became regulatory requirements in many countries. One of the measures for the cooling and stabilization of the molten core is the flooding of the cavity under the reactor vessel where the dropping core melt is caught in a water pool. It is expected that the molten core breaks up into particles of several millimeter sizes and accumulates on the floor, then makes a particle bed, which is hopefully coolable. In this situation, the assessment for the coolability of the particle bed is very important, and a number of experimental and analytical approached suggest that the internal/external structure of the particle bed affects strongly the coolability. The CFD-DEM is a promising candidate for an analysis method that can clarify such details of the particle bed formation.

We tested the applicability of the multiphase CFD-DEM code, an open-source developed by CFDEM®coupling in 2011, for the simulation of particle sedimentation and particle bed formation under water. Our focus in the present work is on the prediction of the particle falling characteristics and particle bed shape. Although the real condition during a severe accident includes very high temperature particles (>1000K) and boiling water (two-phase flow) in a large scale, we started with a simulation of much simple experiment cases without heat transfer or gas phase, with spherical particles only as the first step for the validation of the method. The experimental data with solid particles of various diameters (4 mm, 6 mm) and densities (3600, 6000, 7800 kg/m^3), particle release nozzle diameters (20, 30, 40 mm) were referred to. The results of numerical simulation were compared with the experimental data in terms of particle falling time, maximum particle dispersion angle, particle average falling velocity, particle bed distribution, and so on. The particle falling time was well predicted by the simulation with the maximum deviation 5%. The maximum particle dispersion angle by the simulation significantly deviated from the experimental data probably due to the influences of initial disturbances. The particle average falling velocity was overestimated. The final particle bed profile by the simulation showed agreed with the experimental observation accurately in the peripheral region, but the height of the center part was underestimated. The fluid dynamic drag model and the particle-particle contact model would need improvement for better accuracy. The present work confirmed the validity of CFD-DEM algorithm on the application for the particle sedimentation in a simple liquid pool. The particle-pool interaction including the gas phase will be the next step.

Keywords: CFD-DEM; particle-fluid interaction; particle sedimentation; nuclear safety; severe accident

509440 Application of Extedned Discrete Element Method to the Melting Process of a Packed of Particles

Abstract:

Application of eXtedned Discrete Element Method to the melting process of a packed of particles

Mehdi Baniasadi, Maryam Baniasadi, Bernhard Peters, Xavier Besseron

Université du Luxembourg, Faculté des Sciences, de la Technologie et de la Communication, Luxembourg

e-mail: mehdi.baniasadi@uni.lu, maryam.baniasadi@uni.lu, bernhard.peters@uni.lu, xavier.besseron@uni.lu, web page: http:/www.xdem.de

ABSTRACT

The phenomena of melting takes place in many industrial processes, such as metal processing, environmental engineering and thermal energy storage systems. The problem of modeling of solid-liquid-gas flow including physical and chemical conversions of particles such as melting is the subject of intense research in industrial processes. These systems are difficult to model with numerical analysis due to the complex interaction between the different fluids and granular phases condering heat, mass and momentum excahnge. Basically models used to predict multiphase flows with particles are Eulerian-Eulerian and Lagrangian-Eulerian. The difference between the two models is based on different treatment of the movement of the particles. The particles are tracked individually in Lagrangian-Eulerian approach while they are considered as a continuum medium in Eulerian-Eulerian. Lagrangian-Eulerian is closer to the real physical processes rather than the Eulerian-Eulerian model for granular flows with particleâ€"particle interaction[1].

In this paper, our main focus is on the modelling of the melting process for a packed bed of particles subject to a fluid flow using a numerical technique based on Lagrangian-Eulerian so-called eXtended Discrete Elemnet Method (XDEM)[2]. In this approach, the particles are resolved as discrete phase coupled via heat, mass and momentum transfer to the surrounding fluid phase. The Discrete Element Method (DEM) coupled by heat transfer is used in order to evaluate temperature distribution, melting rate as well as velocity, oriantion and position for each particle individually. In addition the approach allows to be linked with continues numerical approaches such as Computational Fluid Dynamics (CFD) in order to solve fluid fields. The model has been validated by comparing predicted results with existing experimental data for melting of a single ice particle as well as a packed bed of ice particles in the horizontally flowing water. Fluid-particle and particle-particle intractions has also been considered. In addition, the model has the capability to be extended to the packed bed of particles with different size and properties to produce different liquid phases.

REFERENCES

[1] Dierich, F., Nikrityuk, P. A. & Ananiev, S. 2D modeling of moving particles with phase-change effect. *Chem. Eng. Sci.* **66,** 5459â€"5473 (2011).

[2] B. Peters, Thermal Conversion of Solid Fuels (WIT Press, Southampton, 2003).

514248 Real-Time Interpretation of Chaotic Characteristics and Instability of Dense Gas Solid Fluidization

Abstract:

The establishment of instability monitoring system targets at analyzing the chaotic characteristics and evaluating the instability of the fluidization process for dense gas solid fluidization. Experimental pressure fluctuation and pressure drop measurements were interpreted from dense gas solid fluidization system based on chaotic time-series analysis. The pressure fluctuation signals were filtered in terms of attenuating the noise signals. The filtered pressure signals were analyzed with respect to attractor reconstruction. Based on the reconstructed attractor, the maximum Lyapunov exponent and entropy were investigated and estimated under different bed heights and air velocities. Results show that the maximum Lyapunov exponents udner different bed heights and air velocities are positive, which indicates that dense gas solid fluidization exhibits chaotic characteristics. With the increase of bed height, estimated entropy of dense gas solid fluidization increases with the increase of bed height under identical air velocity, while it varies slightly with the increase of air velocity when the air velocity alters from minimum fluidization velocity to 1.36 times of the minimum fluidization velocity.

508510 Glucan Particles As Potential Carriers of Natural Flavonoids for Treatment of Idiopatic Inflammation Diseases

Abstract:

One of the novel methods of pharmaceutical formulation is the loading of the drug into biocompatible microparticles, preferably extracted from natural sources. This approach offers several potential advantages, *e.g.* increased bioavailability (the ratio of the administered drug, which actually enters the bloodstream) compared to conventional drug delivery methods, enabling the option of oral administration even for parenteral drugs. Multiple principles of the loading process are possible – drug encapsulation, adsorption onto the particle surface or precipitation inside the pores a hollow particle. The presented work focuses on precipitation of a drug from an organic solvent inside glucan particles. Glucan microparticles are ellipsoid porous shells, which are obtained from cell walls of *Saccharomyces cerevisiae* (baker´s yeast). The particle size is 2-5 micrometers. Cell walls of glucan particles are mainly composed of β-1,3-glucans. In order to prevent inflammatory reaction in organism, the yeast organelles are removed by a series of alkaline and acidic extraction steps. Subsequently, present lipopolysacharides are extracted using different organic solvents.
Lymphatic cells are mainly present in inflammation area in organism where they are responsible for progress of inflammation. Glucan particles (GPs) interact with lymphatic cell´s receptors on their surface and after that β-1,3-glucans are phagocytosed by the lymphatic cells. Therefore, GPs represent an effective oral delivery system for transporting of biological active substances. Curcumin is an anti-inflammatory flavonoid, which occurs in plants of family *Zingiberacaea.* In our project, curcumin was delivered to an inflammation in organism using GPs. Loading of curcumin into GPs was enabled via precipitation of curcumin from organic solvents. Curcumin dissolved in organic solvents was added to the GPs and through pores penetrated into cavity of glucan shell. Next, the solvent was removed and the curcumin was precipitated in glucan microparticles. GPs with the flavonoid inside were observed using confocal microscopy and SEM (scanning electron microscopy). The achieved content of curcumin was successfully analyzed by other methods, which can not be mentioned directly due to article in preparation. Furthermore, the curcumin-loaded GPs were tested *in vitro* on macrophage like cells and *in vivo* on laboratory mice in order to evaluate their effectivity. The obtained results show that these composites are promising materials for biomedical applications such as efficient treatment of inflammatory diseases (e.g. idiopatic inflammatory bowel disease, rheumatoic arthritis).

508601 Use of Advanced Imaging Techniques in Tablet Disintegration Study

Abstract:

Certified testing of tablet disintegration and dissolution provides us useful information. However, there are many more that are hidden due to techniques' restrictions. Such information is e.g. mechanism of disintegration, change of surface morphology or uniformity of the substances' distribution in the tablet. Therefore, there are many techniques used to obtain the remaining information such as Magnetic Resonance Imaging (MRI), Fourier Transform Infra-Red (FTIR) or Raman spectroscopy and microscopy, Terahertz spectroscopy, Scanning Electron Microscopy (SEM) and other.

Presented work uses tablet made by direct compression of binary mixture of ibuprofen and lactose (based on previous results there was 0.2 - 0.5 w/w of ibuprofen content used) pressed at different compaction pressures (low, medium and high). The tablets were subjected to several solutions and were studied by techniques that have insight into the tablet (MRI) and ones that can see the surface (FTIR and SEM).

MRI experiments provided us with several interesting and several expectable results. As the content of ibuprofen was increasing so was increasing the dissolution time and the wetted layer was decreasing. Similar results were observed when the compaction pressure was increased. When the tablets were submerged to dissolution limiting solutions (saturated lactose solution and solution saturated with both lactose and ibuprofen) the disintegration was severely limited and some tablets began to increase their volume. However, in case of 0.2 w/w ibuprofen tablets it didn't matter whether there was dissolution discriminating solution or not, it disintegrated anyway.

FTIR microscopy showed, that not only the amount, but also the shape and size of the particles from which the tablets were compacted can be determined. These findings were further supplemented by the dry state information obtained by Scanning Electron Microscopy after different dissolution times. Further the FTIR-ATR imaging technique helped to understand the basic mechanisms of the tablet disintegration and to study the dissolution and water penetration into the model ibuprofen-lactose tablets.

508632 Correlation Study between Liquid Penetration and Mechanical Properties of Pharmaceutical Tablets

Abstract:

Main phenomenon influencing treatment of patient in pharmacy is amount of active pharmaceutical ingredient dissolved in medium such as water or blood. Even though this isa key step to achieve desirable medical effect it is not the first one occurring.The first step is governed by surface properties such as wettability and roughness and the first contact point is followed by liquid penetration which is highly dependent not onlyon type of used liquid, ionstrength, liquid density but also on mechanical and structural properties of tablets such as porosity, structure strength or particle bonds.Through many analytical methods we can describe each of thesesteps individually. Aim of this study is to find proper correlation amongseveralmethods such as sessile drop method, texture analysis (TA) and magnetic resonance imaging (MRI) of tablet during dissolution.

For experiments binary mixture of tabletwas used. Tablets consist of lactose (hydrophilic) and ibuprofen (hydrophobic)with variable content (from 0.2 to 0.5 w/w ratio) and were also variable in compaction pressure used on them (from low â€“ 5 kN through 10kN as middle to 18 kN as high).Three different solutions were used according to their saturation with primary substances to prevent the tablet from dissolving. Surface properties were evaluated on contact angle and soaking rate of liquid, both measured by sessile drop method. Texture analysis was performed by recording the material resistance during dry and wet state of tablet. Then the data was processed and trends were obtained.Tablets during the MRI experiments were submerged into dissolution medium and the proceeding wetting front was measured. Later the images were processed via image analysis and trends were obtained.

Results obtained from surface experiments show that higher the ibuprofen content is the more hydrophobic surface appears to be and therefore the contact angle is increasing. Contact angle is independent on manufacturing pressure but soaking rate on the other hand depends strongly.MRI images show that this technique is not suitable for rapidly disintegration tablets, where the calculated error is within experiment conditions. On the other hand, tablet with prolonged dissolution are well measured. Results from MRI experiments also state that local saturation with substances supress dissolution of tablet. Texture analysis of tablets show that they undergo immediately structural weakening after being poured with solution independently on type of solution. Rate of this structural change mainly depends on solution's acidity. Thenthe tablet decreases its resistance in time which can be interpreted as relaxation rate. With increment of ibuprofen or saturation with primary substances this rate decrease, which means that dissolution is suppressed. Correlation between MRI and relaxation rate was done. Results shows that the correlation is mainly dependent on texture analysis measurement. On the other hand, correlation between MRI and soaking rate appears to be more tentative.

Experiment characterising surface, structural and liquid penetration properties were investigated by three different techniques. Obtained results were evaluated and put to mutual correlation. From liquid penetration point of interest there is correlation between MRI and contact angle method. Also, a correlation between MRI and texture analyser was found.

508642 In Situ Drug Amorphisation By Microwave Irradiation Stabilized By Mesoporous Silica

Abstract:

One of the common issues in the pharmaceutical industry is the poor solubility of the newly developed active pharmaceutical ingredients (APIs). The solubility might be significantly enhanced by changing the form of the API from crystalline to amorphous form. The amorphous state is characterized by lower thermodynamic stability and thus better dissolution properties leading to better bioavailability. In addition, the amorphous API in a dosage form (e.g. tablets) must also remain stable and not recrystallize for at least the shelf life of the drug product (usually 2 - 3 years). Various methods are being utilized to prepare and stabilize amorphous formulations in the industry, however they usually involve difficult formulation steps or suffer from poor efficiency, low stability or can be patent obstructed.

In this work a stable amorphous formulation is prepared simply using the standard wet and dry granulation process or direct tableting. Crystalline form of API is used during the initial process. The amorphous form is then prepared *in situ* and â€œon demandâ€ either during manufacturing, or by the end user by applying microwave radiation (MW).

To stabilize the API in an amorphous form after the MW, multi-structured mesoporous silica particles were added to the formulation. The particles were prepared using an emulsion synthesis of octylamine in tetraethyl orthosilicate. Using ethanol as an inhibitor instead of the commonly used nitric acid as a catalyst reduces the silica primary nano-particles nucleation rate. The formed nuclei grow larger and then self-assemble around the octylamine droplets and form hollow micro-particles. The prepared microparticles were characterized using the Scanning electron microscopy (SEM) and the Transition electron microscopy (TEM). To characterize the porous structure, nitrogen sorption (BET) analysis was conducted. The results clearly show the particle assemblies contain macropores (cavities), mesopores and micropores and possess very high specific surface area and also very large pore volume. Together with the thermal and chemical stability of silica these properties make the particles well suitable for preserving the melted API in an amorphous form after the MW.

As a proof of concept, initially powder mixtures and later simple tablets composed of Ibuprofen (model API), silica micro-particles and a filler excipient were experimentally investigated. The *in situ* amorphization of ibuprofen was conducted using a commercial microwave oven. The heating properties of the oven, i.e. the spatial distribution of microwave heating was characterized. Various microwave power input and varying composition of the mixtures (excipient/Ibuprofen ratio) were tested with respect to the melting point of the Ibuprofen. The change from the crystalline to amorphous structure due to MW heating was analyzed by x-ray diffraction (XRD), differential scanning calorimetry (DSC) and FTIR-ATR techniques. The dissolution profiles of ibuprofen from both MW untreated and treated tablets were measured using the High Performance Liquid Chromatography (HPLC).

Based on these initial results, fully formulated tablets were prepared with various API dosage strength using a tablet press. Same compression force and same dwell time was used for all of the tablets. The final tablet composition includes excipient capable of absorption of the MW, which ensures a uniform heating in the whole volume of the tablet. Next it contains the silica microparticles, which by capillary forces absorb the API once melted and stabilize it after cooling down.

Lastly, the tablets contain standard filler, lubricant and disintegrant excipients. After the MW application the tablets change their structure due to the melted API intake into the silica micro-particles, forming visible pores in the tablet structure. This has a notable impact on the dissolution and hardness of the resulting tablets. For the final tablet characterization SEM, Texture Analysis, FTIR and Magnetic Resonance Imaging were used.

508646 Manufacture of Personalised Medicines By API Printing on Porous Tablets

Abstract:

Personalised medicine offers an opportunity to produce tailor-made medicaments for patients. In order to realize this vision, it is necessary to find an alternative approach to the conventional tablet design and production. One that will replace the existing mass production of drugs offering only a few dose options of the active pharmaceutical ingredient (API), while at the same time it must meet the strict reproducibility and a defined relationship between composition and release rate of active substances requirements.

One of such approaches is to print the API on demand on a solid substrate directly when it becomes needed. Appropriate dosing of API and the possibility to even combine several APIs into a single medicament improves the variability of the dosages available for prescription and reduces the overall number of pills the patient has to take. The goal of this work is to find a simple, automated and cost-effective method of preparing personalised medicaments, which could take place directly in the pharmacy shops or at home.

The method in this work consists of a precise dosing of a dissolved API into an already prepared substrate â€“ porous placebo tablets with high sorption capacity. The tablets are prepared using direct compression in a tablet press from a simple physical mixture of a binder, filler, disintegrant and lubricant. As a filler, mesoporous silica particles were used. The high pore volume provides the free space for API sorption into the tablets and the small size of the mesopores ensure the API is deposited in an amorphous form.

The prepared particles used in the formulation were characterized using Transmission electron microscopy (TEM), Dynamic light scattering (DLS) and nitrogen sorption analysis (BET). The prepared tablets hardness and uniformity was characterized both prior and after loading. The tablets after loading were characterized using the X-ray powder diffraction and the Fourier transform infrared spectroscopy to determine the API form and deposition uniformity, followed by dissolution tests to confirm the applied dosage strength and to measure the release profile. So far, several BCS Class II APIs were successfully infused into the substrates using precise on-demand dosing. Impregnation of various doses of API were accurate. Later characterization revealed no crystalline API were present and the dissolution profiles were very satisfactory.

509441 Understanding Phase Transition of Acetaminophen in the Bulk and Surface of Acetaminophen

Abstract:

Pharmaceutical solids processing often lead to the generation of new phases in the bulk and on the surface of the processed drugs. Phase transformation of the API can potentially change the physiochemical properties and influence the drug performances including solubility, dissolution and stability. To detect these changes, analytical tools such as XRD are usually employed. However, these techniques remain suitable for bulk analysis. Whereas, phase transition on the surface remains hard to detect due to the lack of literature, researches in this area and the absence of surface analytical techniques. The aim of this work is to investigate different phases (pure crystalline, partially crystalline partially amorphous and amorphous) of Acetaminophen drug by combining bulk and surface analytical techniques. For bulk analysis, FTIR and XRD have been employed. The former shows no significant differences between Acetaminophen various phases. Whereas the latter confirmed the presence of different phases indicated by a remarkable variation in the absorbance intensity the peak width. Where, high and sharp peaks obtained for the pure crystalline Acetaminophen, fewer peaks observed for semi crystalline Acetaminophen and a single broad peak observed for the amorphous Acetaminophen. For the surface analysis, Nitrogen K near-edge X-ray absorption fine structure (NEXAFS) spectra of a series of NEXAFS indicates that the precise determination of IPs and Ï*f** shape resonance energies is sensitive enough to distinguish between the different phases of Acetaminophen. DFT calculations provides information on the molecular orbitals involved during electronic excitation occurred as a result of NEXAFS experiment.

509599 Treatment of Cystinosis through Vitamin E Modified Silicone Hydrogel

Abstract:

Cystinosis is an orphan disease caused by a genetic mutation that leads to deposition of cystine crystals in many organs including cornea. Ophthalmic manifestation of the disease can be treated with hourly instillation of cysteamine eye drops. The hourly eye drop instillation is tedious to the patients leading to poor compliance and additionally, significant degradation of the drug occurs within one week of opening the bottle, which further complicates this delivery approach. Our work focuses on designing a contact lens to treat the disease with improved efficacy compared to eye drops. Our goal is to design a lens that is safe and that can deliver a daily therapeutic dose of cysteamine to the cornea while retaining drug stability. We show that cysteamine diffuses out rapidly from all lenses due to its small size. The rapid release of cysteamine and other ophthalmic drugs can be prolonged significantly by integration of vitamin E aggregates inside of the polymer matrix of silicone hydrogels. The increase in release duration increases as square of the volume fraction of the loaded vitamin E fraction which leads to high efficacy and suggests that the mechanism of attenuation in drug release is formation of nanosized barriers. The increase in duration depends on the type of the contact lens. The barriers are not effective in single-phase hydrogel lenses while the effect in the bi-phasic silicone hydrogels depends on the type of the contact lens. Vitamin E incorporation increases the release duration of both ACUVUE®OASYS® and ACUVUE® TruEye™ but the effect is more pronounced in TruEye™ likely due to the low solubility of vitamin E in the lens matrix and higher aspect ratio of the barriers. This result along with the high aspect ratio seen through SEM imaging in silicone hydrogels suggests that barriers could be forming at the interface of the silicone and hydrogel phases. The presence of vitamin E has an additional beneficial effect of reduction in the oxidation rates, likely due to a transport barrier between the oxygen diffusing through the silicone channels and drug located in the hydrogel phase. Based on this study, both Acuvue®OASYS® and ACUVUE® TruEye™ can be loaded with vitamin E to design a cysteamine eluting contact lenses for effective therapy of cystinosis. The lenses must be worn for about 4-5 hr. each day, which is less than the typical duration of daily-wear. Vitamin E loaded contact lenses could be very useful for treating many other ophthalmic diseases as well.

517683 Systems Integration for Dry Granulation Based Continuous Pharmaceutical Tableting

Abstract:

Over the past decade, the pharmaceutical industry has seen significant progress in the mechanistic understanding of relevant unit operations, availability of multiple sensor technologies for inline monitoring of critical quality attributes and advances in data analytics to support manufacturing decision-making. These have enabled a systematic realization of continuous manufacturing for pharmaceutical tablets using the direct compaction route. However, the limitations and difficulty in handling of some particulate materials provide the impetus for investigating granulation routes to the continuous manufacturing of tablets.

In this work, the integration of dry granulation into a continuous tableting line is investigated. Specifically, the Process Analytical Technology (PAT) tools required for inline and at-line monitoring of the tableting line, and the semi-mechanistic models for the roller compactor and the tablet press are discussed. While the Kawakita model is used for the tablet press, the roller compaction process model is consisting of an integration of material balance and actuator dynamics from Hsu et. al (2010) with the modified Johansonâ€™s model presented by Liu & Wassgren (2016). The PAT tools, mechanistic equations, and automation system are integrated for measurement validation and data accuracy using data reconciliation in a structured ISA-95 network using DeltaV. The experimental data is presented from the operation of the continuous dry granulation line-based pilot plant at Purdue University.

509248 Modeling Granular Material Mixing and Segregation Using a Multi-Scale Model

Abstract:

A multi-scale model is presented for predicting the magnitude and rate of powder mixing and segregation in blenders. The model combines diffusion and percolation correlations from small-scale discrete element method (DEM) simulations with advective flow field information from large-scale finite element method simulations. The multi-scale model predictions closely match results from both large-scale DEM-only simulations and experiments, but they are much more computationally efficient than DEM-only simulations. Therefore, in contrast to DEM, the multi-scale model is well suited for predicting blending in industrial-scale systems.

Modelling Granular Media using Dynamical Density Functional Theory

T. Hurst, B. Goddard & R. Ocone

Motivation

Granular media is ubiquitous in industrial and physical problems. One of the characteristic qualities of granular media at the microscopic level is inelastic hard body collisions, which drive macroscopic behaviour. Modelling granular media is a complex and interesting problem. Solid-like phenomena make it difficult to approximate a granular system with a continuum model, while chaotic behaviour and inelastic collapse in hard particle systems introduce modelling challenges on the microscopic scale. We investigate uses of dynamical density functional theory (DDFT), augmented with data gained from microscopic models, for granular media simulation.

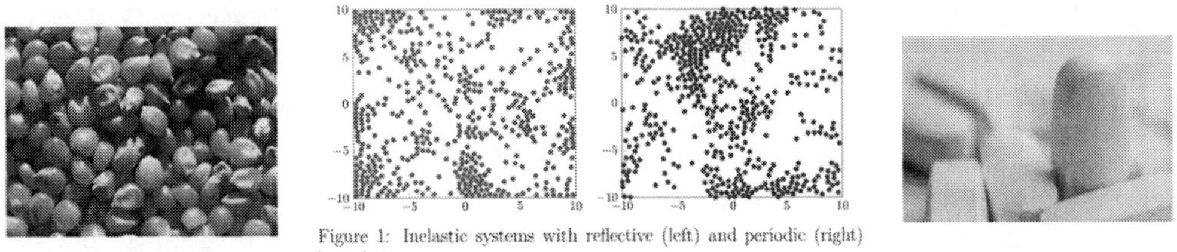

Figure 1: Inelastic systems with reflective (left) and periodic (right) boundaries

Dynamical Density Functional Theory

To derive a DDFT, we start with a set of Langevin equations for N hard particles of diameter σ in d dimensions with positions $\boldsymbol{r}^N$ and momenta $\boldsymbol{p}^N$:

$$\frac{d\boldsymbol{r}^N}{dt} = \frac{\boldsymbol{p}^N}{m}, \quad \frac{d\boldsymbol{p}^N}{dt} = -\nabla_{\boldsymbol{r}^N} V(\boldsymbol{r}^N, t) - \gamma \boldsymbol{p}^N + (m k_b T \gamma)^{1/2} \boldsymbol{G}(t), \tag{1}$$

on the hard sphere domain, $\mathcal{D}_N = \{(\boldsymbol{r}^N, \boldsymbol{p}^N) \in \mathbb{R}^{2dN} : \forall i, j, \|\boldsymbol{r}_i - \boldsymbol{r}_j\| \geq \sigma\}$, where $V(\boldsymbol{r}^N, t)$ includes external and interparticle potentials, and $\boldsymbol{G}(t)$ is an optional stochastic force term. The Kramer's equation is then a partial differential equation for its distribution function $f^{(N)}(\boldsymbol{r}^N, \boldsymbol{p}^N, t)$:

$$\frac{\partial f^{(N)}}{\partial t} + \frac{1}{m} \boldsymbol{p}^N \cdot \nabla_{\boldsymbol{r}^N} f^{(N)} + \nabla_{\boldsymbol{r}^N} V \cdot \nabla_{\boldsymbol{p}^N} f^{(N)} = \gamma \nabla_{\boldsymbol{p}^N} \cdot \boldsymbol{p}^N f^{(N)} + \gamma k_b T \nabla^2_{\boldsymbol{p}^N} f^{(N)}. \tag{2}$$

We integrate out all but one position and momenta in the Kramers equation, to find:

$$\frac{\partial}{\partial t} f^{(1)} = \mathcal{L}_1 f^{(1)} + \mathcal{L}^{2,\ldots,N}_{\text{part}}(f^{(k)}) + \mathcal{L}^{(1)}_{\text{diss}} f^{(1)} + \mathcal{L}_{\text{coll}}(f^{(2)}), \tag{3}$$

where $\mathcal{L}_1$ introduces effects from external potentials, $\mathcal{L}^{(1)}_{\text{diss}}$ effects due to dissipation and friction, $\mathcal{L}^{2,\ldots,N}_{\text{part}}$ effects from interparticle potentials, and

$$\mathcal{L}_{\text{coll}}(f^{(2)}) = -\sigma^{d-1} \int_{\mathbb{R}^d} d\boldsymbol{p}_1 \int_{\mathbb{R}^d} d\boldsymbol{p}_2 \int_{S_1^{d-1}} (\boldsymbol{p}_1 - \boldsymbol{p}_2) \cdot \omega f^{(2)}(\boldsymbol{r}_1, \boldsymbol{r}_1 + \sigma\omega, \boldsymbol{p}_1, \boldsymbol{p}_2, t) d\omega, \tag{4}$$

is a general collision term that arises due to the hard-sphere domain, when assuming only binary collisions. By taking the first two $\boldsymbol{p}$-moments and upon additional assumptions we arrive at a closed set of equations to model the dynamics of the system, a DDFT [1, 2]:

$$\frac{\partial}{\partial t} \rho = -\nabla \cdot (\rho \boldsymbol{v}), \tag{5}$$

$$\left(\frac{\partial \boldsymbol{v}}{\partial t} + \boldsymbol{v} \cdot \nabla \boldsymbol{v}\right) + \gamma \boldsymbol{v} = -\frac{1}{m} \nabla \frac{\delta \mathcal{F}[\rho]}{\delta \rho} - \sigma^{d-1} g_\sigma \int_{S_1^{d-1}} \rho(\boldsymbol{r} + \sigma \boldsymbol{\omega}/2) \rho(\boldsymbol{r} - \sigma \boldsymbol{\omega}/2)$$
$$\times \boldsymbol{\omega} \cdot \left\{ \frac{k_B T}{m} \mathbf{1} + \boldsymbol{v}(\boldsymbol{r} + \sigma\boldsymbol{\omega}/2) \otimes \boldsymbol{v}(\boldsymbol{r} + \sigma\boldsymbol{\omega}) - \boldsymbol{v}(\boldsymbol{r} + \sigma\boldsymbol{\omega}/2) \otimes \boldsymbol{v}(\boldsymbol{r} - \sigma\boldsymbol{\omega}/2) \right\} d\boldsymbol{\omega}, \tag{6}$$

where $\mathcal{F}$ is the well-studied Helmholtz free energy operator, which can include particle-particle interactions and external potentials, ρ is the density, $\boldsymbol{v}$ is the local velocity and g_σ is the value of the correlation function at σ, which must be determined. We choose to model g using statistical data from microscopic simulation.

DDFT Simulations

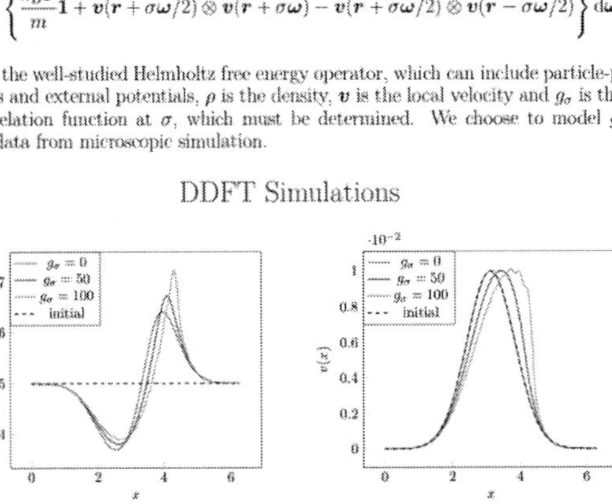

Figure 2: A simple 1D system neglecting all but the collision operator in (5-6), for different values of g. Left: density $\rho(x)$, right: local average velocity $v(x)$.

Data-Driven Modelling of g

A system of N hard spheres of diameter σ in d dimensions with positions $\boldsymbol{x}^N$ and velocities $\boldsymbol{v}^N$ obey

$$\frac{d\boldsymbol{r}^N}{dt} = \boldsymbol{p}^N, \quad \frac{d\boldsymbol{p}^N}{dt} = V(\boldsymbol{r}^N, t), \tag{7}$$

with a pairwise collision condition: if particles i and j are in contact, i.e. $\|\boldsymbol{r}_i - \boldsymbol{r}_j\| = \sigma$ and $(\boldsymbol{r}_i - \boldsymbol{r}_j) \cdot (\boldsymbol{p}_i - \boldsymbol{p}_j) < 0$, we instantaneously change the velocities via the rule:

$$\boldsymbol{p}_i^{out} = \boldsymbol{p}_i^{in} - \frac{1+\alpha}{2} \boldsymbol{\nu}_{ij} \cdot (\boldsymbol{p}_i^{in} - \boldsymbol{p}_j^{in}) \boldsymbol{\nu}_{ij}, \quad \boldsymbol{p}_j^{out} = \boldsymbol{p}_j^{in} + \frac{1+\alpha}{2} \boldsymbol{\eta}_{ij} \cdot (\boldsymbol{p}_i^{in} - \boldsymbol{p}_j^{in}) \boldsymbol{\eta}_{ij}, \tag{8}$$

where $\boldsymbol{\eta}_{ij} = \frac{\boldsymbol{r}_i - \boldsymbol{r}_j}{\|\boldsymbol{r}_i - \boldsymbol{r}_j\|}$ and $\alpha \in [0, 1]$ is the coefficient of restitution. For simple V, the Newton equations can be modelled using Event Driven Particle Dynamics (EDPD). The cell method and bounded increasing priority queues [3] reduce the computational effort to $\mathcal{O}(1)$ per collision. Using statistical methods for synthetic data (e.g. emulation [4]) we gain an understanding of the correlation function g near equilibrium, which we can include in (6-7).

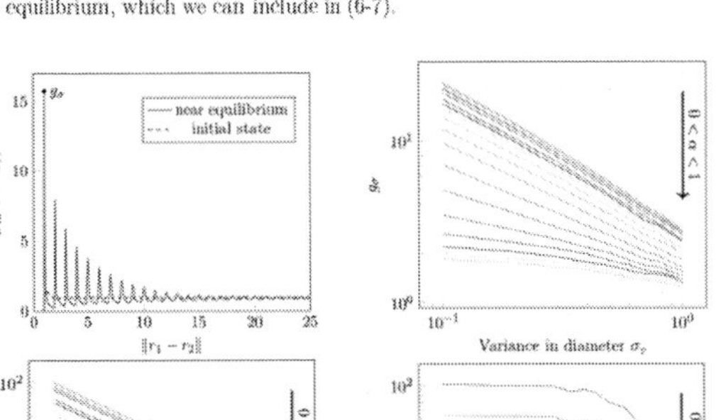

Figure 3: Top left: $g(\|r_1 - r_2\|)$ near equilibrium for a 1D system with $\rho = 0.5$ and $\alpha = 0.5$, using 10000 samples of EDPD simulations. The other three graphs show how g_σ changes for different parameters.

Future Work

We wish to include additional moments in the DDFT derivation, simulate DDFT in higher dimensions including interparticle potentials, and use EDPD in higher dimensions to calculate g.

[1] Archer, A., *J.Chem.Phys.*, **130**, 014509, (2009).
[2] Goddard, B. et. al., *Phys.Rev.Lett*, **109**, 1-5, (2009).
[3] Paul, G., *J.Comp.Phys.*, **221**, 615-625, (2011).
[4] O'Hagan, A., *Reliab.Eng.Syst.Saf.*, **91**, 1290-1300, (2006).

The support from EPSRC (grants EP/N034066/1, EP/L025159/1 and EP/L016508/01) is kindly acknowledged.

509555 A Study on Partially Wetted Particle Collisions with a Wet Wall

Abstract:

For the production of poly-ethylene, a gas-phase polymerization reaction is executed in a fluidized bed reactor. The polymerization reaction produces an excess of heat and the reactor is unable to efficiently remove the excess heat. The introduction of an Inert Cooling Agent (ICA) solves the temperature problem but also has a downside. Introducing the ICA in the reactor influences the particle collisions, leading to substantially different flow behavior. Besides the reduction of the restitution coefficient, the introduction of the liquid can result in adhesion instead of collision. This adhesion results in the formation agglomerates in the reactor and reduces the efficiency of the process. Due to the complex physics of this problem, Computational Fluid Dynamics (CFD) is a suitable method to gain insight in this phenomenon.

In order to study this complex system of partially wetted particle collisions a Direct Numerical Simulation (DNS) is used for incompressible flow. To model the coupling between the solid phases and the fluid phase, a second order accurate Immersed Boundary Method (IBM) is used (Deen et al.,2012). The interface between the fluids is tracked by a Volume Of Fluid (VOF) method (Baltussen et al., 2017). The contact-line dynamics are solved by imposing a contact angle as a boundary condition (Patel et al., 2017).

The simulated results will be compared with dry particle collisions on a wet wall (Tang et al., 2017) and with wet particle collisions on a wet wall.

References:

1. Deen, N.G., Kriebitzsch, S.H., van der Hoef, M.A. and Kuipers, J.A.M., Direct numerical simulation of flow and heat transfer in dense fluidâ€“particle systems. Chemical Engineering Science, 81 (2012): 329-344.

2. Baltussen, M. W., Segers, Q. I. E., Kuipers, J. A. M., and Deen, N. G., Cutting bubbles with a single wire. Chemical Engineering Science 157 (2017): 138-146.

3. Patel, H.V., Das, S., Kuipers, J.A.M., Padding, J.T. and Peters, E.A.J.F., A coupled Volume of Fluid and Immersed Boundary Method for simulating 3D multiphase flows with contact line dynamics in complex geometries. Chemical Engineering Science, 166 (2017): 28-41.

4. Tang, Y., Buck, B., Heinrich, S., Deen, N.G., Kuipers, J.A.M., Interface-resolved simulations of normal collisions of spheres on a wet surface. American Institute of Chemical Engineers, (2017).

510375 Influence of Interparticle Forces and Particle Collision Properties on the Fluidization Behavior at Elevated Temperatures

Abstract:

Recent experiments have shown an influence of temperature on the minimum fluidization conditions, even when the gas phase density and viscosity have been kept constant (Campos Velarde et al., 2016). Correlations that are available in the open literature, for predicting the minimum fluidization velocity and the bed voidage at minimum fluidization conditions at elevated temperatures, fail to accurately describe the experimental data, in particular how the bed porosity at incipient fluidization conditions is changing with temperature.

It is believed that at higher temperatures inter-particle forces become more influential. Inter-particle forces are already known to be important in the fluidization of very fine powders, and may incur detrimental effects in industrial processes such as the formation of particle agglomerates and reduced particle mixing with a strong influence on the solids circulation patterns. The experimental results by Campos Velarde (2016) indicate that inter-particle forces may also become important for larger particles at increased temperatures. In this work, we characterize the effects of the inter-particle forces using numerical simulations with a Discrete Particle Model (DPM). Moreover, it is known from the literature that in addition to inter-particle forces, also particle collision properties can change with temperature. Different materials behave differently with changing temperature, and in this work we will also show how the change of particle collision properties can influence the fluidization behaviour.

An in-house DPM has been used to perform investigations into the minimum fluidization velocity (U_{mf}), bed porosity ($_{mf}$) and average particle velocities for different fluidization conditions. DPM is an Euler-Lagrange type model with a discrete description of the solids phase and a continuous description of the gas phase. Particle-particle collisions are dealt with deterministically, using a soft-sphere collision model which allows multiple simultaneous contacts between several pairs of particles. The motion of each individual particle is tracked and described with Newton's second law. The additional force that represent the inter-particle interactions is derived from the Lennard-Jones potential and is described with two constants, one for attractive part (Van der Waals) and one for the repulsive part.

By comparing simulation results for fluidization at different temperatures, with different fluidization gas properties and different particle collisional properties, with unique experimental data of high-temperature fluidization by Campos Velarde (2016), we investigate the influence of the inter-particle forces and the particle collisional properties on U_{mf}, $_{mf}$, flow fields and bubble properties to further elucidate the dominating phenomena prevailing at high-temperature fluidization. REFERENCES

VELARDE, I.C, GRIM, R., GALLUCCI, F., VAN SINT ANNALAND, M., (2016), "Influence of Temperature on Minimum Fluidization Properties of Gas-Solid Fluidized Beds", *Powder Technology*, Submitted.

DELIMONT, JACOB M. "Experimental Investigation of Temperature Effects on Microparticle Sand Rebound Characteristics at Gas Turbine Representative Conditions." (Doctoral dissertation) 2014

MIHAJLOVIC, M. ROGHAIR, I. VAN SINT ANNALAND, M. "High temperature fluidization - influence of inter-particle forces on fluidization behaviour" In J. E. Olsen, & S. T. Johansen (Eds.), Progress in Applied CFD – CFD2017 (pp. 107-114). Oslo: SINTEF Academic Press, 2017

507717 Wet Stirred Media Milling of Organic Crystals: Interactions between Product Formulation, Grinding Media Wear and Colloidal Stability

Abstract:

Wet stirred media milling of organic crystals: Interactions between product formulation, grinding media wear and colloidal stability

F. Flach, S. Breitung-Faes, A. Kwade

Institute for Particle Technology, Technische Universität Braunschweig, Volkmaroder Str. 5, 38104 Braunschweig, Germany, f.flach@tu-bs.de

Wet stirred media milling is a common technology to produce fine particles. Recent developments are focused on the production of organic nanoparticles for different applications, e.g. pharmaceuticals, high quality printing inks or fine chemicals. General specifications of those products are well defined particle size distributions and high purity standards. In particular, contaminations, like mill and grinding media wear, represent a critical quality attribute which determines product quality and production costs.

Latest investigations on fine grinding processes of organic particles revealed new insights into the mechanisms of grinding media wear generation. In contrast to former investigations on "hard" inorganic products (minerals, ceramics, ...) other specific dependencies were identified. Machine parameters influence wear characteristics, a lower hardness ratio between product and grinding media is less advantageous and especially the young´s modulus of the grinding media has a significant impact [1]. Additionally, the product particles and the used steric stabilization additives provide much potential to reduce wear rates by damping of grinding media collisions. This effect can be enhanced by controlling particle-particle interactions in order to adjust a certain degree of agglomeration. By stressing agglomerated particles more collision energy is transferred into the deformation and breakage of agglomerates and the applied stress intensity is assumed to be closer to the optimum one which is linked to minimum contamination values [2].

Wear particles do not only represent a contamination of the product suspension, they may also lead to the agglomeration of product particles. Investigations have shown that due to different surface charges the formation of heteroagglomerates between product and wear particles is possible. Heteroagglomeration can be controlled by formulation parameters i.e. pH-adjustment or stabilizer choice as well as by adapting the grinding media material [3].

Overall, the results show that especially formulation parameters have great potential for the reduction of product contamination in fine grinding processes of organic crystals. In addition the targeted adjustment of the product formulation enables also the control of interactions between product and wear particles.

[1] F. Flach, C. Konnerth, C. Peppersack, J. Schmidt, C. Damm, S. Breitung-Faes, W. Peukert, and A. Kwade, "Impact of formulation and operating parameters on particle size and grinding media wear in wet media milling of organic compounds – A case study for pyrene," *Adv. Powder Technol.*, 2016.
[2] S. Breitung-Faes and A. Kwade, "Use of an enhanced stress model for the optimization of wet stirred media milling processes," *Chem. Eng. Technol.*, vol. 37, pp. 819-826, 2014.

[3] F. Flach, S. Breitung-Faes, and A. Kwade, "Grinding media wear induced heteroagglomeration of electrosteric stabilized particles," *Colloids Surf., A*, vol. 522, pp. 140-151, 2017.

508042 Hydriphobic Attraction between Silanated Silica Surfaces of Different Hydrophobicity

Abstract:

Despite the recent progress in understanding hydrophobic attraction, the origin and nature of the attraction have not been completely clarified so far. Particularly, the relationship between surface hydrophobicity and the forces acting on surfaces has not been fully understood.

In this study, we have conducted force measurements between hydrophobic surfaces of different hydrophobicity using atomic force microscopy in order to investigate how surface hydrophobicity affects the resulting interaction forces. Silica particles and silicon wafers were hydrophobised with octadecyltrichlorosilane. Interactions between a hydrophobic particle with a contact angle of 108° and a silicon wafer with various hydrophobicity were evaluated. The force curves obtained between the hydrophobic particle and a completely hydrophilic silicon wafer fitted well to DLVO theoretical predictions, indicating that no additional forces were present between the surfaces. On the other hand, when the silicon wafer was slightly hydrophobised, with a contact angle of 40°, an additional attractive force which was longer-ranged than van der Waals attraction was observed between the surfaces. This suggests that, when one surface is highly hydrophobic, the hydrophobic attraction can occur even when the other surface is weakly hydrophobic. We also found that the range of the additional attractive force depends weakly on the contact angle, whereas the adhesion force steeply increases with increasing hydrophobicity of the wafer.

508131 Determination of Particle Surface Properties Using Hansen Dispersibility Parameters Demonstrated By Means of Carbon Black and Functionalized ZnO Nanoparticles

Abstract:

Keywords: Formulation, analytical centrifugation, dispersibility, particle functionalization, colloidal stability

The performance of small nanoparticles in various fields ranging from optoelectronic devices like solar cells or light emitting diodes (LEDs), pigments but also medical applications, largely depends on their surface chemistry and how well these particles can be embedded in an appropriate continuous phase [1-3]. Thus, formulation is absolutely decisive in terms of the later product performance [4-7]. However, to date no holistic approach to formulation of nanoparticles does exist which includes thermodynamics and kinetics of ligand binding on the molecular level, dispersibility and colloidal stability on the mesoscale, as well as product properties like optoelectronic device performance on the macroscopic level.

To overcome this limitation and to establish a novel approach for knowledge-based formulation, we introduce a new, standardized scoring method for HSP [8] determination of colloidal systems using analytical centrifugation (AC). AC quantifies optically in a direct and accelerated way the behavior of dispersed particles by concentration profiling over the entire sample height of up to 12 samples simultaneously [9, 10]. However, as for particles not dissolution but dispersion is in focus, we propose to use the term Hansen Dispersibility Parameters (HDP) instead of HSP whenever dispersibility and stability of particles against agglomeration are to be discussed.

First, we implemented a defined dispersion routine for the well-known, industrially highly relevant pigment carbon black (CB). Then, a standardized method for the evaluation of measured AC profiles and appropriate ranking of NPs in different liquids was developed. Noteworthy, in contrast to other approaches the final discrimination of "good" and "poor" liquids is not based on preset arguments but derived automatically within the HDP evaluation procedure. This is important to establish HDP as material property that can be analyzed and compared throughout different laboratories.

Then, we demonstrate outstanding reproducibility of our procedure by comparing HDP derived for the same CB material from independent experiments performed at two different affiliations by different experimenters. For further validation, we prove the predictive power of HDP and the accuracy of our approach by evaluating the dispersibility of CB in additional liquids and mixtures of so-called "good" and "poor" liquids crossing the border from stable to unstable conditions.

Finally, motivated by the enormous potential of AC to determine the HDP of colloidal systems using a standardized and non-subjective method to access particle interactions and colloidal stability, we applied the technique to surface functionalized, quantum confined ZnO semiconductor nanoparticles (quantum dots, QDs). Aim was to link mesoscopic findings on dispersibility and colloidal stability with interface properties at the molecular level. Noteworthy, for the molecular characterization, a well-validated toolbox of different methods that has been established by our group was applied [11, 12]. Thereby gained information on binding energy and in particular on surface coverage could be successfully linked to the particle level by dispersibility parameters, δ_p (polar contributions), δ_d (disperse contribution) and δ_h (ability to exchange electrons).

In conclusion, our hierarchical multiscale approach in combination with standardized HDP analysis allows establishing structure-property relationships between molecular features of individual ligand molecules, ligand binding to colloidal particles at the molecular level and dispersibility at the mesoscopic level. This is seen as a major step forward in the whole field of interface characterization and formulation as it paves the way from empiricism towards knowledge-based approaches.

Acknowledgements

The authors want to thank the funding of Deutsche Forschungsgemeinschaft (DFG) through the Cluster of Excellence "Engineering of Advanced Materials". Moreover, we acknowledge the Federal Ministry of Economic Affairs through the Arbeitsgemeinschaft industrieller Forschungsvereinigungen "Otto von Guericke" e.V. (AiF, project no. KF 2347922UW4).

Literature

[1] Q. Sun, Y.A. Wang, L.S. Li, D. Wang, T. Zhu, J. Xu, C. Yang, and Y. Li, Nat. Photonics, 1 (2007) 717-722

[2] X. Michalat, Science, 307 (2005) 538-544

[3] Y.L. Kong, et al., Nano Lett., 14 (2014) 7017-7023

[4] M. Magdassi, A. Bassa, Y. Vinetsky, and A. Kamyshny, Chem. Mater., 15 (2003) 2208-2217

[5] I. Brigger, C. Dubernet, and P. Couvreur, Adv. Drug Del. Rev., 54 (2002) 631-651

[6] S.E. Cross, M.S. Roberts, T. Tsuzuki, T.A. Robertson, and P. McCormick, Skin Pharmacol. Physiol., 20 (2007) 148-154

[7] J. Jain, S. Arora, J.M. Rajwade, P. Omray, S. Khandelwal, and K.M. Paknikar, Mol. Pharmaceutics, 6 (2009) 1388-1401

[8] C.M. Hansen, *Hansen solubility parameters: a user's handbook* 2nd edition ed2007, Boca Raton, FL, USA: CRC Press.

[9] D. Lerche and T. Sobisch, Coll. Surf. A: Physicochem. Aspects, 440 (2014) 122-130

[10] D. Lerche, S. Horvat, and T. Sobisch, Dispersion Letters, 6 (2015) 13-18

[11] W. Lin, M. Haderlein, J. Walter, W. Peukert, and D. Segets, Angew. Chem. Int. Ed., 55 (2016) 932-935

[12] W. Lin, J. Walter, A. Burger, H. Maid, A. Hirsch, W. Peukert, and D. Segets, Chem. Mater., 27 (2015) 358-369

508492 Collisional Dissipation Rate of Flexible Rods Measured Using Driven and Non-Driven DEM Simulations

Abstract:

Particle flexibility is a complex phenomenon that has a significant effect on common particle systems (e.g. biomass refining). Current large-scale simulation techniques are far from being able to describe flexible particles, but discrete particle method (DEM) simulations offer the opportunity to investigate important particle-particle interactions. These interactions include the collisional dissipation rate, which describes the decrease in energy over time due to particle-particle collisions.

Homogeneous cooling system (HCS) and shear-flow (SF) simulations are used to investigate the effect of particle flexibility on the dissipation rate. As particles transition from rigid to flexible, new vibrational and potential energy degrees of freedom are introduced and accounted for in the energy loss due to collisions. As a particle becomes more flexible (decrease in Youngâ€™s modulus) the collision frequency decreases due to increased contact time. Thus, in systems where no energy is lost due to flexibility, the collisional dissipation rate decreases. The bond damping coefficient, describes energy loss due to flexibility, is found to be a more significant contribution to the collisional dissipation rate compared to the Youngâ€™s modulus and coefficient of restitution. As the bond damping coefficient increases, the collisional dissipation rate increases until it reaches a plateau. The effect of the coefficient of restitution and bond damping coefficient on the dissipation rate can be isolated and summed to create a model that is applicable for two-fluid simulations.

508629 PickeringÂ foam Formulation By Wet Nano-Milling

Abstract:

Pickering foams or emulsions are very interesting formulations because they are stabilized by solid particles instead of surfactants. We observed during formulation of a nano suspension of a hydrophobic API by wet ball milling spontaneous undesirable formation of a viscous foam, which was stabilized by solid nanoparticles dispersed in water. We have prepared a Pickering foam by the milling of a hydrophobic API suspension in a flow-through ball mill Netsch LabStar by zirconium dioxide balls as the milling media. We have characterized the particle size distribution (PSD) of the API particles in the foam by static light scattering and Scanning Electron Microscopy (SEM). We have studied the 3D structure of the foam and calculated its porosity using micro CT and confocal optical microscopy. A Pickering foam produced by the ball nano-milling process is stabilized by the solid particles of the hydrophobic API. The foam formation time is dependent on the initial surfactant concentrations used in the milling process. Mechanical properties such as viscosity are dependent on the milling time and increase because of increasing number of stabilizing particles and their surface area. Solid nanoparticles can be used as emulsifiers for the stabilisation of the Pickering foams or emulsions, which have advantages compared to surfactant stabilized formulations. Sometimes the formation of a Pickering foam can be a spontaneous and unwanted phenomenon, which causes a change of mechanical properties of the material and makes the technological process more challenging. To avoid undesired complications of the technological process, it is important to know what causes the changes and to know how to avoid them. This was achieved by studying the undesirable Pickering foam produced by the formulation of a nanosuspension. The foam creation is significantly dependent on the surfactant concentration relative to the specific surface area of the hydrophobic API.

REFERENCES:

1. Pyotr M. at al. Advances in colloids and interfaze science 165, 108-116 (2011)

2. Chevalier Y. & Bolzinger M. Colloids and surfaces A: Physicochemical and engineering aspects, 439, 23-34 (2013)

509860 Electrostatics of Dry Powder Aerosols for Inhalation

Abstract:

Electrostatics of pharmaceutical aerosols for inhalation is a relatively new research area. Particle charges have been proposed to potentially influence deposition in the lungs but no conclusive in vivo data have been reported. Investigation of the relationship between the formulation and aerosol charging is required to further explore this area. The modified electrical low pressure impactor (ELPI) and bipolar charge analyser (BOLAR) have been used to measure the size and electrostatic charge of particles generated from dry powder inhalers (DPIs). DPIs are commonly used to deliver drugs to the lungs. The powder is dispersed by the inhalation effort of the patient. Net charges are generated through triboelectrication during aerosolisation. These charges were found to be formulation- and humidity-dependent [1, 2], as well as bipolar [3, 4]. The knowledge gained may be useful in the development of pharmaceutical aerosols and pulmonary drug delivery in the future.

References

1. Philip Chi Lip Kwok, Hak-Kim Chan (2008) Effect of relative humidity on the electrostatic charge properties of dry powder inhaler aerosols. Pharmaceutical Research, 25(2): 277-288.
2. Jennifer Wong, Philip Chi Lip Kwok, Tim Noakes, Ali Fathi, Fariba Dehghani, Hak-Kim Chan (2014) Effect of crystallinity on electrostatic charging in dry powder inhaler formulations. Pharmaceutical Research 31(7): 1656-1664.
3. Jennifer Wong, Yu-Wei Lin, Philip Chi Lip Kwok, Ville NiemelÃ¤, John Crapper, Hak-Kim Chan (2015) Measuring bipolar charge and mass distributions of powder aerosols by a novel tool (BOLAR). Molecular Pharmaceutics 12(9): 3433-3440.
4. Jennifer Wong, Philip Chi Lip Kwok, Ville NiemelÃ¤, Desmond Heng, John Crapper, Hak-Kim Chan (2016) Bipolar electrostatic charge and mass distributions of powder aerosols â€“ Effects of inhaler design and inhaler material. Journal of Aerosol Science 95: 104-117.

507600 Electrification and Dispersion of Particles Using Mesh Electrode

Abstract:

Electrostatic forces cause spontaneous movement in every charged particle. Thus, electrostatic technologies are attracting attention in the actual processes of powder handling, i.e., separation, classification, and collection. When conductive particles come in contact with any one of the inside walls of parallel electrodes, the particles are charged to the same polarity as that of the electrode by induction. If Coulomb forces acting on the charged particles are larger than the adhesion forces between the contact surfaces, the particles will levitate from the electrode. After the particles move and reach the counter electrode, the polarity of the particles inverts, and then the particles levitate again. As a result, the particles oscillate between the parallel electrodes. Replacing the counter electrode with a mesh electrode allows the passage of the charged particles through the mesh openings; then, the particles with the same polarity as the first electrode are dispersed by mutual electrostatic repulsion. However, since the particles are polarized in the external electric field, chain agglomerates can be formed; thereby the particle behavior becomes more complicated. In powder handling processes, dielectric particles are often used. Such particles are accumulated and form multi-layers on the electrodes; this affects the phenomena of particle charging and levitation. The phenomena for single dielectric particles have been reported, but there are few studies focusing on the induction charging and levitation for particle layers.

We present the charging, levitation, and relevant behaviors for dielectric particle layers, which are obtained using parallel electrodes consisting of a lower plate electrode and an upper mesh electrode. First, multi-layers of particles are formed on the lower electrode. Next, DC voltage is applied to the lower electrode and the upper electrode is grounded to vertically form a high-intensity electric field. Then, the particles are charged and levitated in the electric field. The particle motion around the lower electrode is observed via a high-speed microscope camera, showing that agglomerated as well as single particles levitate from the surface of the particle layers, i.e., the particles are polarized in the electric field and form chain agglomerates by mutual electrostatic interactions. When the specific charge, i.e., charge-to-mass ratio of particles is small, single particles cannot levitate because of small Coulomb forces. However, chain agglomerates can levitate because of large Coulomb forces even though the specific charge is small. The Coulomb forces of the chain agglomerates can exceed the adhesion forces and gravitational forces acting on primary particles in the agglomerates.

The specific charge of each levitating particle is estimated by fitting experimentally obtained particle trajectories with calculated ones in simulated electric fields, showing that as the number of primary particles of the chain agglomerate increases, their specific charge decreases. In addition, it is found that as the adhesion force of each primary particle increases, the number of primary particles of the chain agglomerate increases.

Furthermore, a unique phenomenon is observed around the upper electrode. Fig. 1 shows a chain agglomerate that was dispersed. Agglomerates can be rotated in the dispersion process. The charges of the agglomerates during the dispersion process are estimated using a simulated non-uniform electric field around the upper electrode and the effect of the particle charge on the dispersion process is studied.

Fig. 1. Dispersion of chain agglomerate by electrostatic repulsion around upper mesh electrode.

NOVEL ELECTROSTATIC FIELD METER USING ROLLING SECTOR

(FIRST PAPER)

Dr. Kwangseok Choi, National Institute of Occupational Safety and Health, Japan, Tokyo, Japan
Dr. Teruo Suzuki, Kasuga Denki Inc., Kanagawa, Japan

Abstract

This paper investigated experimentally the fundamental characteristics of the novel electrostatic field meter in order to measure electrostatic charge levels. The novel meter consists mainly of a grounded plate with a sensing rolling sector, the electric motor, a signal cable, an enclosure, and two air supply lines. The grounded plate had four 0.9 mm holes that had a trench shape to satisfy the international standard. The rolling sector was rotating at 1100 rpm. The enclosure was supplied with compressed air (air purge) to prevent combustible gases or powders from getting inside. As the experimental demonstrated, (1) the electrostatic field strength of the simulated charge material and detected signal were clearly in direct correlation, and its validity was confirmed; (2) the signal of the novel meter strongly depends on the thickness of the grounded metal plate; and (3) the performance of the novel meter was not affected by air purge in this study.

Keywords: *electrostatic charges, electrostatic discharges, electrostatic field, powder, explosion*

1. Introduction

Electrostatic charges and discharges generated in industry can cause gas and dust explosions and damage electronic components. Especially, unwanted electrostatic charges and discharges can cause major damage such as fires in buildings, damage to sensitive electrical equipment, or injury to humans, including death. Many papers have reported major accidents caused by electrostatics in the industrial process [1–4].

This work deals with the experimental investigation of a novel electrostatic field meter (hereafter, the novel meter) based on its performance on a laboratory scale as a first paper.

2. Experimental apparatus and method

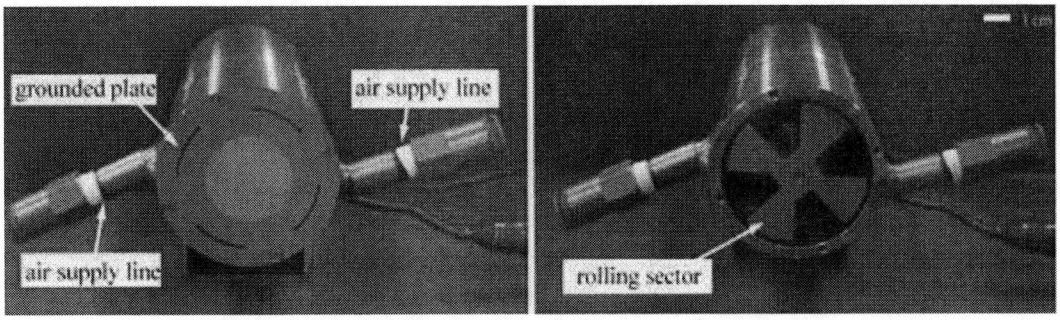

Fig. 1 Novel electrostatic field strength sensor

Figure 1 shows that the novel meter consists mainly of a grounded plate with a sensing rolling sector, electric motor, signal cable, enclosure, and two air supply lines. The grounded plate was attached to the front of the enclosure. It had four 0.9 mm holes that had a trench shape to satisfy the requirement of the International Electrotechnical Commission (IEC)-Ingress Protection (IP) code [5]. The sensing rolling sector (signal-detecting electrode) was placed 1 mm behind the grounded plate. Using the electric motor, the signal-detecting electrode rotated at 1100 rpm. To measure signals, the rolling sector was separated from the motor's axis by an insulator. The enclosure was supplied with compressed air (air purge) to prevent combustible gases or powders from getting inside. In this study, a laboratory test was conducted to assess the performance of the novel meter. The experimental setup consisted of the novel meter, a rotating sector controller, an HV power source, a charged metal plate (500 mm × 500 mm in area, 2 mm thick), a digital oscilloscope, an air compressor, an automatic air dryer, an electrometer, a mass flow meter, and other auxiliary devices. The charged metal plate in this study was assumed to be the charging material and was attached to an HV DC power source to control its charge level. The novel meter was fixed toward the center of the charged metal plate. The novel sensor is a device based on electrostatic induction. A 100 M□ was installed inside the electric circuit to convert into voltage in order to measure the signal. The effects of the electrostatic field E [kV/cm], the thickness t [mm] of the grounded plate, and the air purge P [MPa] on signal V [mV] were examined in this study.

3. Experimental results and discussion

Figure 2 shows the effect of the voltage applied to the charged metal plate on the signal detected by the novel meter without purging air. The applied voltage means the potential difference between the novel meter and the charged metal plate. Because the distance between the novel meter and the charged metal plate was 10 mm in this study, the values of the applied voltage and the electrostatic field were the same. The range of applied voltage to the metal plate was in 0 to10 kV in bipolar. It should be noted here that the signal of the novel meter was almost zero when high voltage was not applied to the metal plate. As for the results, the applied voltage had a positive correlation to the detection signal. The positive and negative values were very similar, only with different signs, showing a fairly linear graph when plotted. This shows that, as voltage increased, the detected signal increased as well. That is to say, electrostatic field E of the simulated charging material (metal plate) and detected signal V were clearly in direct correlation, and its validity was confirmed.

Figure 3 shows the effect of the thickness, t, of the grounded metal plate on the signal detected by the novel meter when V and d were 10 kV and 10 mm, respectively, also without the air purge. The thickness was in a range of 0.5 mm to 2.0 mm, every 0.5 mm. Results showed that thickness had a negative effect on the detection signal. For example, detected signals V for t : 0.5 mm, 1.0 mm, and 1.5 mm were 26.6 m, 5.87 mV, and 1.53 mV, respectively. When the thickness was 2.0 mm, the detected signal was almost zero. This can be explained by the fact that the electric field generated by the charged metal plate could not fully reach the rotating sector because it headed toward the grounded metal plate.

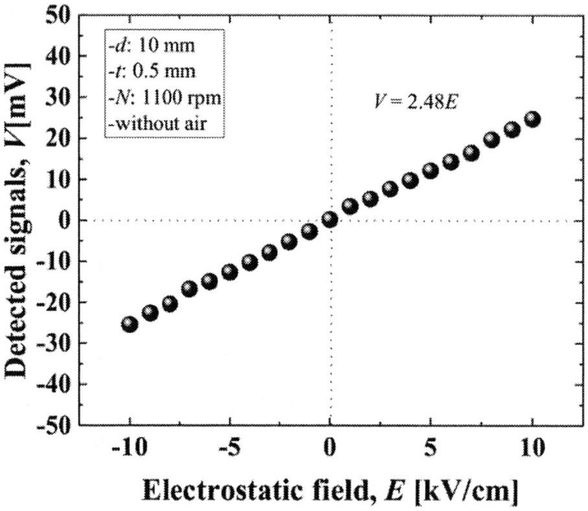

Fig. 2 Relationship between the electrostatic field and the signal of the novel meter

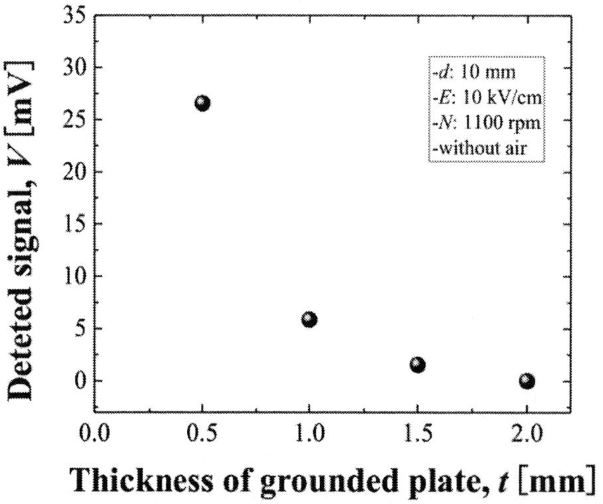

Fig. 3 The signal of the novel meter as a function of the thickness of the grounded metal

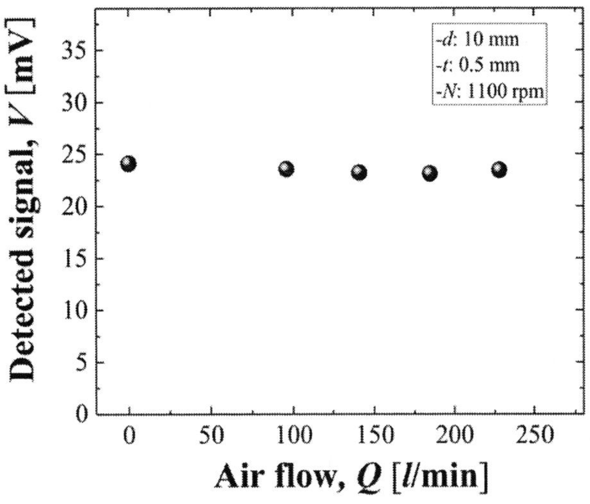

Fig. 4 Relationship between the detected signal and the air purge of the novel meter

Figure 4 shows the effect of the air purge on the signal detected by the novel meter when E was10 kV. The flow of air for the purge, Q, in the novel meter ranged between 0 and 230 l/min. The pressure used for purging in the novel meter was also measured. Results show that the performance of the novel meter was almost unaffected by the purge of air supplied to protect combustible gases or powders from getting inside. It is notable that even in processes with combustible materials, the novel meter can still give an accurate signal. The pressure in the novel meter due to the purge of air increased as the air flow increased. When the air flow was 96 l/min, 141 l/min, 185 l/min, or 230 l/min, the pressure was 212 Pa, 443 Pa, 734 Pa, and 1020 Pa, respectively. Air flow of the novel meter at or above 96 l/min can satisfy the requirement for an internal pressure considered explosion-proof according to the IEC standard, which is 50 Pa [6].

4. Conclusion

This paper investigated experimentally the effect of the electrostatic field, the thickness of the grounded plate, and the air purge pressure on its performance in a laboratory setting. A summary of the experimental results follows:

1) The tests confirmed the linearity of the relationship between the detected signal and the electrostatic field.

2) The detection signal decreased as the thickness of the grounded plate increased. When the thickness was 2.0 mm, the detected signal was almost zero.

3) The air flow (max. 230 l/min.) for purging the novel meter did not have an effect on the detected signals.

References

1. T. H.Pratt (2000): Electrostatic ignitions of fires and explosions, Center for chemial process safety, pp.132-142.
2. F. Riedewald (2012): A fire in a secondary pharmaceutical powder transfer operation, Process safety progress, 31, 4, pp. 390-392.

3. M. Glor (2003): Ignition hazard due to static electricity in particulate processes, Powder technology, 135, 136, pp. 223-233.
4. N. Gibson: Static electricity- an industrial hazard under control, J. electrostatics 40, 41, pp.21-30.
5. IEC 60529 (2004), Degrees of protection provided by enclosures (IP code).
6. IEC 60079-2 (2015), Explosive atmospheres- part 1: Equipment protection by flame proof enclosures.

507317 Investigation of Granular Surface Roughness Effect on Electrostatic Charge Generation

Abstract:

Abstractï¼šElectrostatic charge generation is a multivariable and complex process whose working mechanism has never been fully understood. The objective of this study is to investigate the effect of granule surface roughness on electrostatic charge generation. Two kinds of granule material, namely Polyvinyl chloride (PVC) and polypropylene (PP) were used with the granule size of 4mm in diameter and 2mm in height and in the shape of either cylinder or semi-cylinder. The working surfaces were grounded and the corresponding roughness ranged from 0.140 to 8.600 Î¼m. Uneven surfaces were found to give rise to voids between two solids, where air stored in the voids was able to accelerate discharging. With the same roughness, PVC tended to generate more electrostatic charge than PP by one order of magnitude. For both materials, electrostatic charge generation first increased with surface roughness and then decreased. The maximum electrostatic charge generation was found to occur when the effects of interaction, contact area and voids discharging were at equilibrium. With the combined effect of humidity, surface roughness and contact area, highest electrostatics generation occurred near the mid-roughness point tested in this work. Humidity had more effect on electrostatic charge generation as the granule working surface had lower roughness.

Keywords: Electrostatic charge generation; Surface roughness, Granule.

DIFREX Reactor and Technology Solutions for Many Particle Types and Sizes in Catalytic and Non-Catalytic Reaction Systems

Subhash Dutta, Stephen Arnold, Aashish Gaurav, Jim Brenner
DIFREX LLC., 8900 Astronaut Blvd. Cape Canaveral, FL 32920 USA

Full understanding of particle flow and hydrodynamic behavior, characterization of the properties of these materials, and of their effects on the operation and performance of industrial reactors, is essential for successful industrial reactor design and scale-up. This discussion will focus on gas-solid reactor systems and will use the term "particle" or "solid" to represent both particulates and particles. Where the terms are distinguished, "particulates" are defined as fine particles usually suspended in gas and in dynamic and/or flow mode with a bulk gas flow stream, and are the solid types used in fluidized bed reactors of various kinds and in pneumatic transport lines. The term "particles", on the other hand, is used for relatively larger particles that are usually utilized in fixed or packed bed reactors and remain static or in flow mode in moving bed reactors. Schematics of the four basic industrial solid-gas reactor types are shown in **Figure 1**: packed bed, tubular fixed bed, bubbling/turbulent fluid bed, and circulating fluid bed.

Reactor Types and Particle Technology

Common industrial solid-gas reactors are of two types: a packed/fixed bed reactor (PKB/FXB) or a fluidized bed reactor, with variations of each. A fluidized bed reactor uses solid particulates. Two prominent versions of the fluidized bed are a bubbling/turbulent fluidized bed (BFB) and a circulating fluidized bed (CFB). Impacts of solid-gas interactions, and effects of particulate flow and hydrodynamic behaviors of fluidized beds, on heat and mass transfer characteristics of reaction systems are discussed with examples, including designs and performances of the reactors.

Packed/Fixed Bed Reactors, which employ larger particles in a static bed, are traditionally viewed as the most obvious choices in industry due to easy design and scale-up. For a multi-tubular FXB reactor (TFXB) development and testing, testing in a full-size single tube is ordinarily sufficient for the design of an industrial size reactor of virtually any capacity – the industrial size simply having multiple tubes (hundreds or thousands of the single tube). An FXB reactor also eliminates many potential issues typically associated with a fluidized bed reactor, like solid attrition, entrainment loss and dust emission. However, a multi-tubular FXB reactor suffers from many serious drawbacks that include the following:

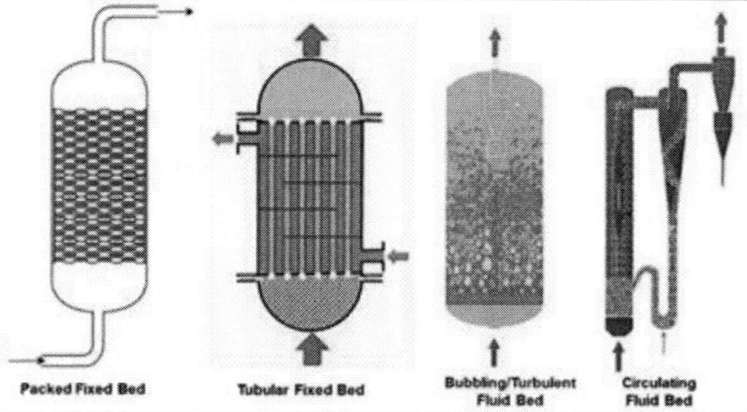

Figure 1: Basic Commercial solid-gas reactor types

1. High heat and mass transfer resistances leading to a steep axial temperature profile in the reactor.
2. Poor catalyst utilization or low catalyst effectiveness factor, due to its use of relatively large particles.
3. Possible maldistribution of gas between all tubes (including impact of catalyst not loaded equally).
4. No continuous solid (catalyst or material) regeneration and replacement capability.
5. No load-following capability.
6. Hot spot formations possibly leading to catalyst activity/selectivity loss and even catalyst/material vitrification.
7. Higher pressure drop.
8. Restricted window of allowable feed composition, temperature and/or pressure to avoid possible temperature runaway for highly exothermic reactions.

Fluidized Bed Reactors, both Bubbling/Turbulent (BFB) and Circulating (CFB), do not suffer from the drawbacks of PKB/FXB reactors indicated above. The smaller particle sizes in these particulate systems lead to better heat and mass transfer rates, allowing operation at uniform bed temperature, which is normally preferred. Close to 100% catalyst utilization can be achieved, with no gas maldistribution. These reactors also provide continuous catalyst/material regeneration and replacement capability due to fluidizability of the chosen particulates that can be continuously discharged and recharged after regeneration and/or feeding a make-up stream. These reactors also have load-following capability, allowing operation across a broad window of production capacity as desired. They can be operated with lower pressure drops and are usually also free of hot spots. The windows of allowable feed composition, pressure and temperature are much wider compared with a fixed bed reactor, with less chance of reactor temperature runaway. On the other hand, the key drawbacks of a fluidized bed reactor include the following:

1. Potential dust emission with possible environmental concern.
2. Loss of expensive catalyst by entrainment, to the extent that fines are not fully captured and returned.
3. Particulate attrition causing excessive fines buildup, leading to excessive solid loss or depletion of fines that are needed for good fluidization, and other potential problems like line plugging and agglomeration.
4. Potential erosion of lines or internal heat transfer tubes.
5. Poorer solid-gas mass and heat transfer rates for BFB through the interface films of gas bubbles or voids.
6. Potential neglect of safety concerns (even unacceptable bypassing of safety instrumentation and procedures) if operators do not fully understand, and become complacent with respect to the high flame-arresting safety benefits of a BFB, e.g., for partial oxidation (POX) reactions, including knowing when or where they do not apply.

Circulating Fluidized Bed Reactors (CFBs) are usually designed for fast reactions and for reactor-regenerator dual-bed systems. In the dual-bed system, the solid particulates, catalyst or non-catalytic material, can be regenerated in the regenerator bed, or sometimes in a transfer line, and be returned back to the reactor in a continuous circulation of the solid. (This is also done with some BFBs.) The primary section of a CFB system operates without gas bubbles or voids, and has minimal solid-gas heat and mass transfer resistances. However, it suffers from the following drawbacks compared with the other systems:

1. The bed voidage throughout these reactors is usually very high (in the range of 88-98%), thus, solid concentration per unit bed volume very low, resulting in much lower reaction rate per unit bed volume.
2. These reactors operate at relatively much higher gas velocity and therefore are applicable to fast reactions only.
3. At the high gas velocities, which includes the catalyst movement, they demand much tougher or more attrition-resistant catalyst or material, unless the material is desired to be broken down (or reacted away) and to pass once-through as in gasifiers or combustors.
4. The reactors need to be much taller (in the order of 30 meters or even more) for adequate reaction residence time to achieve satisfactory conversions at their high gas velocities.

DIFREX Reactor Solutions with Particle Technology

DIFREX software, services and solutions integrate chemical reaction engineering with particle technology to provide the right solutions for design, development, scale-up and successful commercialization of gas-solid reactors, enabling one to promptly establish the design and performance of virtually any particle or solid-gas reactor type for reaction systems of different complexities. Since it has fundamental basis in the two engineering sciences (reaction and particle) and avoids much of the empiricism of traditional process development, especially in the HC processing industry, it is better suited for new designs and also revamp or upgrade of an existing technology/ reactor/ process, particularly for a new feed or product stream. This refers to both catalytic and non-catalytic solid-gas reactions, including the use of a catalyst (active components with or without carrier) as the solid for catalytic processes or a non-catalytic solid particle like coal, biomass or a mineral (FeO or NiO, for example) for combustors, gasifiers and metal extraction/roasting processes, and also solid particles as adsorbents to transfer atoms and other molecular species from one stream or vessel to another.

Figure 2 illustrates the extensive and convenient framework and easy use of the DIFREX software as represented as a ***dial-a-reactor***. It incorporates detailed flow and hydrodynamic behavior for many different reactor types, allowing prompt and easy calculation switching (dialing) from one reactor type to another, for the same chemistry and kinetics of the reaction system. Better, faster initial determination of the best conceptual/ preliminary design for a reaction system, catalyst or process can therefore ***now*** be achieved by comparing the design and performance details projected by the software at the earliest project stages. The best choice could be a packed bed (PKB), a tubular fixed bed (TFXB), a bubbling/turbulent fluid bed (BFB), a circulating fluid bed (CFB), all with or without quench and/or recycle, or a hybrid reactor (CSTR-CFB-BFB) as in a typical FCC reactor or other system. The best **Reactor Of Choice (ROC)** can now be determined along with the **Catalyst or Material Of Choice (COC/MOC)** and much earlier – even while the solid catalyst or material is being developed. Final determination of the best ROC, however, would be determined by **consideration of other criteria** that include cost, particulate handling/recycle, strength and attrition resistance of solid catalyst, and miscellaneous other characteristics like particulate fluidizability and agglomeration problems for fluidized bed systems. However, the software allows one to reach each **GO/NO-GO decision gate** much sooner, leading to substantial savings in both cost and time for catalyst, process or technology development.

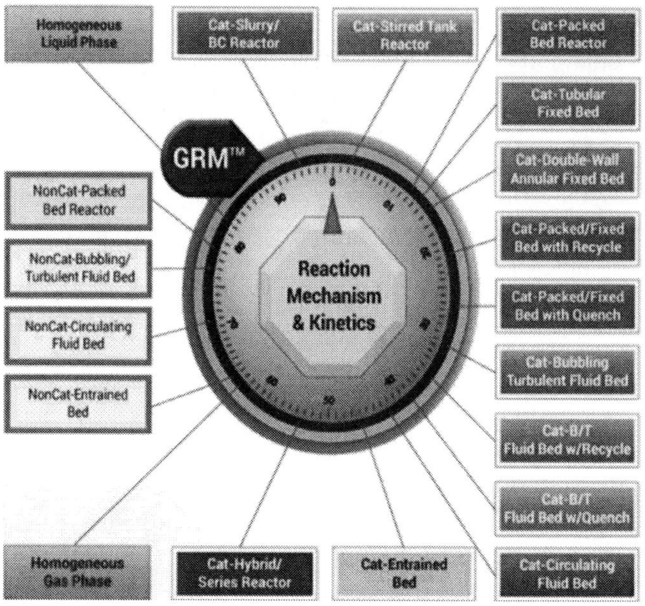

Figure 2: DIFREX: Selection for Reactor of Choice

Current DIFREX team members were instrumental in successful commercialization of several new process technologies, involving all four basic solid-gas reactor types: PKB, TFXB, BFB and CFB reactors and their variations. A prime example is the ALMA process, a fluidized bed POX process for production of maleic anhydride from n-butane and air. These designs were products of seamless integration of particle technology (including particulate flow and hydrodynamic behaviors) with chemical reaction engineering principles. Results of some of these works are available in a series of publications by Dutta, Arnold and other authors[1-20]. Many of the theories and procedures used in these works can be found in the two classical books on Fluidization Engineering and Chemical Reaction Engineering by Kunii and Levenspiel[21-22]. The following discussion provides information on some of these reactor designs and the results achieved.

Packed/Fixed Bed Reactors

The Fischer-Tropsch process is of great industrial significance. Projected axial profiles for temperature, conversions and pressure drop of a typical Fischer-Tropsch reaction system are shown in **Figure 3**, for syngas conversion to paraffins, olefins and alcohols, carried out in a steam-cooled Tubular Fixed Bed (TFXB) reactor system. PR and SR refer to primary reactant CO and secondary reactant H_2. There is a steep temperature rise near the reactor entrance, due to high reaction rate at the concentrated inlet composition, plus relatively high heat transfer resistance typical of this solid-gas fixed-bed reactor system. Temperature runaway with inadequate temperature control, hot spot formation, catalyst selectivity loss, deactivation and possibly even vitrification are potential problems of this system. **Figure 3** also shows the profiles of primary, secondary and undesirable product (PP, SP, UDP) selectivity, per-tube productivity, catalyst productivity (kg/day of PP per kg catalyst), and catalyst effectiveness factor (% utilization) for the primary reaction.

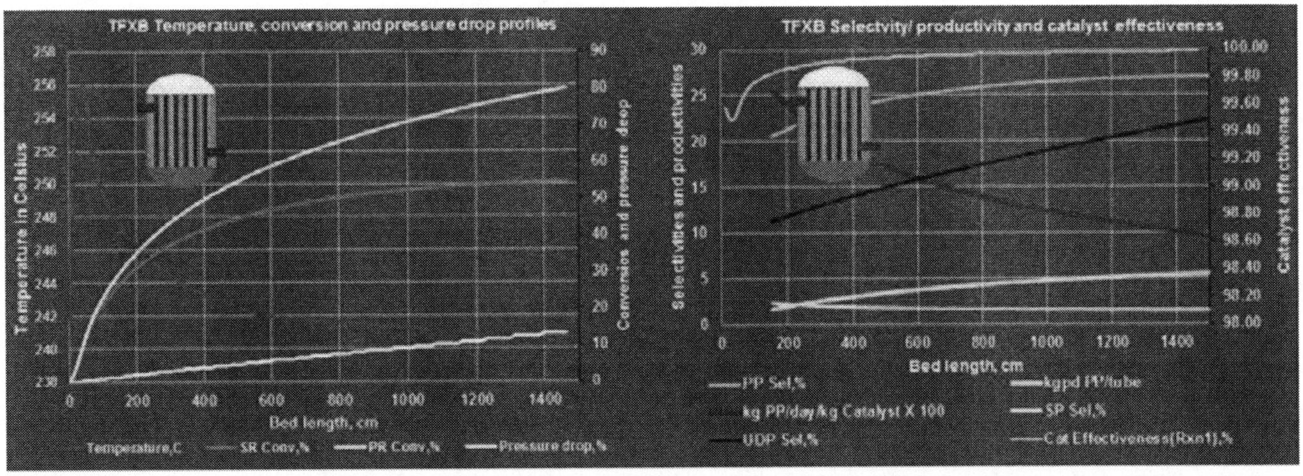

Figure 3: DIFREX capabilities in detailed reactor design studies: Case of FT synthesis in a TFXB reactor

Bubbling/Turbulent Fluidized Bed Reactors

The conversion of n-butane to maleic anhydride (B→M) presents a unique example where extensive performance tests simultaneously in both tubular fixed bed (TFXB) and bubbling or turbulent fluidized bed (BFB) reactors, accompanied by modeling and scale-up, led to successful commercialization of the world's leading fluidized bed process for this reaction system, called the ALMA Process. Unification of particulate technology principles, flow and hydrodynamic behavior with chemical reaction engineering formed the core of the reactor model that was developed. The model was also applied to simulation and scale-up analysis of two other catalytic reaction systems (N→P and P→A) that vary considerably in mechanism and kinetics. The three reaction systems are demonstrated in **Figure 4**. The data for the N→P system were obtained from published data of a commercial reactor.

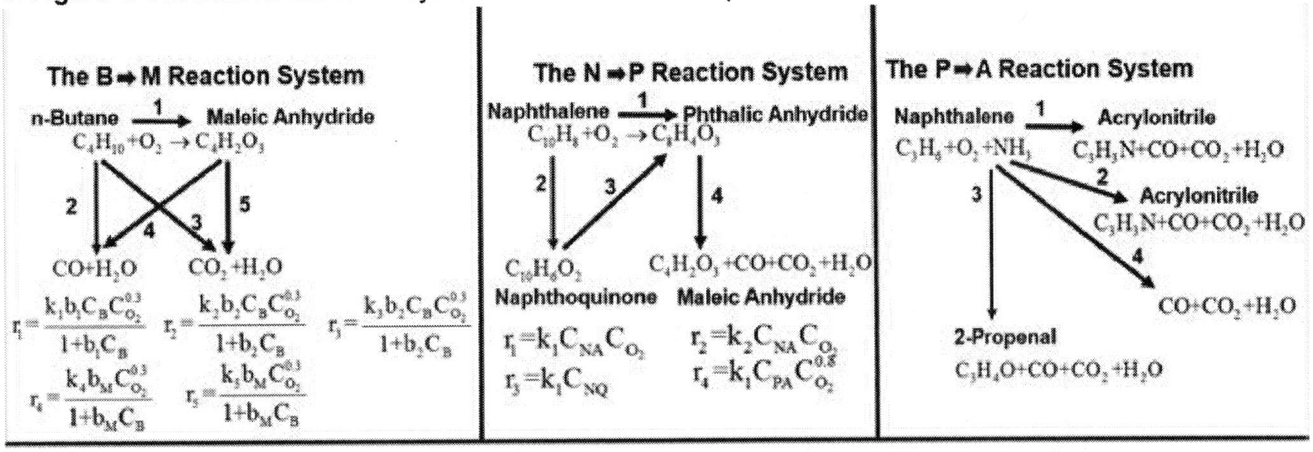

Figure 4: Catalytic reaction systems modelled and studied in TFXB and BFB reactors

In the unified model, the entire height of the BFB is viewed to consist of 5 possible zones: 1) ***grid zone*** where bubbles form and grow continuously until they become slugs or reach a constant average size, 2) ***bubble zone*** where bubbles of a constant average size (maximum stable size) rise up through surrounding solid-gas emulsion phase, 3) ***slug zone*** where slugs of constant size rise, filling up the vessel diameter (mostly in smaller-diameter lab or bench-scale reactors), 4) ***splash zone*** where the bubbles burst and splash the particles above the dense bed below and then the catalyst density decreases with height in the dilute phase, and 5) ***post-TDH zone*** where the density remains constant in the dilute phase ("freeboard") above the Transport Disengagement Height. The DIFREX model accounts for particulate flow and dynamic behavior in each zone, and how it affects the reaction system and reactor performance, including conversions and selectivities of each reaction. Details of these treatments are available in the publications of Dutta et al.[14-15].

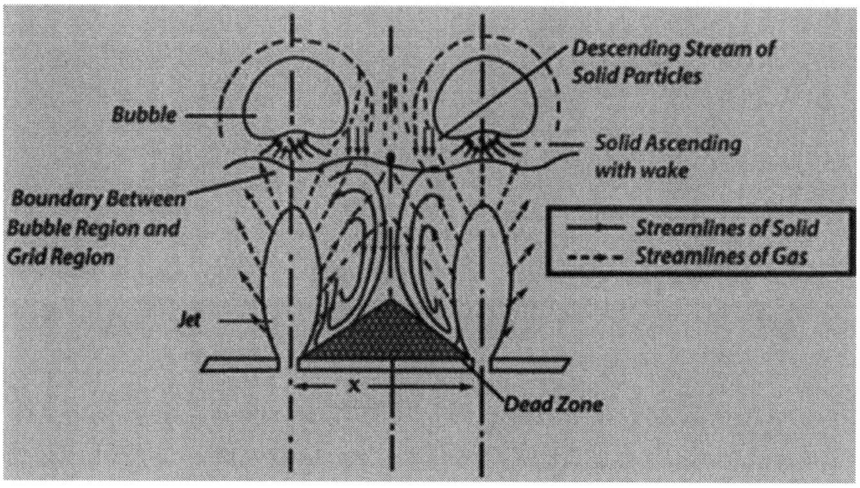

Figure 5: Schematic of gas and solids flow patterns and dead zone formation in the grid region

Unlike fixed bed (FXB), the particulate system behavior is quite complex in BFB and may change dramatically from reactor bottom to the top. Proper account of these changes and their effect on the reaction system is necessary for developing a successful BFB scale-up model. A schematic representation is shown in **Figure 5** of gas and solid flow patterns immediately above the grid at the very bottom of a commercial BFB. Due to intense solid-gas mixing, turbulence and smaller bubbles in this zone, plus the higher feed concentration, as much as 60% of the total conversion of a commercial BFB can happen in the ***grid zone***.

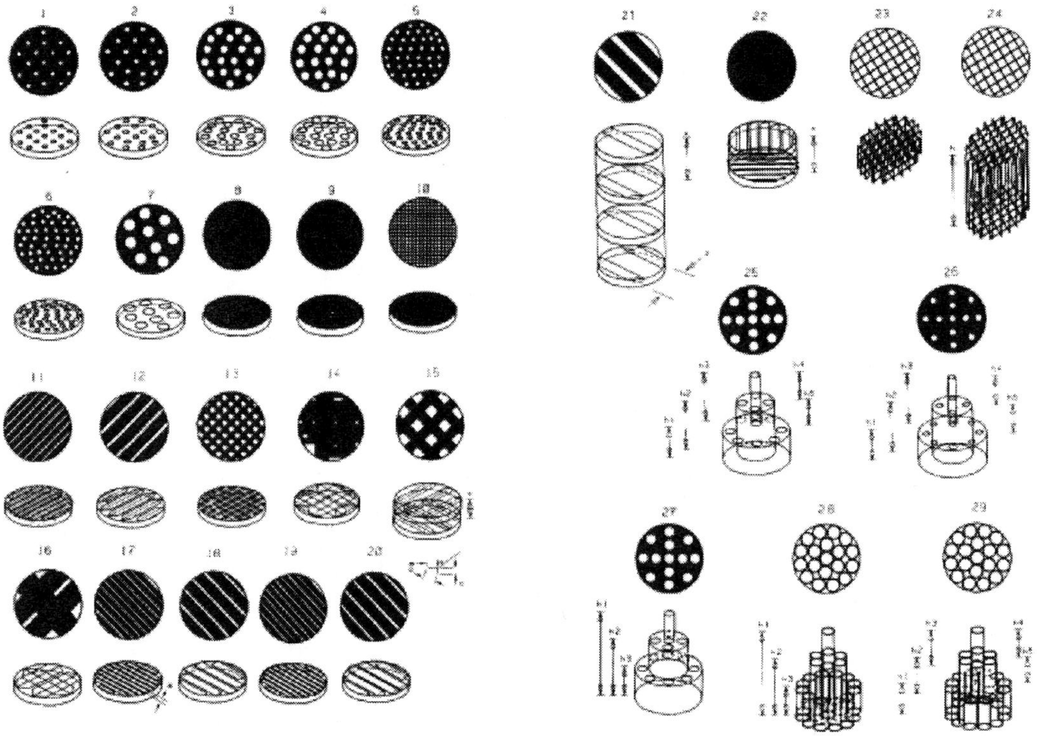

Figure 6: Top views of experimental baffles [Adapted from Dutta and Suciu[10]]

Grid design of a BFB is important and a challenging task since several objectives need to be achieved in this design. These include support strength, optimal pressure drop, minimum dead zone, minimum jet attrition of particles, and avoidance of particle slippage through the grid openings. Similar considerations affect the design of internal baffles, etc. A study of designs for the grids and other internals, with different open areas, layers, slot angles, egg-crate shapes, etc., was conducted for the ALMA process as exemplified in **Figure 6**. DIFREX team members have extensive experience in design and commissioning of such internals.

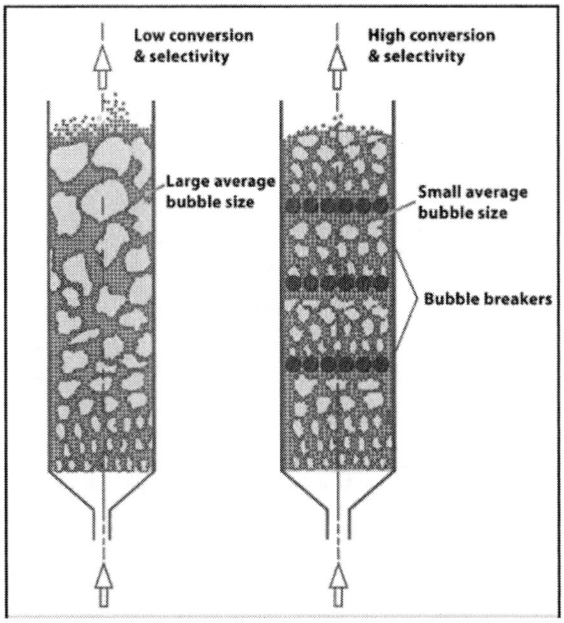

Figure 7: Bubble size control in BFB

Bubble size control above the grid zone is also an important consideration for higher reactor conversion, because smaller bubbles mean both higher and more selective conversion, while bigger bubbles mean lower conversion and poor selectivity. This is due to higher mass transfer resistance for the latter case, which slows the exchange (or mass transfer) rate between the rising bubbles with the bulk of the gas and the surrounding emulsion phase containing most of the catalyst. Proper design of bubble breakers therefore can also be an important part of a BFB design. DIFREX can provide help in this area as well. The concept of bubble size control is shown schematically in **Figure 7**.

Applicability of three BFB models were examined and compared during the ALMA process development, as described in a publication by Dutta, Arnold, Suciu and Verde[16]. Schematics of these models[21, 23, 24] are shown in **Figure 8**. Modification of the Kunii-Levenspiel (K-L) model[21] to account for growing bubbles and also the ***splash zone*** as proposed by Hatano et al.[25] are three important features of the DIFREX BFB reactor model. More details are available in Dutta and Suciu[14].

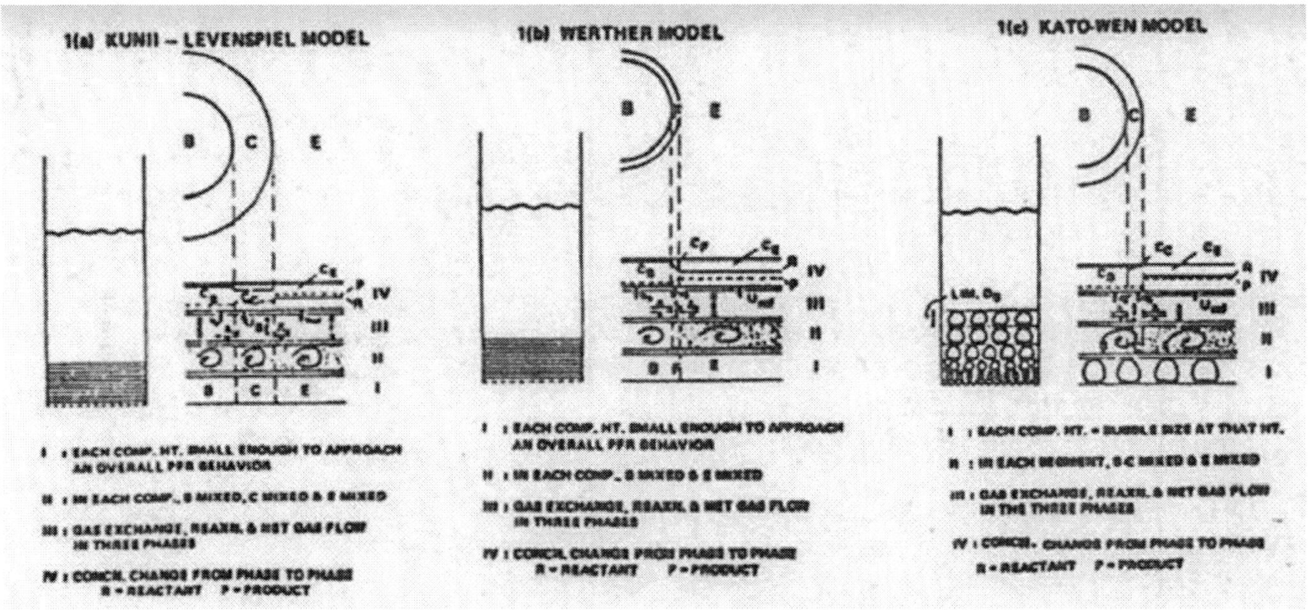

Figure 8: Schematic representation of three BFB models [Adapted from Dutta et.al [16]]

Projected BFB reactor performances and the key particulate hydrodynamic behaviors affecting these for the same FT reaction system used in the FXB reactor (those projections shown in Figure 3) are shown in **Figure 9**. The figure shows bubble size growth and expanded ***dense phase*** - the two critical properties of the particulate system that primarily influence the performance of a BFB reactor. The case shown is for a reactor with bubble size control to a maximum size of 10 cm. This figure also shows solid loadings and dilute phase voidage up to the TDH in the dilute phase region, where the declining solid loading (shown by the dotted curve) characterizes the ***splash zone***. The model estimates the TDH, the information necessary to decide the minimum height necessary for the dilute phase of the reactor before the exit gas nozzle location (or entry into internal cyclones, if used). The BFB temperature is ordinarily constant within the bed, and so the temperature profile is not shown in the figure. No significant reaction is noticed in the ***dilute phase*** for this particular case and reaction system used. However, depending on the design and operating conditions, reaction system, relative volume and temperature of the two zones (dilute phase vs. dense phase), significant reaction may occur in the dilute phase, too.

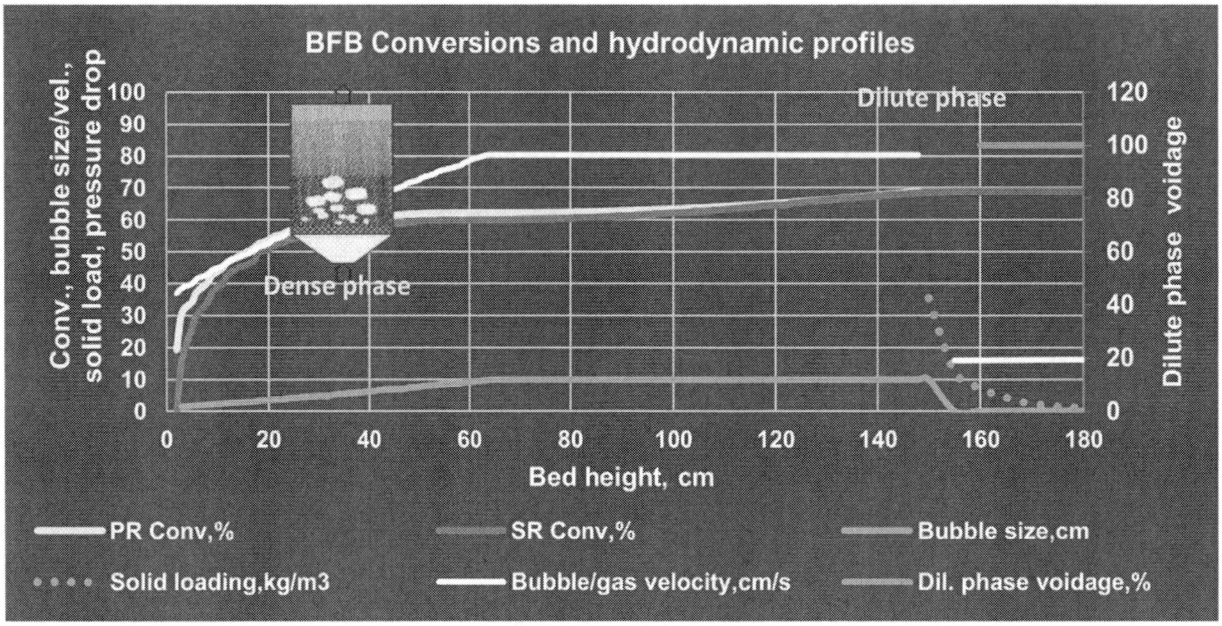

Figure 9: Projected BFB reactor performances and key particulate hydrodynamic effects

The DIFREX reactor design and model package GRM™ enables quick generation of the complete design and operating conditions and performance comparison of alternative reactors (FXB, BFB or CFB) if the information required for complete definition of the reaction system involved can be provided or is generated with DIFREX help. **Figure 10** shows one such comparison of complete design and operating conditions of three different solid-gas reactor types - a TFXB (Tubular Fixed Bed), a BFB and a QTFXB (Quenched Tubular Fixed Bed), for 100 TPD production of the primary product (PPX) or product group (like alcohols, paraffins or olefins) from syngas for an FT (Fischer-Tropsch) reaction system (as in Figures 3 and 9). **Figure 11** shows the performance comparison of the three reactors. The best Reactor Of Choice (ROC) can be quickly determined based on the criteria of the user that usually include cost, mechanical and other considerations. In **Figure 11**, the best choice shown is based solely on the highest yield.

Uncertainty/ fear of scale-up of a BFB reactor from bench to commercial scale can now be drastically reduced if proper account of advances made in particulate technology is incorporated in the design and scale-up of this reactor system.

Circulating Fluidized Bed Reactors

Successful design and operation of a circulating fluidized bed (CFB) reactor also require proper interplay of particulate flow and hydrodynamics behavior with reaction chemistry and kinetics and heat and mass transfers. The two key advantages of this reactor are: 1) It offers an elegant means of continuous regeneration of spent solid (catalyst or non-catalytic/ reactive particles) in a recirculation loop that removes the solid after only a short exposure (residence) time in the reactor, and subsequently returns the regenerated material back to near the reactor bottom. It is therefore particularly suited for fast-deactivating catalyst and for regenerable catalysts as in an FCC reactor. It is also a prime candidate for applications in **coal/ biomass/ tar sand combustion and gasification and minerals processing**, fast reactions where particle ***attrition*** is not an issue of concern. 2) CFB is also a ***bubble-less*** reactor above the bottom mixing zone, unlike BFB, thereby enjoying minimum heat and mass transfer resistances when compared with both FXB and BFB reactors.

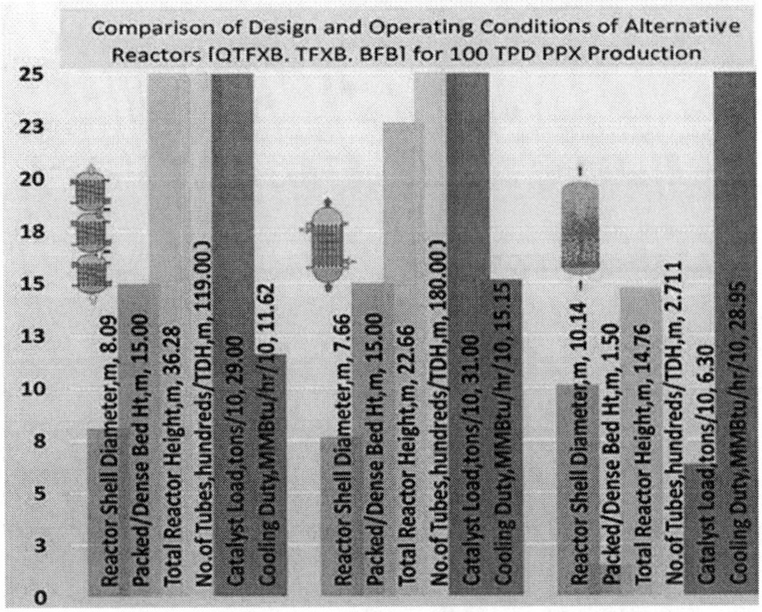

Figure 10: Reactor design and operating conditions for 3 different solid-gas reactor types

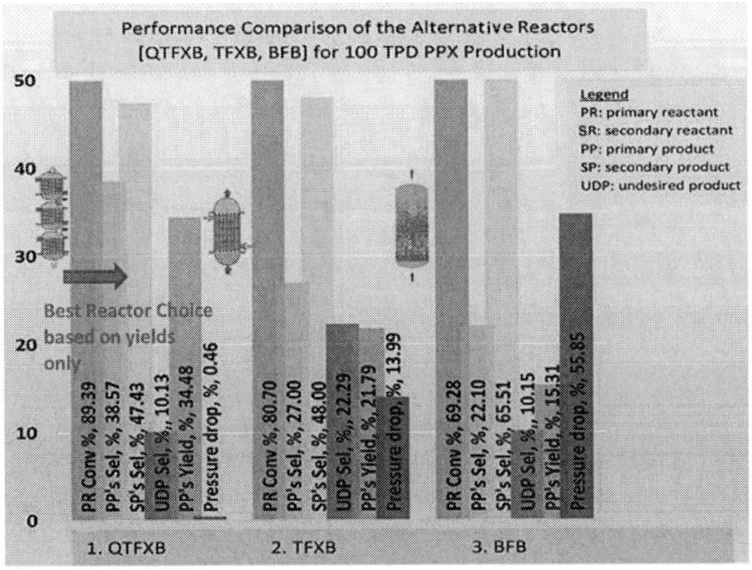

Figure 11: Performance Comparison for 3 solid-gas reactors

The two most important particulate system parameters used in the design of a CFB are reactor voidage (**ε**) profile and solid loading or "flux" of the up-flow solid-gas stream. Voidage profile correlations proposed by Kwauk et al.[26] used in the CFB reactor design and model have been first tested for valid simulation of voidage profiles of common solids of commercial interest and found to be satisfactory, as shown in **Figure 12.** For the CFB, reactor pressure drop was calculated by using four major pressure drop components: bed hydrostatic pressure, solids acceleration, solids friction and gas friction, as proposed by Yang and Keairns[27]. More details of these treatments are available in Dutta et al[13]. The DIFREX CFB design and modeling tool has been applied to both catalytic and non-catalytic reactor designs that include ***acrylonitrile production from propylene, FCC catalyst regeneration***[13], ***high-grade crude extraction from oil shale, iron ore reduction*** **and** ***hot gas desulfurization by zinc titanate***[6]. Application of particulate technology and principles is integral to the design of these reactors.

Temperature, conversions and bed/ particulate hydrodynamics for a propylene-to-acrylonitrile CFB system is shown in **Figure 13**, where PR and LR denote primary reactant propylene (C_3H_6) and limiting reactant ammonia (NH_3). In this reactor design, the bed bottom gas-solid mixing zone has a smaller diameter than the top/main bed. Profiles of temperature, reactor voidage, local gas velocity and solid loading of the up-flowing gas are shown.

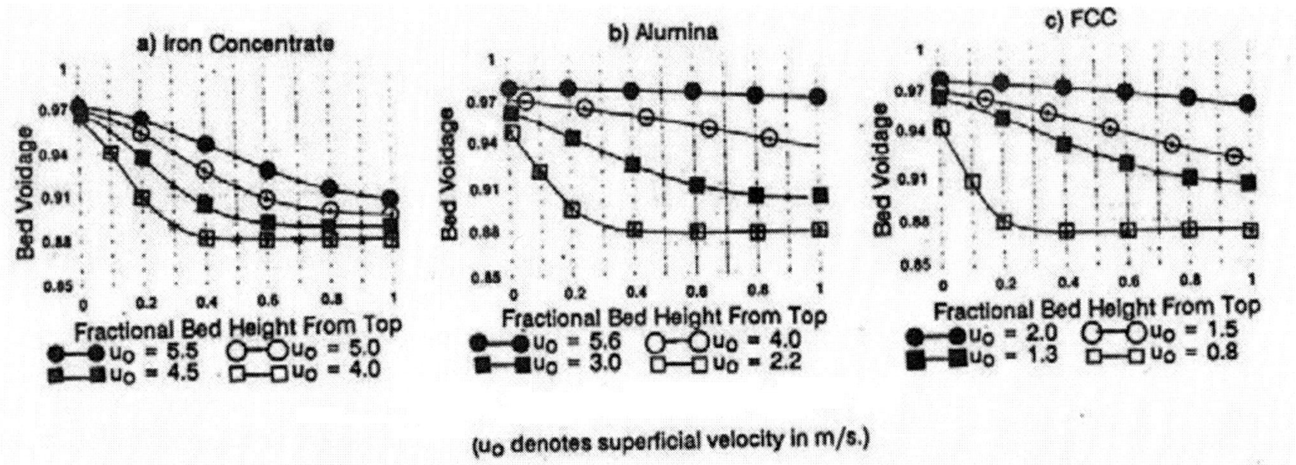

Figure 12: Simulation of CFB model based on voidage profile correlations proposed by Kwauk et al.[26]

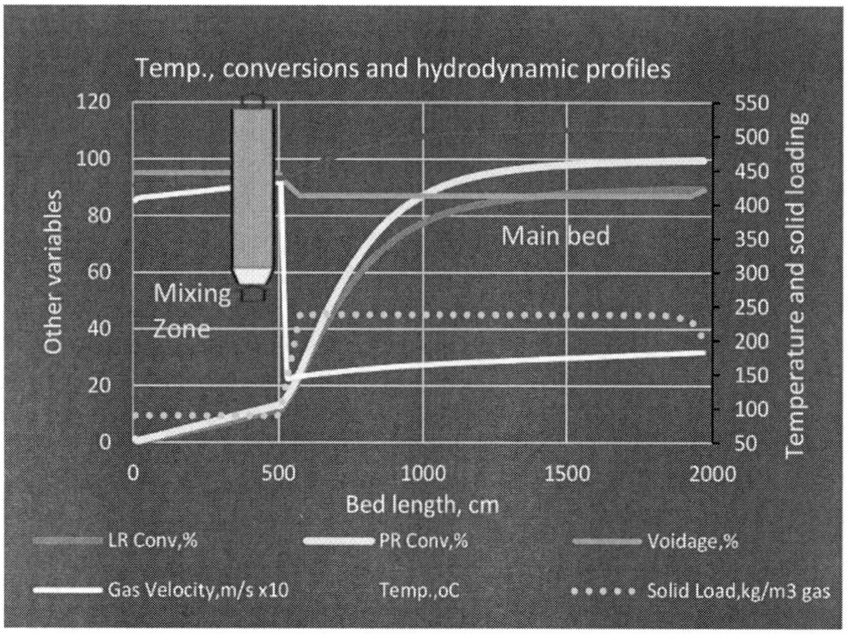

Figure 13: Profiles (Temperature, conversion and hydrodynamics) for a CFB system

An Interesting Observation

An interesting observation while studying the effect of CFB hydrodynamics on a reaction system is worth mentioning. A *CFB* can be viewed as a suspended and moving fixed bed (*FXB*) of particulates (fine particles) but with a very high bed voidage and with significant solids slippage (falling backward in the annulus along the wall). However, due to the intimate best heat and mass transfer rates possible, the bed temperature (rise) profile (for an exothermic reaction) in an adiabatic CFB reactor does not show as steep a temperature rise or spike as can be problematic in an FXB. Also, the bed temperature rise can be controlled by both gas flow and solid circulation rate.

Due the various impacts, the upper portion of a CFB that has a higher average bed voidage (a CFB "suspended bed" with lower solid/catalyst concentration per unit volume) may achieve higher conversion than a CFB "suspended bed" with lower bed voidage. Lower solid loading in this more diluted suspended bed allows a higher temperature rise due to lower volumetric heat capacity because of its lower solid content. The effect of rising temperature in the CFB may then more than compensate for the effect of lower solid/catalyst concentration. This is illustrated in **Figure 14**. Higher temperature and higher conversions are observed for the case with higher voidage and lower solid loading (kg/m^3 gas) in spite of the low solid concentration. On the other hand, since the inlet temperature is the same for both cases and the average and upper zone temperatures are higher for the case with higher bed voidage (due to extra temperature rise), the catalyst deactivation is likely to be faster and the selectivity is also likely to be different (for several reasons).

It is important to understand that, unlike the dense bed of a BFB, a CFB is not isothermal even though it is more isothermal than an FXB.

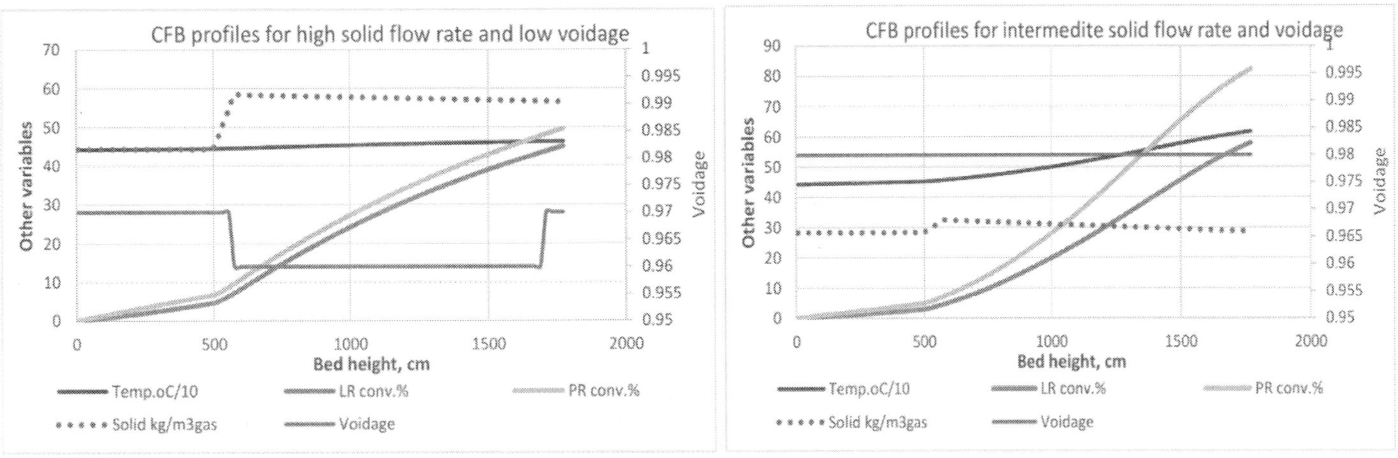

Figure 14: Effect of solid flow rate and bed voidage in a CFB case study

Hybrid Particulate Systems

A prime application of particulate technology is in FCC reactors. There are a variety of configurations of these reactors. In many installations, the FCC reactor is a "riser" type, with high gas velocity carrying all the catalyst upward in a dilute phase, with minimal residence time, etc. For this riser kind of FCC reactor, it can be represented as a hybrid/ combination of three particulate-gas contact types: a bottom regenerated solid-feed gas mixing (CSTR or BFB) zone, a middle which is the main CFB (riser) section, and a gas-solid disengaging zone at the top (which may create a further bed that can be simulated as a BFB). This representation is shown in **Figure 15.**

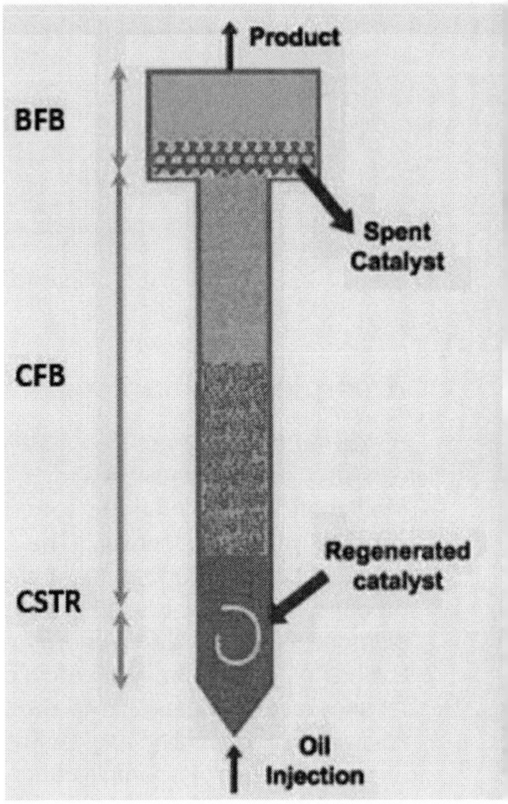

Figure 15: 3-zone simulation of an FCC performance

Using 1) a 5-lump reaction scheme with gasoil feed as shown in **Figure 16**, 2) surrogate compounds to represent the components involved, and 3) particulate system behavior in the three zones, a simple FCC reactor model was developed by Dutta and Gualy[8] that simulated the observed performance data reasonably well. The simulation results are shown in **Figure 17(a-d)**.

Reaction system: FCC 5-Lump reactions with GASOIL feed

GASOIL	GASOLINE	Rxn 1
GASOIL	DRYGAS	Rxn 2
GASOIL	LPG	Rxn 3
GASOIL	COKE	Rxn 4
GASOLINE	DRYGAS	Rxn 5
GASOLINE	LPG	Rxn 6
GASOLINE	COKE	Rxn 7
LPG	DRYGAS	Rxn 8

Figure 16: FCC 5-lump reactions with GASOIL feed

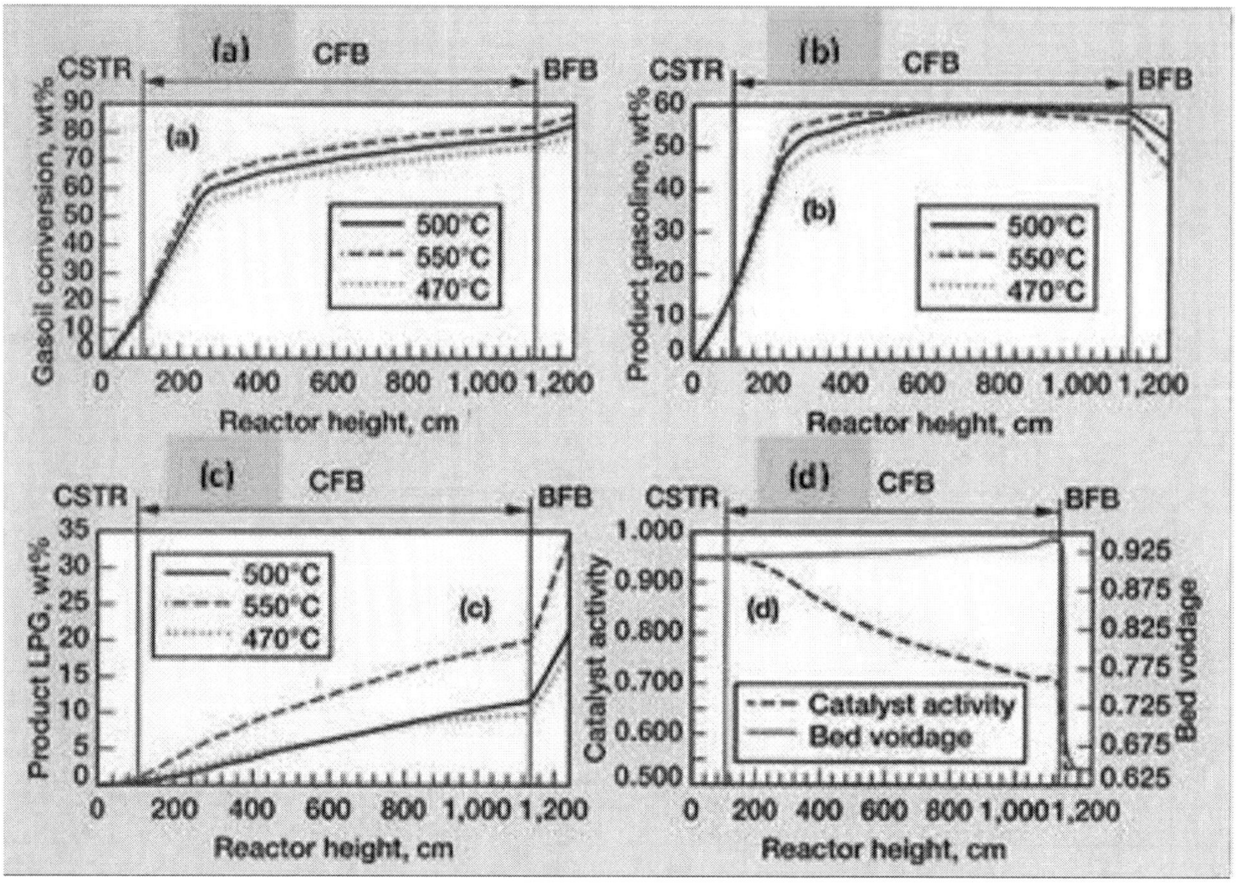

Figure 17: FCC simulation using a 5-lump reaction scheme

Emerging Technologies and Particulate Systems

Is ***Methanol*** the fuel of the future? There is a worldwide push to capture CO_2 from atmosphere and power plant exhausts and convert this and H_2O into methanol. One of the emerging technologies is ***PTM*** [(solar) Power to Methanol] that may lead to zero carbon footprint – as solar energy seems to be becoming increasingly more competitive and economical. The particulate technology can play a vital role in the PTM process, where the CO_2 could be captured and concentrated ***economically*** by using a regenerable particulate sorbent in a fluidized bed process. The other is the conventional route of ***syngas-to-methanol*** through syngas production from coal or biomass – the ***renewable energy route***. In both of these technologies, the particulate technology plays a vital role. Syngas production by gasification of coal or biomass in a fluidized bed reactor is one of the most attractive means for the second route. Due to a rather unique and perhaps the most diverse expertise and experience in both fluidization and reaction engineering, and also in coal combustion and gasification technology[28-31], DIFREX can play an instrumental role in implementing a rapid and successful move to ***the fuel of the future***. An integrated CTSM/ BTSM (Coal-to-Syngas-Methanol/ Biomass-to-Syngas-Methanol) plant can be designed and built promptly with help from DIFREX. This will be further facilitated by DIFREX already having an advanced and proprietary syngas to methanol (S2M) reactor design available.

References

1. Dutta, S. and R. Gualy, "Build Robust Reactor Models", Chemical Engg. Progress, 96(10), October 2000, p. 37.
2. Dutta, S. and R. Gualy, "Super-modeling Reactors for Profitability", Chem. Engg., 107(6), 2000, p. 72.
3. Dutta, S. and R. Gualy, "Modernize Process Reactors", Hydrocarbon Processing, November, 1999, p.91.
4. Dutta, S. and R. Gualy, "Overhaul Process Reactors", Hydrocarbon Processing, September 1999, p. 43.
5. Dutta, S. and R. Gualy, "General Reactor Model Improves HPI Applications", Hydrocarbon Processing, July 1999, p. 45.
6. Dutta, S., "Circulating Versus Bubbling Fluidized Bed for Hot Gas Desulfurization by Zinc Titanate", Proc. 5th Int'l. Circulating Fluidized Bed Conf., Beijing, PRC, May, 1996, p. pr-7.
7. Dutta, S., "Reactor Model: Usefulness of a Lesser Known Tool in Process Plants", paper presented at SABIC Tech. Meeting, May, 1996.
8. Dutta, S., "Fluidized and Moving Bed Desulfurization by Zinc Titanate", AIChE Symp. Ser., 301(90), 1994, p. 157.
9. Dutta, S. and B. Jazayeri, "Alternate Reactor Concepts for Oxidative Coupling of Methane", Proc. 7th Int'l. Fluidization Conf., ed. Potter and Nicklin, Brisbane, Australia, May, 1992, p. 445.
10. Dutta, S. and G. D. Suciu., "An Experimental Investigation on the Effectiveness of Baffles in Breaking Bubbles in Fluid Beds", J. Chem. Engg. Japan, 25(3), 1992, p. 345.
11. Dutta, S., "A Novel CFB Application: A Chemical Compressor", Proc. 3rd Int'l. Circulating Fluidized Bed Conf., ed. Basu, Horio and Hasatani, Nagoya, Japan, Oct., 1990, p. 465.
12. Dutta, S., "A Novel Circulating Bed Metal Hydride Compressor", paper presented at the 8th World Hydrogen Conf., Honolulu, Hawaii, 1989.
13.Dutta, S., B. R. Christian and M. F. Ratterman, "Circulating Fluid Bed Reactor Model Developed for FCC/ART Catalyst Regenerator", Proc. 6th Int'l. Fluidization Conf., ed. Grace, Shemilt and Bergougnou, Banff, Canada, May, 1989, p. 9.
14. Dutta, S. and G. D. Suciu, "Unified Model Applied to the Scale-up of Catalytic Fluid Bed Reactors of Commercial Importance", Proc. 6th Int'l. Fluidization Conf., ed. Grace, Shemilt and Bergougnou, Banff,Canada,May,1989, p.311.
15. Dutta, S., "A Unified Approach to the Modeling of Catalytic Fluidized Bed Reactors", Proc. World Cong. III Chem. Reaction Engg., Tokyo, 1986, p. 287.
16. Dutta, S., S. C. Arnold, G. D. Suciu and L. Verde, "Scale-Up of a Commercial Catalytic Fluid Bed Reactor Involving Complex Kinetics", Inst. Chem. Engg. Symp. Ser. 87, 1984, p. 517.
17. Dutta, S., "Measurement of Bubble Characteristics by Optical Probe", paper presented at the AIChE Annual Meeting, Chicago, Nov., 1980.
18. Dutta, S., "A Simple Probe for Fluidized Bed Measurements", Cana. J. Chem. Engg., 57, 1979, p. 115.
19. Dutta, S., "Dead-zone Heights near the Grids of Fluidized Beds", Fluidization, Cambridge University Press, 1978, p. 32.
20. Wen, C. Y. and S. Dutta, "Research Needs for Analysis, Design and Scale-up of Fluidized Bed Reactors", AIChE Symp. Ser., 73(161), 1977, p.1.
21. Fluidization Engineering, 2nd edition, by Daizo Kunii and Octave Levenspiel, Butterworth-Heinemann, 1991.
22. Chemical Reaction Engineering, 3rd edition, by Octave Levenspiel, John Wiley & Sons, 1999.
23. Werther, J., Chem. Engg. Sci., 35, 1980, p. 372.
24. Kato, K., and C. Y. Wen, Chem. Engg. Sci., 24, 1969, p. 1351.
25. Hatano, H., T. Okuma and M. Ishida, J. Chem. Engg. Japan, 17, 1984, p. 491.
26. Kwauk, M. N., N. Wang, Y. Li, B. Chen and Z. Shen, Circulating Fluidized Bed Technology, ed. P. Basu, Pergamon Press, 1986, p. 33.
27. Yang, W. C. and D. L. Keairns, AIChE Symp. Ser., 141(70), 1974, p.27.
28 Wen, C.Y. and S. Dutta, Book Chapter "Kinetics of Coal Pyrolysis and Gasification Reactions", Ch.2 Coal Conversion Technology, Addison-Wesley Publishing Company, 1979.
29. Wen, C.Y. and S. Dutta, "Solid-Gas Reactions in Coal Conversion Processes", Coal Processing Technology, CEP Tech. Manual, IV, 1978, p. 40.
30. Wen, C.Y. and S. Dutta, "Reactivity of Coal and Char: Part I – In Carbon Dioxide Atmosphere", I&EC Process Design & Development, 16, 1977, p.20.
31. Wen, C.Y. and S. Dutta, "Reactivity of Coal and Char: Part II – In Oxygen-Nitrogen Atmosphere", I&EC Process Design & Development, 16, 1977, p.31.
32. www.difrex.com
33. http://difrex.com/pdf/BrochureContents.pdf.
34 Dutta, Subhash, Stephen Arnold, Aashish Gaurav and Jim Brenner, "DIFREX Nexgen Expands and Advances Reactor & Technology Solutions", paper presented at AIChE Spring Meeting, Orlando, April, 2018.

508152 Erosion and Mobilisation of Highly ACTIVE Simulant Suspensions with Impinging Vertical Jets

Abstract:

Based on the nuclear industry in the U.K., storage of nuclear waste poses a particularly complex risk. Part of the waste during storage before being disposed of begins to form colloidal particles. These particles begin to settle out, and can compromise the integrity of their storage container. Using an Impinging vertical liquid jet is the means to prevent this. However the operational effectiveness of these jets is unknown due to the nature of the containers. Reduced scale modelling is required to assess their efficacy, and understand the dimensionless results as they allow systems of scale to be compared. This is of particular interest due to the wide reaching applications to all industries that may require utilisation of impinging jets, or require mass transfer, or even heat transfer.

An experimental rig of 0.5 m^3 has been used to model the central jet of the storage tank system. A Honite simulant (SiO_2, glass spherical particles) of varying size distributions has been used for the initial non-cohesive particle testing, and modelling. Each experiment uses 10 different jet diameters. Data is collected with an Ultrasonic Velocity Profiler, which scans the simulant bed in real time under impingement utilising an automated traverse. Acoustic backscatter data was recorded to yield a dynamic scour depth, and dynamic crater radius to allow dimensionless analysis of the erosion for each size distribution. Dimensionless analysis takes the form of the Erosion Parameter, E_c, which allows comparison for systems of scale and a means of correlating multiple simulants against one another, so patterns are easily identifiable. Part of the dimensionless analysis was also to deduce the Shieldâ€™s Parameter from vane viscometry to allow for a comparison of erosion behaviours taking into account particle properties, flow properties and the overall influence these have on the erosion profile.

505818 REAL WORLD Examples of Influences of Particles and Cake Formation on SOLID-Liquid Separation Technology Operation

Abstract:

The paper begins with a general discussion of the bench-top laboratory tests that are conducted for solid-liquid separation technology problem analysis, process development, technology selection and scale-up. The tests include pressure, filter media, filter aids, and similar process parameters.

Solid-liquid separation technologies are explained for continuous pressure filtration and batch pressure filtration. This provides an overview for the testing and troubleshooting.

Three different case histories are presented illustrating different scenarios of the impact of particles, particle size distribution (PSD) and cake formation on the separation technology operations. The first is a specialty chemical application for a continuous rotary pressure filter where the particles changed from lab scale to production scale. The second case is a clarification application where the system was running successfully for over one-year but inexplicitly, the performance changed drastically and the filter media began plugging very quickly during the cycle. The final application looks at lab testing and pilot scale testing where the cake formation changed resulting is a unique operating technique.

Sherlock Holmes and Dr. John Watson are fictional characters of Sir Arthur Conan Doyle. Process engineers who live in the real-world can learn many things from the two of them for solving process filtration problems. One example that Holmes proves time and again is that there is no benefit to â€œjumping to conclusions.â€ As discussed in the illustrations, there are many choices of filtration technologies to achieve the necessary level of quality with a cost-effective and a reliable process. Also, in terms of troubleshooting, there are many causes of filtration symptoms. With careful analysis and taking a creative and "non-conventional approach" by looking at what is behind the process data and the operating parameters, the optimum solution will be realized.

Influence of electrostatic charges on the particle concentration in wall-bounded turbulent flows

H. Grosshans[a,*], L. Villafañe[b], A. Banko[b], M.V. Papalexandris[c]

[a]*Physikalisch-Technische Bundesanstalt (PTB), Bundesallee 100, 38116 Braunschweig, Germany*
[b]*Department of Mechanical Engineering, Stanford University, CA*
[c]*Institute of Mechanics, Materials and Civil Engineering, Université catholique de Louvain, 1348 Louvain-la-Neuve, Belgium*

Abstract

Preferential concentration of inertial particles in wall-bounded turbulent flows is of paramount importance and, thus, subject of fundamental research. In practical applications of confined particle-laden flows the particles experience frequent collisions with the piping system which may result in an unwanted electric charge separation through triboelectric effects. The consequential occurrence of electrostatic forces alters the particle trajectories and, thus, impairs the general validity of the measured results to an unknown extent. In this work the influence of triboelectric charging on the preferential concentration of inertial particles in a fully developed turbulent duct flow was investigated by means of a combined numerical and experimental approach. The order of magnitude of the potential charge accumulation was estimated and imposed in a parametric study to the particulate phase in the simulations. The simulations demonstrate that the concentration profiles are for the most part independent of the prescribed charge. However, the peaks of the particle number density at the walls caused by turbophoresis are strongly reduced through the local increase of repelling electrostatic forces. The comparison of numerical with experimental data indicates that the particles in the experimental setup are affected by a surface charge density of the order of 40 $\mu C/m^2$. The presented results aim to elucidate the impact of electrostatic forces in the particle distribution in wall-bounded particle-laden flows.

Keywords: electrostatics, particle dynamics, particulate flows, experiment, simulation

Introduction

Preferential concentration of inertial particles in fully-developed wall-bounded turbulent flows is of relevance to many industrial applications, such as pneumatic powder conveying [1], pulverized coal combustion [2] or particle-based solar receivers [3]. For this reason, this type of flows is subject to experimental research since decades. Noteworthy classical works include those by Tsuji and Morikawa [4], Tsuji et al. [5] who collected a large set of data concerning the mean velocity

*Corresponding author
Email address: holger.grosshans@ptb.de (H. Grosshans)

of powders in vertical and horizontal flows through circular pipes. This kind of experimental study requires the usage of a long feed-line in order to obtain a fully-developed flow at the test section and to ensure that the particles are independent of their initial conditions. However, as the particles are advected along the piping system they may experience frequent collisions with the walls and between each other. If the powder and the piping system are of a different kind of material their contact results in an electric charge separation through triboelectric effects [6]. Under those circumstances a considerable amount of electric charge might be accumulated on the surface of the particles or on the structural components of the facility.

One easily observable consequence is the deposition of particles on the surfaces of the test section [7, 8, 9]. If no adequate measure is taken, such as the usage of an antistatic agent (as done by Tsuji and Morikawa [4], Tsuji et al. [5]), particle deposition might alter near-wall particle concentrations. Also, deposition layers may induce detrimental effects on the experiments, i.e., constrain optical access and light transmission. Another, less obvious, implication of electrostatic forces is the modulation of the particle trajectories of airborne particles. In an experimental setting, the exact influence of electrostatics on the measured particle concentration statistics is unknown. Moreover, in an industrial context the formation of deposits can impair the functionality of the facilities by reducing the effective cross-sectional area of pipes. Also, excessive local charge accumulation can effectuate sparks which challenges the operational safety of a plant [10, 11, 12].

Experimental measurements in the near-wall region in particle-laden flows are challenging due to the wall interference with optical measurement techniques, and also because of the temporal and spatial resolution required to capture the small scales of the particle and flow structures of interest. Limited data is available of gas and particle phase statistics near walls [13]. Measurements are often restricted to a certain parameter range due to experimental limitations, i.e. relatively large small flow scales to be able to measure all fluid fluctuations, or particle diameters larger than the dominant near-wall eddies [14]. Conversely, numerical simulations have been extensively used to study near wall phenomena. A remarkable early contribution was the ability of capturing the accumulation of particles in the viscous sublayer through *turbophoresis*, i.e. the tendency of particles to migrate in the direction of decreasing turbulent intensity, by means of direct numerical simulations (DNS) [15, 16]. More recent examples include the explanation of the mechanisms for particle transfer in the turbulent boundary layer by Marchioli and Soldati [17] and the computation of the inertial migration of particles in laminar square duct flow by Tabaei Kazerooni et al. [18]. Nonetheless, simulations rarely consider the possible effect of arising electrostatic forces which are expected in real particle-laden flows.

Previous numerical works proposing approaches to predict the charge build-up process of solid-fluid mixtures incorporated different levels of simplifications. For example, the simulations by Hogue et al. [19] described the charging of particles but neglected the influence of aerodynamic forces on their trajectories. Watano et al. [20] included drag forces in the numerical framework but they assumed the gaseous phase to follow a predefined velocity profile, i.e. the interaction between particle dynamics and turbulence was not accounted for. The same counts for the studies of Kolniak and Kuczynski [21] and Tanoue et al. [22, 23] who solved the Reynolds-Averaged Navier-Stokes equations for the turbulent gas flow. The fluid-particle interaction was computed through large-eddy simulations (LES) by Lim et al. [24]. However, they focused on the influence of an external electric field on already charged particles. The authors of the present paper implemented dynamic models to predict the contact charging of particles when colliding with a solid surface or with each other into a computational fluid dynamics solver. Employing this computational framework generic geometries such as channel and pipe flows were studied while accounting by DNS [25] or LES [26] for the turbulence of the carrier fluid. Albeit electrical forces were included in the set of equations, the size of the flown-through vessels was small, i.e. the initial charge build-up was evaluated. Therefore, the accumulated charge was rather low and did not significantly influence particle dynamics.

To sum up, the influence of accumulated electrostatic charge on the preferential concentration of particles in wall-bounded flows in large-scale facilities is not known to date. While experimental results may be affected by electrostatic forces, quantification of the precise impact is at present lacking. In order to evaluate its contribution we investigated the influence of triboelectric charging on inertial particles in a fully-developed, vertical, turbulent duct flow by comparison of numerical and experimental results. In the following section of this paper the experimental facility is presented followed by an outline of the numerical methodology. Results are discussed in the fourth section followed by the main conclusions from the present study.

Experimental methods

Experimental data was obtained in a square section duct flow facility at Stanford University, see Villafañe et al. [27]. A 5.4 m long smooth aluminum duct with a 40 mm wide cross section ensures a fully developed turbulent air flow at the glass test section where measurements are

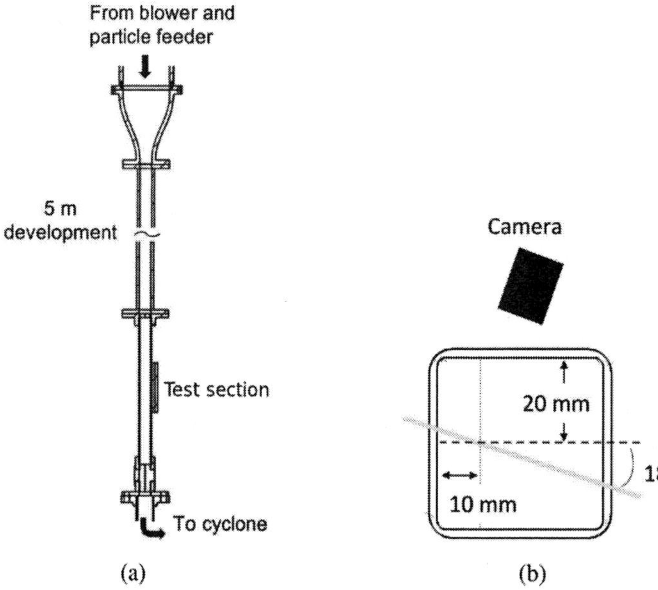

Figure 1: Experimental apparatus: (a) layout of the particle-laden duct flow facility, (b) laser sheet configurations used for measurements of mean particle concentration [27].

reported. An schematic of the vertical rig is shown in figure 1(a). A variable speed centrifugal pressure blower controls the inlet air flow to sustain a constant Reynolds number during the tests. A volumetric screw feeder is used to feed the particles into a flow conditioning section with a cross section 16 times larger than that of the main flow duct. Several grids placed downstream of the particle injection help to laterally disperse the particles prior to the contraction between the flow conditioning unit and the development duct section. The test section consist on a 425 mm long borosilicate glass duct with a nominal internal cross-section of 40 mm × 40 mm, and 2 mm thick walls. The test section was manufactured by extrusion to eliminate the presence of joints. The manufacturing process is responsible for the 2 mm internal corner radii in the test section cross section that limits the optical accessibility to the near wall region along the wall normal direction. Figure 1(b) illustrates the test section cross section and the optical set-up configuration used to measure particle concentration profiles in the vicinity to the wall. Downstream of the test section the particles are separated from the air and collected at a cyclone, and the air flow is exhausted to the atmosphere.

Planar measurements of mean particle concentration were performed at a Reynolds number based in the duct width and bulk flow velocity of $Re = 10\,000 \pm 0.7\%$. The nickel particles used have a number-based mean diameter of 11.8 ± 0.5 μm. The particle-size distribution shown in figure 3 is characterized by low probability tails covering a range from 6 μm to 20 μm. Particle size distributions before and after the tests were measured by a Coulter Counter and confirmed that particles were not selectively filtered by the rig. An average Stokes number of $\mathrm{St}_\eta = 5$ corresponds to the experimental particle and flow parameters. The average Stokes number is based on the mean particle diameter and the Kolmogorov flow time scale derived from the channel-averaged dissipation rate. A ratio of particle to gas mass loading of 0.01 was imposed in the experiments, with a maximum uncertainty of 0.004.

Particle velocity and concentration measurements were performed using planar optical techniques, i.e. a collimated laser sheet from a pulsed NdYag laser crossing the test section with a constant thickness and width, and a high resolution CCD camera normal to it. A set of 4 cylindrical lenses were used to form a laser sheet with a $1/e^2$ Gaussian beam thickness of 1.6 mm and a width of 25 mm. The angled laser sheet configuration shown in figure 1(b) was adopted to measure the concentration distribution near the wall. The laser sheet is inclined 18° with respect to the wall-normal and crosses the central plane of the test section at 10 mm from the wall. The high resolution camera synchronized with the pulsed laser acquired time independent images of the particles within the laser sheet. Image processing algorithms were used to identify individual particles in 2000 independent images, and extract the two-dimensional coordinates of particle centroids that were used to compute concentration profiles.

Mathematical model and numerical methods

This section summarizes the CFD solver which was employed in the present study. This solver utilizes a four-way coupled approach where the continuous gaseous phase is described in Eulerian and the dispersed particulate phase in Lagrangian framework [28]. In the gaseous phase model given by Grosshans and Papalexandris [26, 25] the flow is considered dilute so the volume fraction of the gaseous phase can assumed to be unity. Thus, the mass and momentum balance laws are given by the Navier-Stokes equations with constant diffusivities, namely

$$\nabla \cdot \boldsymbol{u} = 0 \tag{1a}$$

$$\frac{\partial \boldsymbol{u}}{\partial t} + (\boldsymbol{u} \cdot \nabla)\boldsymbol{u} = -\frac{1}{\rho}\nabla p + \nu\nabla^2\boldsymbol{u} + \boldsymbol{F}_{\mathrm{s}} + \boldsymbol{F}_{\mathrm{f}}\,. \tag{1b}$$

In the above equations $\boldsymbol{u}$ denotes the gas velocity, p its pressure, ρ its density and ν its kinematic viscosity. A source term, $\boldsymbol{F}_{\mathrm{s}}$, accounting for the momentum transfer from the particulate to the gaseous phase is also introduced. The flow is forced in positive x-direction by adding a constant pressure gradient, $\boldsymbol{F}_{\mathrm{f}}$.

The governing equations of LES are obtained by applying a spatial filter operator to equations (1a) and (1b). Thus, the large-scale turbulent motions are directly resolved on the computational grid. On the other hand, the small sub-filter scales, which exhibit universal characteristics, are considered through an appropriate turbulence model. To this end, the implicit model by Boris et al. [29] is applied. The above Eulerian equations are discretized by the Finite Difference Method (FDM). In particular, the convective terms are approximated by an up to fifth-order Weighted Essentially Non Oscillatory (WENO) scheme [30]. The diffusive and pressure terms are approximated by fourth-order central differences and the time derivatives by an implicit second-order backward scheme. For further details concerning the numerical implementation the reader is referred to Gullbrand et al. [31].

The particulate phase is treated according the numerical methodology proposed by Grosshans and Papalexandris [32] and implemented by [33]. This methodology is extended to facilitate the treatment of a large amount of particles as outlined in the following. In the current setting, the particle probability density function is given by

$$\zeta = \zeta(\boldsymbol{u}_{\mathrm{p}}, r_{\mathrm{p}}, Q_{\mathrm{p}}; \boldsymbol{x}, t)\,. \tag{2}$$

This expression gives the probable number of particles having a velocity of $\boldsymbol{u}_{\mathrm{p}}$, a radius of r_{p} and carrying an electrical charge of Q_{p} at a certain location, $\boldsymbol{x}$, and time instance t. In order to handle the large amount of particles the stochastic parcel method [34] is adopted to approximate ζ. Accordingly, the above function is discretized in computational parcels, each parcel representing $\zeta = 10$ particles, which each is tracked in a Lagrangian framework.

The trajectory of a parcel is altered by accelerations due to aerodynamic, $\boldsymbol{f}_{\mathrm{ad}}$, collisional, $\boldsymbol{f}_{\mathrm{coll}}$, gravitational, $\boldsymbol{f}_{\mathrm{g}}$, and electric field, $\boldsymbol{f}_{\mathrm{el}}$ forces. As regards the aerodynamic acceleration the particle drag coefficient is derived as a function of the particle Reynolds number following the classical correlation of Schiller and Naumann [35]. Furthermore, gravitation accelerates the particles towards the positive x-direction.

With respect to acceleration during collisions, binary particle collisions and elastic reflection of the particles off the walls is accounted for. To enhance computational efficiency, a statistical technique that yields a collision frequency is employed [36]. More specifically, it is assumed that the probability P_n that the particles of parcel n collide with the particles of parcel m during the time increment $\mathrm{d}t$ follows a Poisson distribution, namely

$$P_n = 1 - \mathrm{e}^{-\omega_{nm}\,\mathrm{d}t}\,. \tag{3}$$

In this equation, the time-averaged collision frequency ω_{nm} is defined as

$$\omega_{nm} = \pi \frac{\zeta_m \left(r_{\mathrm{p},n} + r_{\mathrm{p},m}\right)^2 |\boldsymbol{u}_{\mathrm{p,n}} - \boldsymbol{u}_{\mathrm{p,m}}|}{(\boldsymbol{e}_x \cdot \boldsymbol{\delta})(\boldsymbol{e}_y \cdot \boldsymbol{\delta})(\boldsymbol{e}_z \cdot \boldsymbol{\delta})} \tag{4}$$

where $\boldsymbol{\delta}$ is a vector pointing from the location of one parcel to the other, $\boldsymbol{e}_x$, $\boldsymbol{e}_y$ and $\boldsymbol{e}_z$ denote the unit vectors of the Cartesian coordinate system and $\cdot$ the scalar product.

As regards electrostatic forces, the specific force acting on a particle due to the electric field is calculated as

$$\boldsymbol{f}_{\mathrm{el}} = \frac{Q_{\mathrm{p}} \boldsymbol{E}}{m_{\mathrm{p}}} \tag{5}$$

where m_{p} is the mass of the particle. Concerning the computation of the electric field strength, $\boldsymbol{E}$, there are two different approaches available, namely Gauss' and Coulomb's law. In the present study, a computational efficient and accurate hybrid scheme which combines both approaches which was proposed by Grosshans and Papalexandris [32] is applied. Therein, the interaction of an individual particle with the space charge, i.e. with the electric field originating from the particle cloud, is calculated by Gauss' law. Furthermore, the derivation of the forces between particles located close to each other (here: present in the same grid cell) is based on Coulomb's law.

According to Gauss' law, the divergence of the electric field is related to the electric charge density per unit volume, ρ_{el}, by

$$\nabla \cdot \boldsymbol{E} = \frac{\rho_{\mathrm{el}}}{\varepsilon}. \tag{6}$$

In the above equation $\varepsilon = 8.854$ F/m denotes the electrical permittivity of the vacuum and ρ_{el} is defined as

$$\rho_{\mathrm{el}} = \left((\boldsymbol{e}_x \cdot \boldsymbol{v})(\boldsymbol{e}_y \cdot \boldsymbol{v})(\boldsymbol{e}_z \cdot \boldsymbol{v})\right)^{-1} \int_{\boldsymbol{v}} \zeta Q \, \mathrm{d}\boldsymbol{x} \tag{7}$$

where $\boldsymbol{v}$ is the unit volume.

For the case of charged particles, alternatively Coulomb's law can be used, i.e.

$$\boldsymbol{E}(\boldsymbol{r}) = \frac{\zeta_m Q_m \boldsymbol{r}}{4 \pi \varepsilon |\boldsymbol{r}|^3}. \tag{8}$$

Therein, $\boldsymbol{E}(\boldsymbol{r})$ is the electric field a particle is subjected to due to the presence of another parcel at a distance of $\boldsymbol{r}$ consisting of ζ_m particles carrying the charge Q.

Whereas Gauss' and Coulomb's law are mathematically equivalent, the numerical methods for solving them when dealing with a large number of particles may differ considerably in terms of accuracy and computational costs. For the proof of their equivalence and a detailed discussion concerning the advantages and disadvantages of each approach, the reader is referred to Grosshans and Papalexandris [32].

As regards the numerical set-up, in streamwise direction, the computational domain extends 15 H. Further, periodic boundary conditions are chosen in streamwise direction in order to mimic an infinite duct length and no-slip is assumed for the gaseous phase at the sidewalls. Consistent to the experiments, a duct flow of $Re = 10\,000$ was considered. Moreover, the particle size distribution, which is detailed in the following section, replicates the utilized powder. The duct is defined to be bounded and manufactured of a conductive material. This choice implies that an electrical charge which might be brought to the ducts surface upon contact with a particle, vanishes almost instantly towards ground. Nonetheless, a so-called image charge is accounted for at the surface if a charged particle residing in its vicinity. Aiming to obtain efficiently results for long averaging times, a rather coarse numerical grid was applied. Namely, the first grid point was placed at a distance of $y^+ = 6$ from the wall which results in a total number of 1.2 million points.

Results and discussion

Code validation

The above Eulerian-Lagrangian solver has been applied previously to simulate a wide range of dispersed flows such as LES and DNS of pneumatic powder conveying [26, 37, 38, 39, 9, 25]. The mentioned references include a large amount of validation studies of various aspects of the solver, realized e.g. through grid resolution studies and comparison with experiments. In particular, the implementation of the dynamics of charged particles in the influence of an electric field was validated in Ref. [32].

Specifically for the current set-up the implementation of the Eulerian solver for the fluid phase has been validated by Grosshans [33] through comparison of the solution for a single-phase flow with the DNS data of Zhu et al. [40] which was also performed for a duct flow of $Re = 10\,000$ and the empirical law of the wall [41] and log-law [42]. The profiles of the mean streamwise velocity component, i.e. averaged in time and in the homogeneous streamwise direction and normalized to the friction velocity u_τ, compare very well with the earlier DNS and to an expected degree with the empirical laws.

Further, the profiles of the root-mean-square (r.m.s.) fluctuations of the three velocity components are plotted in figure 2. These and the following results are given in terms of wall units,

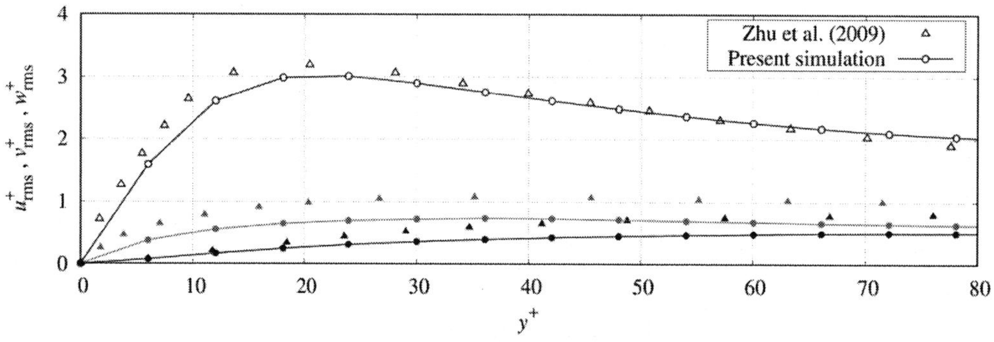

Figure 2: Profiles of r.m.s. velocity fluctuations for a single-phase flow. Comparison of our numerical results with the DNS data of Zhu et al. [40]. The empty symbols denote the streamwise velocity component u^+_{rms}, the black symbols the wall-normal component v^+_{rms}, and the grey symbols the spanwise velocity component w^+_{rms} [33].

$y^+ = u_\tau y/\nu$. In general, good agreement to the DNS data of Zhu et al. [40] is ascertained. However, in comparison the turbulent fluctuations predicted by our simulations are of a lower amplitude. This can be explained by the relative coarse grid employed for the computations which impedes the resolution of the smallest flow scales in the viscous wall region ($y^+ < 50$) where the production of turbulent kinetic peaks [43]. Nevertheless, the curves agree sufficiently to provide qualitatively reliable results.

Initial particle charge

The expected charge of the particles at the glass test section inlet is imposed as an initial condition in the simulations and varied systematically. By doing so, the effect of electrostatic charge accidentally aggregated by the particles on their preferential concentrations in the experimental apparatus is evaluated. For this purpose first the order of magnitude of the potential amount of charge carried by the particles when reaching the test section is estimated. This estimation serves as a basis for the decision on the parameter range to be investigated. With respect to the methodology for the evaluation, we compare the maximum charge predicted by the charging model implemented by

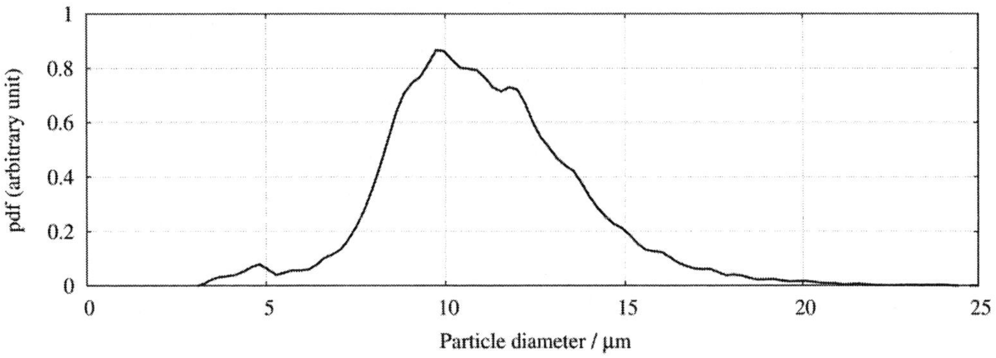

Figure 3: Particle size distribution measured with a Coulter Counter and replicated in the simulations.

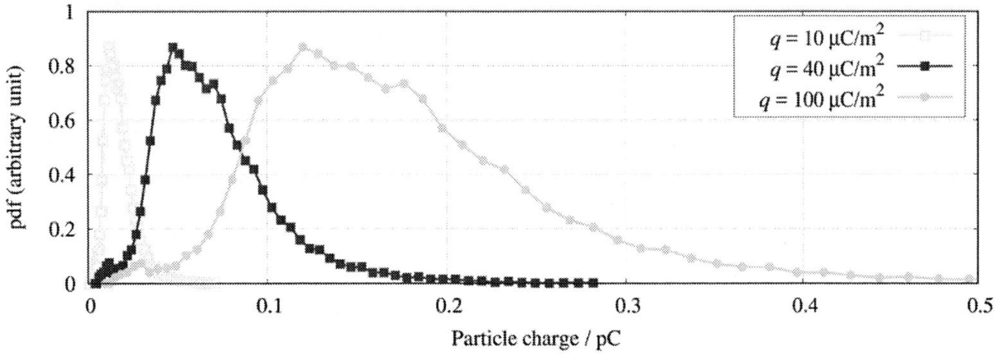

Figure 4: Charge distributions assigned to the particles in the simulations.

Grosshans and Papalexandris [25] with the outcome of the charge relaxation model by Matsuyama [46].

As regards the former approach, it is assumed that the charge originates from triboelectric effects during collisions of the particles with structural components such as the blower, feeder or duct walls (cf. figure 1(a)) prior to the test section. Whereas the amount of collisions taking place is unknown one can utilize the dynamic model implemented by Grosshans and Papalexandris [25] to predict an order of magnitude of the maximum charge a particle can obtain. This charge value is also termed equilibrium charge, since for further collisions the amount of pre-charge transferred from the particle to the surface equals the triboelectric charge separation, i.e. the net charge transfer is zero. Following this model, even though the total equilibrium charge depends on the particle size, the equilibrium surface charge density, q_{eq} is independent. More precisely, for a theoretically infinite number of collisions the model predicts an asymptotic limit for q of

$$q_{\mathrm{eq}} = \frac{\varepsilon U}{h} . \tag{9}$$

In this equation U denotes the contact potential and h the particle-surface separation distance. Consulting the well-known triboelectric series [e.g. 44] one may assume a value of the order of 1 V for the contact potential. Concerning h, a commonly used estimate relates to the order of the range of repulsive molecular forces due to surface irregularities, namely $h = 10^{-9}$m [45, 21, 26]. These assumptions lead to a maximum charge density of $q_{\mathrm{eq}} = 8.85 \times 10^{-3}$ C/m^2.

Nonetheless, as recently discussed by Matsuyama [46] such a high value cannot be obtained in reality since a charge relaxation takes place immediately after the particle-surface contact. This relaxation occurs due to discharge at the contact gap when the potential difference equals the gaseous breakdown limit potential. Thus, this mechanism limits the maximum charge a particle can carry. Matsuyama [46] compared in his work the prediction of the breakdown limit potential by Paschen's law with their impact charging experiments of single particles [cf. 47, 48, 49]. The compared data set indicate that a particle of the size of 12 μm, which corresponds to the mean particle size in the present study, is able to carry a maximum charge of about 0.5 pC or, in other words, a surface charge density of about $q = 10$ μC/m^2.

Considering the discussion provided above, we assigned a constant charge density to the particles during each simulation. First, for comparison purposes one simulation with an uncharged particulate phase was ran. Further, four cases with different charge density values, namely $q = 10$ μC/m^2, $q = 40$ μC/m^2 and $q = 100$ μC/m^2, were computed. In other words, a parameter space which includes uncharged particles, particles carrying a charge which corresponds to the limit predicted by the charge relaxation model and up to ten times the respective value was explored. This choice aims to give an overview over the flow pattern to be expected in this type of flows. However, it is noted that several underlying assumptions which may not coincide with the realistic conveying situation. These include the assumption of homogeneously distributed charge on the surface of each particle which might not be reachable considering the finite number of collisions of the particle with the test rig. Further, the charge of particulate phase might be dependent on the distance to the wall, i.e. a particle close to the wall is expected to accumulate more charge than a particle in the bulk of the flow.

A consequence of the decision to assign a constant surface charge density is the dependence of the total particle charge on its size. The particle size distribution measured before the experiment and replicated in the simulations is plotted in figure 3. The resulting particle charge distributions when applying the surface charge densities in the simulations are given in figure 4.

Particle concentrations

In the simulations an initially fully-developed turbulent flow field was considered. The particles were introduced at random positions whereas their initial velocity was assumed to equal the local velocity of the surrounding fluid. The resulting mean concentration profiles along the laser sheet centerline normalized by the mean concentration at $y = 0.05\ H$ are plotted for the near-wall region in figure 5. Since the planar laser sheet was inclined by 18° with respect to the wall-normal (cf. figure 1(b)) the z^+ position varies with the y^+ coordinate.

The differences in the presented numerical results reveal the potential influence of electrostatic charges on preferential particle concentration. Apparently, the effect of the electrostatic charges on the particle concentrations is most pronounced in the vicinity of the wall, i.e. in the region $y^+ < 15$. Comparing the different numerical data suggests that the case corresponding to $q = 40\ \mu C/m^2$ replicates the closest the experimental measurements. The proximity of both curves indicates that the powder utilized in the experiment might be subjected to a charge of this respective order of magnitude. However, a full compliance between experiments and simulations cannot be expected from the present study. On the one hand this discrepancies can be attributed to inaccuracies in the numerical methodology, namely the rather crude assumptions underlying the charge distribution which have been discussed in the previous subsection and the limited resolution of the turbulent flow scales in the simulations (cf. figure 2). Also the boundary conditions to the electric field may play a role: while in the simulations the wall is considered to be conductive and grounded, the corresponding components in the experimental facility consist of a mix of materials including

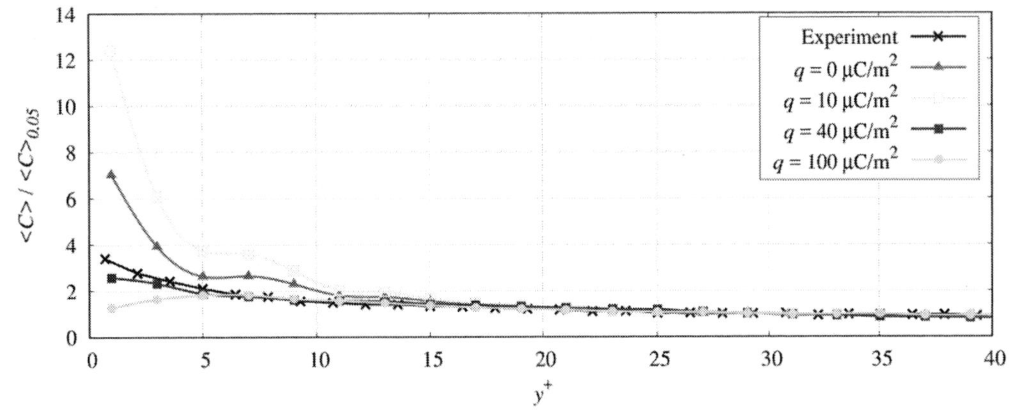

Figure 5: Comparison of the measured and simulated mean particle concentration distribution in the near-wall region. The local concentration is normalized by the mean concentration at $y = 0.05\ H$. Note that the plotted data is collected along the laser beam (cf. figure 1(b)), i.e. the respective z^+ position varies with the y^+ coordinate.

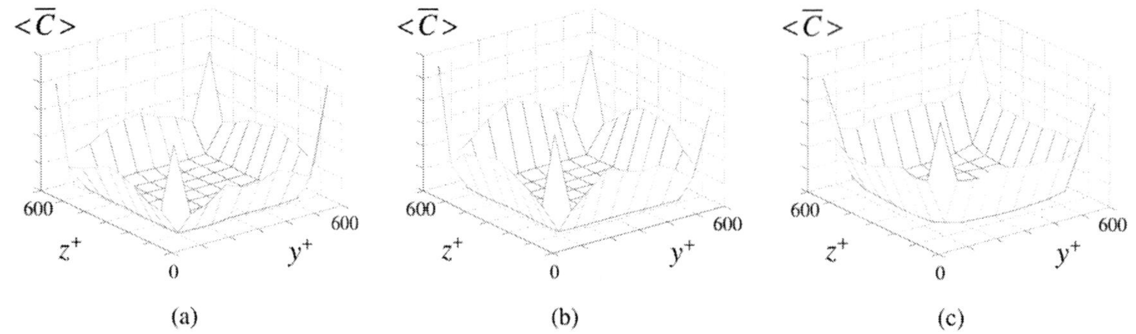

Figure 6: Normalized mean particle concentration in arbitrary units, $<\overline{C}>$, over the duct cross-section. The particles carry a surface charge density q of (a) 0 μC/m^2, (b) 10 μC/m^2 and (c) 40 μC/m^2

grounded metal parts but also glass. On the other hand there are few differences in the data acquisition. In the numerical post-processing procedure all particles are counted whose center point resides within the thickness of the laser sheet. In contrast, the laser beam in the experiments is of a Gaussian intensity profile which decreases the probability to detect small particles at the beam flanks. Further, due to the laser thickness being much larger than the particle diameter, particles may overlap and not be detected.

All the profiles presented in figure 5 indicate the effect of turbophoresis, i.e. particle accumulation in the near wall region. Comparing the numerical data related to the uncharged case with q = 40 μC/m^2 and q = 100 μC/m^2 shows that the peak at the wall reduces with increasing particle charges. This behavior was observed before [33] and attributed to the increasing electrostatic forces between particles. These forces are most prominent in regions of a high particle number density since there the average spacing between particles is the lowest. Consequently, particles are pushed out of the near-wall region towards the bulk of the flow. In other words, electrostatic charges tend to smoothen the particle concentration profiles.

Nonetheless, the profile related to q = 10 μC/m^2 in figure 5 evolves contradictory to this argumentation. Surprisingly, the concentration of this case peak at the wall higher than for the uncharged particles. To get further insight the mean particle concentrations over the complete duct cross-section are depicted in figure 6. On the whole, the particle accumulation close to the walls can be observed for all cases. In particular, the distributions relating to the uncharged particles (figure 6(a)) exhibit the strongest peaks at the four corners of the duct. Additional maxima are formed at the centerlines of the four sidewalls. These characteristic peaks in the ducts corner and sidewalls centerlines are qualitative agreement with the observations of Noorani et al. [50] and Tabaei Kazerooni et al. [18].

When the charge density is increased to q = 10 μC/m^2 (figure 6(b)) the particles are forced to migrate from the regions of highest accumulation, i.e. the duct corners. Nevertheless, instead of migrating to the bulk of the flow they travel along the sidewalls and feed the peaks at their centerline. Since the profiles in figure 5 depict locations close to the sidewall centerlines, this motion explains the rise of the particle concentration for q = 10 μC/m^2. If the charge density is further increased to $q = 40$ μC/m^2 (figure 6(c)), the forces between the particles residing in these regions becomes high and, thus, the peaks are smoothed out. Consequently, the maximum of the related profile in figure 5 at the wall reduces. For a further increase to $q = 100$ μC/m^2, the particle behavior is qualitatively similar to the case of $q = 40$ μC/m^2, but the electrostatic forces affect an even stronger smoothing of the concentrations. Therefore, for $q = 100$ μC/m^2 only a slight particle accumulations is detectable in the region $0 < y^+ < 10$.

Conclusions

We examined the role of accidentally accumulated electrostatic charge in the mean particle concentration distribution and in particular in the near wall increase of concentration due to turbophoresis. Numerical simulations of particles of equal polarity in a fully developed duct flow were performed considering a homogeneous charge density on the particles' surface. Whereas the simulations reproduced the particle size distribution measured in the experiments the charge density was varied systematically. The resulting concentration profiles were compared to experimental data for which the particle charge is unknown. The numerical parametric study reveals that electrostatic charges play an important role in the particle redistribution, especially in the near wall region. In general, the simulations corroborate that the electrical forces tend to smoothen the particle concentration profiles, i.e. counteract particle accumulation through turbophoresis. The numerical results for a surface charge density of 40 μC/m^2 are the closest to the experimental data. A future elaboration of this topic requires the simulation of the charge build-up process in order to get a more realistic picture of the charge distribution of the powder, as well as higher resolution of the small flow scales.

Acknowledgments

HG gratefully acknowledges the financial support from the Max Buchner Research Foundation. LV and AB gratefully acknowledge the funding by the Advanced Simulation and Computing (ASC) program of the US Department of Energys National Nuclear Security Administration via the PSAAP-II Center at Stanford, Grant No. DE-NA0002373-1. MP gratefully acknowledges the financial support of the National Research Fund of Belgium (FNRS) under the GRANMIX Projet de Recherche grant.

References

[1] S. S. Grossel. Design and Operating Practices for Safe Conveying of Particulate Solids. *J. Loss Prev. Process Ind.*, 25:848–852, 2012.

[2] J. G. M. Kuerten, C. W. M. van der Geld, and B. J. Geurts. Turbulence modification and heat transfer enhancement by inertial particles in turbulent channel flow. *Phys. Fluids*, 23(12): 123301, 2011.

[3] Andrej Lenert and Evelyn N. Wang. Optimization of nanofluid volumetric receivers for solar thermal energy conversion. *Sol. Energy*, 86(1):253 – 265, 2012.

[4] Y. Tsuji and Y. Morikawa. LDV measurements of an air-solid two-phase flow in a horizontal pipe. *J. Fluid Mech.*, 120:385–409, 1982.

[5] Y. Tsuji, Y. Morikawa, and H. Shiomi. LDV measurements of an air-solid two-phase flow in a vertical pipe. *J. Fluid Mech.*, 139:417–434, 1984.

[6] S. L. Soo. Dynamics of Charged Suspensions. In *Topics in Current Aerosol Research*, pages 71–73. Pergamon Press, 1971.

[7] I. Adhiwidjaja, S. Matsusaka, S. Yabe, and H. Masuda. Simultaneous phenomenon of particle deposition and reentrainment in charged aerosol flow – effects of particle charge and external electric field on the deposition layer. *Adv. Powder Technol.*, 11(2):221–233, 2000.

[8] J. Guardiola, V. Rojo, and G. Ramos. Influence of particle size, fluidization velocity and relative humidity on fluidized bed electrostatics. *J. Electrostat.*, 37(1–2):1–20, 1996.

[9] N. Schwindt, U. von Pidoll, D. Markus, U. Klausmeyer, M. V. Papalexandris, and H. Grosshans. Measurement of electrostatic charging during pneumatic conveying of powders. *J. Loss Prev. Process Ind.*, 49:461–471, 2017.

[10] M. Glor. Ignition hazard due to static electricity in particulate processes. *Powder Technol.*, 135–136:223–233, 2003.

[11] W. Fath, C. Blum, M. Glor, and C.-D. Walther. Electrostatic ignition hazards due to pneumatic transport of flammable powders through insulating or dissipative tubes and hoses – new experiments and calculations. *J. Electrostat.*, 71(3):377–382, 2013.

[12] S. Egan. Learning lessons from five electrostatic incidents. *J. Electrostatics*, 88:183–189, 2017.

[13] J. K. Eaton and J. R. Fessler. Preferential concentration of particles by turbulence. *Int. J. Multiphase Flows*, 20:169–209, 1994.

[14] B. M. Sumer and R. Deigaard. Particle motions near the bottom in turbulent flow in an open channel. Part 2. *J. Fluid Mech.*, 109:311–337, 1981.

[15] G. A. Kallio and M. W. Reeks. A numerical simulation of particle deposition in turbulent boundary layers. *Int. J. Multiphase Flows*, 15:433–446, 1989.

[16] J. B. McLaughlin. Aerosol particle deposition in numerically simulated channel flow. *Phys. Fluids*, 1:1211–1224, 1989.

[17] C. Marchioli and A. Soldati. Mechanisms for particle transfer and segregation in a turbulent boundary layer. *J. Fluid Mech.*, 48:283–315, 2002.

[18] H. Tabaei Kazerooni, W. Fornari, J. Hussong, and L. Brandt. Inertial migration in dilute and semidilute suspensions of rigid particles in laminar square duct flow. *Phys. Rev. Fluids*, 2: 1–23, 2017.

[19] M. Hogue, C. I. Calle, P. S. Weitzman, and D. R. Curry. Calculating the trajectories of triboelectrically charged particles using discrete element modeling (dem). *J. Electrostat.*, 66: 32–38, 2008.

[20] S. Watano, S. Saito, and T. Suzuki. Numerical simulation of electrostatic charge in powder pneumatic conveying process. *Powder Technol.*, 135–136:112–117, 2003.

[21] P. Z. Kolniak and R. Kuczynski. Numerical modeling of powder electrification in pneumatic transport. *J. Electrostat.*, 23:421–430, 1989.

[22] K. Tanoue, A. Ema, and H. Masuda. Effect of material transfer and work hardening of metal surface on the current generated by impact of particle. *J. Chem. Eng. Jpn.*, 32:544–548, 1999.

[23] K. Tanoue, H. Tanaka, H. Kitano, and H. Masuda. Numerical simulation of triboelectrification of particles in a gas-solid two-phase flow. *Powder Technol.*, 118:121–129, 2001.

[24] E. W. C. Lim, J. Yao, and Y. Zhao. Pneumatic transport of granular materials with electrostatic effects. *AIChE J.*, 58(4):1040–1059, 2012.

[25] H. Grosshans and M. V. Papalexandris. Direct numerical simulation of triboelectric charging in a particle-laden turbulent channel flow. *J. Fluid Mech.*, 818:465–491, 2017.

[26] H. Grosshans and M. V. Papalexandris. Large eddy simulation of triboelectric charging in pneumatic powder transport. *Powder Technol.*, 301:1008–1015, 2016.

[27] L. Villafañe, A. Banko, C. Elkins, and J. K. Eaton. Gas heating by radiation absorbing inertial particles in a turbulent duct flow. In *Annual Research Briefs*, Center for Turbulence Research, Stanford University, USA, 2017.

[28] S. Elgobashi. On predicting particle-laden turbulent flows. *Appl. Sci. Res.*, 52:309–329, 1994.

[29] J. P. Boris, F. F. Grinstein, E. S. Oran, and R. L. Kolbe. New insights into large eddy simulation. *Fluid Dyn. Res.*, 100:199–228, 1992.

[30] G. S. Jang and C. W. Shu. Efficient implementation of weighted ENO schemes. *J. Comput. Phys.*, 126:202–228, 1996.

[31] J. Gullbrand, X. S. Bai, and L. Fuchs. High-order Cartesian grid method for calculation of incompressible turbulent flows. *Int. J. Num. Meth. Fluids*, 36:687–709, 2001.

[32] H. Grosshans and M. V. Papalexandris. On the accuracy of the numerical computation of the electrostatic forces between charged particles. *Powder Technol.*, 322:185–194, 2017.

[33] H. Grosshans. Modulation of particle-laden flows by electrostatic charges. In *12th International Symposium on Hazards, Prevention and Mitigation of Industrial Explosions*, Kansas City, MO, USA, 2018. submitted.

[34] F. V. Bracco. Modeling of Engine Sprays. *SAE Tech. Pap.*, 850394, 1985.

[35] L. Schiller and A. Z. Naumann. A drag coefficient correlation. *Z. Ver. Dtsch. Ing.*, 77: 318–320, 1933.

[36] G. Stiesch. *Modeling Engine Spray and Combustion Processes*. Springer Verlag, Berlin Heidelberg New York, 2003.

[37] H. Grosshans and M. V. Papalexandris. Evaluation of the parameters influencing electrostatic charging of powder in a pipe flow. *J. Loss Prev. Process Ind.*, 43:83–91, 2016.

[38] H. Grosshans and M. V. Papalexandris. A model for the non-uniform contact charging of particles. *Powder Technol.*, 305:518–527, 2016.

[39] H. Grosshans and M. V. Papalexandris. Numerical study of the influence of the powder and pipe properties on electrical charging during pneumatic conveying. *Powder Technol.*, 315: 227–235, 2017.

[40] Z. Zhu, H. Yang, and T. Chen. Direct numerical simulation of turbulent flow in a straight square duct at Reynolds number 600. *J. Hydrodyn.*, 21:600–607, 2009.

[41] L. Prandtl. Bericht über die Entstehung der Turbulenz. *Z. Angew. Math. Mech.*, 5:136–139, 1925.

[42] T. von Kármán. Mechanische Ähnlichkeit und Turbulenz. In *Proc. 3rd Int. Congr. Applied Mechanics*, pages 85–105, Stockholm, 1930.

[43] J. Kim, P. Moin, and R. Moser. Turbulence statistics in fully developed channel flow at low Reynolds number. *J. Fluid Mech.*, 177:133–166, 1987.
[44] R. Brück. Chemische Konstitution und elektrostatische Eigenschaften von Polymeren. *Kunststoffe*, 71:308–312, 1981.
[45] W. John, G. Reischl, and W. Devor. Charge transfer to metal surfaces from bouncing aerosol particles. *J. Aerosol Sci.*, 11(2):115–138, 1980.
[46] T. Matsuyama. A discussion on maximum charge held by a single particle due to gas discharge limitation. In *AIP Conference Proceedings*, volume 1927. American Institute of Physics, 2018.
[47] T. Matsuyama and H. Yamamoto. Charge relaxation process dominates contact charging of a particle in atmospheric conditions. *J. Phys. D: Appl. Phys.*, 28:2418–2423, 1995.
[48] T. Matsuyama and H. Yamamoto. Charge relaxation process dominates contact charging of a particle in atmospheric conditions: II. The general model. *J. Phys. D: Appl. Phys.*, 30: 2170–22175, 1997.
[49] T. Matsuyama, M. Ogu, H. Yamamoto, J. C. M. Marijnissen, and B. Scarlett. Impact charging experiments with single particles of hundred micrometre size. *Powder Technol.*, 135–136: 14–22, 2003.
[50] A. Noorani, R. Vinuesa, L. Brandt, and P. Schlatter. Aspect ratio effect on particle transport in turbulent duct flows. *Phys. Fluids*, 28(11):115103, 2016.

507384 Powder Electrification Due to Pneumatic Conveying

Abstract:

Electrostatic charging is an unavoidable and in some processes, an undesired phenomenon which occurs during particle handling and processing due to inter-particle and particle-wall collisions. Particle agglomeration, electrostatic discharge, and adhesion of the particles to the wall surface are some of the adverse effects of electrostatic charge generation. Catalytic gas-solid fluidized bed reactors which are used for polyethylene production, are examples in which electrostatic charge buildup has a negative impact on overall process performance. Charged polyethylene and catalyst particles can adhere to the reactor wall, a problem known as "sheeting" which requires regular reactor shut down for cleaning purpose. One source potentially capable of introducing electrostatic effects in gas-solid polymerization fluidized bed reactors is anticipated to be through the catalyst injection into the reactor via pneumatic conveying. Fresh catalyst, which is transferred and fed into the fluidized bed through narrow pneumatic conveying lines, can become highly charged due to continuous collisions among particles and the conveying tube. Upon injection of these highly charged particles into the reactor, image forces could form leading to the particles migration and adhesion on the reactor wall. In an attempt to better understand the charging of powders through pneumatic conveying, this study illustrates the effect of transport tube length, conveying gas velocity and particle size distribution on triboelectrification of fine glass beads as well as a catalyst powder. In addition, the generated electric current over a part of the transport tube as the result of particle movement was measured and compared with the cumulative electrostatic charge of that particle captured in a Faraday cage.

508098 Modeling of Electrostatic Charging of Particles in a Shaker

Abstract:

Multiple spherical particles made of polymer are loaded in a metal container, and shaken for a time. Due to impacts on the metal wall, particles develop electrostatic charge. The development of the charge generation was recorded experimentally in our previous work. In this paper we try to develop a physical model on this. A feature in the experiment was that the saturated amount of charge per particle decreased with an increase of the number of particles loaded in the container. This was understood as space charge effect, by which a particle in contact with a wall is affected on its charge transfer due to the external electric field from other charged particles, or charged cloud. How to involve the space charge effect into the charging model will be discussed in details.

508495 Parametric Study for the Development of a Particle-Particle Collisional Charging Model for Use in the CFD Simulation of Electrostatic Effects in Gas-Solid Fluidized Beds

Abstract:

An operational challenge faced in some gas-solid fluidized beds is the generation of electrostatic charge, which is an inherent and unavoidable issue in this field. Triboelectric charging in gas-solid fluidized beds occur due to the continuous particle-wall and particle-particle contacts during the fluidization process. For instance, in polyethylene reactors, polyethylene and catalyst particles can become highly charged, which results in them adhering to the reactor wall and forming large fused polyethylene sheets. These sheets can break off the reactor walls and obstruct the distributor plate, forcing a reactor shut-down even within a few hours of operation for clean-up. The downtime due to the clean-up, which can last up to 30 days, can result in significant economic loss due to reduced production.

A good understanding of triboelectric charging and its effects in gas-solid fluidized beds allows the development of techniques to reduce or control the charging in the bed. It also gives rise to the development of electrostatic charging models for use in computer simulations. Simulations are a very cost-effective alternative to running experiments, especially for industrial-scale test runs. Thus, it would be beneficial to build an effective Computational Fluid Dynamic (CFD) simulation model to study the electrostatic effects of charge generation in industrial-scale gas-solid fluidized beds. Since triboelectric charging in gas-solid fluidized beds occurs due to both particle-wall and particle-particle charging, the charging model must carefully consider the mechanisms for both types of charging within the fluidized bed. The overall focus of this work is specifically on particle-particle charging, which has received very limited attention.

One important charging phenomenon observed in gas-solid fluidized beds is bipolar charging. Bipolar charging occurs when particles of the same material, but different sizes, become oppositely charged during fluidization. It has been observed that the wall fouling thickness in a fluidized bed depends on both the magnitude of charge in the bed and the presence of both positively and negatively charged particles [1]. Once a layer of charged particles adheres to the wall, the oppositely charged particles are attracted to it to form a second layer. Multiple layers can then form depending on the magnitude of the electrostatic force in the fluidized bed [2]. The mechanism to describe how bipolar charging occurs is still unclear, since the studies on poly-sized particle systems have shown conflicting results; some noted that the smaller particles charged negatively and the larger particles charged positively [3], while others have reported the opposite [4]. By identifying the mechanism for bipolar charging, and the impact of various parameters on this phenomenon, it would be possible to develop a charge transfer model that reflects bipolar charging due to particle-particle collisions. This model can be used to simulate the change in charge of particles in a poly-dispersed gas-solid fluidized bed system. The successful development of this model would bring this research group one step closer to the final objective of this project: to develop a CFD model to simulate triboelectric charging and its effects on an industrial-scale gas-solid fluidized bed.

Thus, the specific objective of this work is to propose a mechanism for bipolar charging by performing systematic particle-particle collision experiments. A novel apparatus enabling particle-particle collisions was designed and fabricated in this work. The apparatus consisted of two Faraday cups situated above a wind-box. Single particles were dropped through the two Faraday Cups to measure their individual initial charges. The particles then went through the wind-box, which was designed to use dry air to push the two particles towards each other and cause a collision. After the collision, the two particles dropped into another set of separate Faraday Cups, located below the wind box, to measure their final charges. The collision and charge measurements were conducted in this way to minimize any external influence on the two particles; the charge transferred was solely due to the collision between the two particles. Since every test run may not experience a successful collision in the wind-box, the collisions were observed using a video recording device linked to a particle-tracking program. Parametric studies of this provide some conclusion on whether larger particles charge negatively or positively when colliding with smaller particles of the same material. Some parameters studied, which may affect the direction and magnitude of charge transfer, included impact velocity, particle size and/or size difference, particle material, and initial charge.

[1] A. Sowinski, A. Mayne, P. Mehrani, Effect of fluidizing particle size on electrostatic charge generation and reactor wall fouling in gas–solid fluidized beds, Chem. Eng. Sci. 71 (2012) 552–563.

[2] D. Song, P. Mehrani, Mechanism of particle build-up on gas-solid fluidization column wall due to electrostatic charge generation, Powder Technol. 316 (2017) 166–170.

[3] K.M. Forward, D.J. Lacks, R. Mohan Sankaran, Methodology for studying particle–particle triboelectrification in granular materials, J. Electrostat. 67 (2009) 178–183.

[4] A. Sowinski, L. Miller, P. Mehrani, Investigation of electrostatic charge distribution in gas–solid fluidized beds, Chem. Eng. Sci. 65 (2010) 2771–2781.

AUTHOR INDEX

AUTHOR INDEX

AUTHOR INDEX

AUTHOR INDEX